Experimental Aspects of Quantum Computing

Experimental Aspects of Quantum Computing

Edited by

Henry O. Everitt
Senior Research Scientist
Army Research Office
Research Triangle Park, North Carolina

Springer

ISBN: 0-387-23045-9

Printed in the United States of America. (BS/DH)

9 8 7 6 5 4 3 2 1

springeronline.com

Contents

Special Issue on Experimental Aspects of Quantum Computing 1
Henry Everitt

Invited Articles

Progress in Quantum Algorithms 5
Peter W. Shor

NMR Quantum Information Processing 15
Chandrasekhar Ramanathan, Nicolas Boulant, Zhiying Chen, David G. Cory, Isaac Chuang, and Matthias Steffen

Quantum Computing with Trapped Ion Hyperfine Qubits 45
B. B. Blinov, D. Leibfried, C. Monroe, and D. J. Wineland

Ion Trap Quantum Computing with Ca^+ Ions 61
R. Blatt, H. Häffner, C. F. Roos, C. Becher, and F. Schmidt-Kaler

Quantum Information Processing in Cavity-QED 75
S. J. van Enk, H. J. Kimble, and H. Mabuchi

Quantum Information Processing with Trapped Neutral Atoms 91
P. S. Jessen, I. H. Deutsch, and R. Stock

The Road to a Silicon Quantum Computer 105
J. R. Tucker and T.-C. Shen

Controlling Spin Qubits in Quantum Dots 115
Hans-Andreas Engel, L. P. Kouwenhoven, Daniel Loss, and C. M. Marcus

Spin-based Quantum Dot Quantum Computing in Silicon 133
Mark A. Eriksson, Mark Friesen, Susan N. Coppersmith, Robert Joynt, Levente J. Klein, Keith Slinker, Charles Tahan, P. M. Mooney, J. O. Chu, and S. J. Koester

Optically Driven Quantum Computing Devices Based on Semiconductor Quantum Dots 147
Xiaoqin Li, Duncan Steel, Daniel Gammon, and L. J. Sham

Implementing Qubits with Superconducting Integrated Circuits 163
Michel H. Devoret and John M. Martinis

Towards Scalable Linear-Optical Quantum Computers 205
J. P. Dowling, J. D. Franson, H. Lee, and G. J. Milburn

Photonic Technologies for Quantum Information Processing 215
Prem Kumar, Paul Kwiat, Alan Migdall, Sae Woo Nam, Jelena Vuckovic, and Franco N. C. Wong

Contributed Articles

Quantum Computer Development with Single Ion Implantation. 233
A. Persaud, S. J. Park, J. A. Liddle, I. W. Rangelow, J. Bokor, R. Keller, F. I. Allen, D. H. Schneider, and T. Schenkel

Bang–Bang Refocusing of a Qubit Exposed to Telegraph Noise 247
Henryk Gutmann, Frank K. Wilhelm, William M. Kaminsky, and Seth Lloyd

Quantum Computing and Information Extraction for Dynamical Quantum Systems . . . 273
Giuliano Benenti, Giulio Casati, and Simone Montangero

One-Dimensional Continuous-Time Quantum Walks 295
D. ben-Avraham, E. M. Bollt, and C. Tamon

Introduction

This year marks the tenth anniversary of the algorithms Peter Shor wrote for factoring and computing discrete logarithms on a quantum computer. It is no understatement to say that those algorithms have revolutionized our thinking about information processing and computability. By showing that there are certain, meaningful problems that are better solved on a quantum computer than on a classical computer, they inspired us to try to tame the weird world of quantum phenomena in order to reap these revolutionary benefits. Spurred by the importance and promise of this fundamentally new form of information processing, worldwide interest in research related to quantum information processing has skyrocketed in the intervening years. One measure of the remarkable impact of Shor's algorithms is seen in the United States' investment in quantum information, which rose from under \$5 M in 1994 to more than \$100 M in 2004.

Nevertheless, practical quantum computing still seems more than a decade away. Researchers have not even identified what the best physical implementation of a quantum bit will be. There is a real need in the scientific literature for a dialog on the topic of lessons learned and looming roadblocks. In order to (1) highlight the lessons learned over the last 10 years, and (2) outline the challenges we will face over the next 10 years, I have organized a special issue of the new journal *Quantum Information Processing* dedicated to the experimental aspects of quantum computing. The special issue includes a series of invited articles that discuss the most promising physical implementations of quantum computing. The invited articles were to draw grand conclusions about the past and speculate about the future, not just report results from the present. Of particular interest were insights that are universal or practical in nature. To provide a unifying theme and structure to these invited articles, the authors were asked to address the following topics:

Historical Review

A brief historical review is followed by an overview of current experimental approaches. By discussing relevant issues regarding materials, fabrication, control, measurements, analysis, phenomenology, etc, each article offers insight into the advantages offered and challenges faced by a given specific physical implementation.

Lessons Learned

Lessons learned and universal conclusions reached over the last 10 years are summarized, both those that can be applied to a specific physical implementation and those that may apply to the entire quantum computing community. Of primary interest were lessons learned about fabrication, control, and scalability, but authors were encouraged to make grander observations about trends and experimental truths.

Future Research Goals

Research roadblocks facing a specific physical implementation are listed, particularly those challenges regarding the control, fidelity, and scalability of quantum bit fabrication and logic operations. Authors were asked to identify research demonstrations that would be recognized as significant steps forward and to pose challenges to the experimental and theoretical communities.

The invited articles are listed below:

Progress in Quantum Algorithms,
Peter W. Shor (MIT)

Nuclear Magnetic Resonance Quantum Computing
NMR Quantum Information Processing, Chandrasekhar Ramanathan, Nicolas Boulant, Zhiying Chen, David G. Cory, Isaac Chuang, and Matthias Steffen (MIT)

Ion Trap Quantum Computing
Quantum Computing with Trapped Ion Hyperfine Qubits, B. B. Blinov, D. Leibfried, C. Monroe, (University of Michigan), D. J. Wineland (NIST)

Ion Trap Quantum Computing with Ca^+ Ions, R. Blatt, H. Haffner, C. F. Roos, C. Becher, F. Schmidt-Kaler (University of Innsbruck)

Neutral Atom Quantum Computing
Quantum Information Processing in Cavity-QED, S. J. van Enk, H. J. Kimble, and H. Mabuchi (CalTech)

Quantum Information Processing with Trapped Neutral Atoms, P. S. Jessen (University of Arizona), I. H. Deutsch, and R. Stock (University of New Mexico)

The Road to a Silicon Quantum Computer, J. R. Tucker (University of Illinois), and T.-C. Shen (Utah State University)

Semiconductor Quantum Dot Quantum Computing
Controlling Spin Qubits in Quantum Dots, Hans-Andreas Engel, (University of Basel), L. P. Kouwenhoven (Delft University of Technology), Daniel Loss (University of Basel), and C. M. Marcus (Harvard)

Spin-based Quantum Dot Quantum Computing in Silicon, Mark A. Eriksson, Mark Friesen, Susan N. Coppersmith, Robert Joynt, Levente J. Klein, Keith Slinker, Charles Tahan (University of Wisconsin), P. M. Mooney, J. O. Chu, and S. J. Koester (IBM T. J. Watson Research Center)

Optically Driven Quantum Computing Devices based on Semiconductor Quantum Dots, Xiaoqin Li, Duncan Steel (University of Michigan), Daniel Gammon (NRL), and L. J. Sham (UC-San Diego)

Superconductor Quantum Computing
Implementing Qubits with Superconducting Integrated Circuits, Michel H. Devoret (Yale) and John M. Martinis (NIST)

Photonic Quantum Computing
Towards Scalable Linear-Optical Quantum Computers, J. P. Dowling (NASA, JPL), J. D. Franson (Johns Hopkins University), H. Lee (NASA, JPL), and G. J. Milburn (University of Queensland)

Photonic Technologies for Quantum Information Processing, Prem Kumar (Northwestern University), Paul Kwiat (University of Illinois), Alan Migdall and Sae Woo Nam (NIST), Jelena Vuckovic (Stanford), Franco N. C. Wong (MIT)

In addition, several contributed articles were received in response to the call for the special issue. These contributed articles appear after the invited ones.

It has been a pleasure and an honor to work with and support this community during these early years. The surprising finding has consistently been that roadblocks may be overcome with conceptual innovation, improved materials, clever designs, and high fidelity controls. It is with great optimism for the future that this special issue series is presented, not only to stimulate new research but also to provide a look back on how far we have come in such a short time.

Guest Editor
Henry Everitt
Senior Research Scientist
U.S. Army Research Office
and
Physics Department
Duke University
May 12, 2004

Progress in Quantum Algorithms

Peter W. Shor[1]

We discuss the progress (or lack of it) that has been made in discovering algorithms for computation on a quantum computer. Some possible reasons are given for the paucity of quantum algorithms so far discovered, and a short survey is given of the state of the field.

KEY WORDS: quantum algorithms; NP-complete.

PACS: 03.67.Lx.

1. INTRODUCTION

It has now been 10 years since I discovered the quantum factoring algorithm.(1) This discovery caused great excitement; although some quantum algorithms had previously been discovered, this was the first algorithm that gave a substantial speedup over a classical algorithm for a well-studied and interesting problem. Many people expected a succession of other interesting quantum algorithms to quickly follow. Lov Grover indeed discovered his quantum searching algorithm shortly thereafter,(2) but the progress since has been disappointing, especially compared with the progress the rest of the field of quantum information processing has been making. Physicists have been proposing and experimenters have been exploring possible physical implementations of quantum computers at a pace I believe is faster than what anybody, but the most optimistic people expected; these developments are covered in the rest of this issue. Quantum cryptography is coming of age, with several theoretical proofs of its security recently discovered, and commercial quantum cryptography systems now on the market. The field of quantum information theory and quantum computational

[1]Department of Mathematics, Massachusetts Institute of Technology, Cambridge, MA 02139, USA. E-mail: shor@mit.edu

complexity have both been quite active, with a succession of interesting and important theoretical results. Meanwhile, the development of quantum algorithms appears to have lagged behind, with what seem like barely any significant new algorithms having been discovered. We will speculate on why more quantum algorithms have not been found, and survey the progress that has been made. This is an expansion and update of my paper[3] which also discusses this issue.

2. THOUGHTS ON QUANTUM ALGORITHMS

One thing I am often asked is why so few new quantum algorithms for solving classical problems have been discovered. It has not been for lack of effort; people have looked quite hard for new quantum algorithms. I can think of two reasons that quantum algorithms might be difficult to discover. The first is that there might really be only a few problems for which quantum computers can offer a substantial speed-up over classical computers; in the most pessimistic scenario, we have already discovered most of the important algorithms. The second is that quantum computers operate in a manner so non-intuitive, and so different from classical computers, that all the experience of the last 50 years in discovering classical algorithms offers little insight into how to go about finding quantum algorithms, so that while efficient quantum algorithms for many more problems exist, they are very hard to find. It appears impossible to tell which of these two cases is the actuality.

Another thing that I am often asked is what kind of problems are susceptible to attack by a quantum computer. Unfortunately, even the classical analog of this question: *What kind of problems are can be solved in polynomial time by a digital computer?* does not have a satisfactory answer. Computer scientists have a plethora of techniques they can try to apply to a problem: linear programming, divide-and-conquer, dynamic programming, Monte Carlo methods, semidefinite programming, and so forth. However, deciding which of these methods is likely to work for a given problem, and how to apply it, remains more of an art than a science, and there is no good way known to characterize the class of problems having polynomial-time algorithms. Characterizing the class of problems having polynomial-time quantum algorithms appears equally, if not more, difficult, one of the main additional difficulties being that we have so far discovered very few algorithmic techniques.

One of the things that has made it difficult to find new quantum algorithms that perform better than classical algorithms is the remarkable job that computer scientists have done over the last 50 years in finding good classical algorithms for problems. For the most part, researchers have been looking for quantum algorithms that efficiently solve problems which are not known to be solvable classically in polynomial time. These would yield the most impressive advances, and are also very likely to be the first problems for which, when and if quantum computers are developed, the quantum algorithms will give a practical advantage in the real world. To find such a problem, if we make the assumption that quantum computers cannot solve NP-complete problems in faster than exponential time, we would need to find a problem which is neither in P nor is NP-hard. Remarkably, in part because of the success of the classical theory of algorithms, there are relatively few natural problems which fit this criterion.

I now give a brief digression on complexity theory. The complexity class P consists of those problems which can be solved using algorithms running in time bounded by a polynomial in the length of the input. The class of problems with probabilistic polynomial-time algorithms is called BPP, and the class with quantum probabilitistic polynomial-time algorithms is called BQP (quantum algorithms are in general inherently probabilistic, and so the class BQP should most fairly be compared with the class BPP rather than P). Polynomial running times are considered to be efficient by theoretical computer scientists. This isn't strictly true—nobody would call an algorithm that runs in n^{100} steps efficient in practice, where n is the length of the input, but this definition has proven to be a good compromise between theory and practice; it appears to be the case that most natural problems in P have algorithms with running time a relatively small power of n. The class NP consists of those problems for which a solution can be verified in polynomial time; this class contains P, and the containment is generally thought to be strict.

Computer scientists have identified a subclass of NP comprising the hardest problems in NP; these are called NP-complete problems,(4,5,6) and a polynomial-time algorithm for any of these problems would imply a polynomial-time algorithm for all problems in NP, showing that P = NP. Remarkably, a large number of NP-complete problems have been identified.(7) When theoretical computer scientists consider a new problem, one of their first goals is to either show that it is NP-complete, or to find a polynomial-time algorithm for it. While these are mutually exclusive outcomes, it is not guaranteed that a problem in NP will either be NP-complete or in P; remarkably, however, the vast majority of problems studied seem to fall in one of these two classes.

Why might we suspect that quantum computers cannot solve NP-complete problems? Let us consider the classical analog of that question: why do computer scientists believe that classical computers cannot solve NP-complete problems efficiently? This is the celebrated P vs. NP question. (See Ref. 8,7,9 for the history of this problem.) The class NP is the class of problems for which, once a solution has been found, it can be verified in polynomial time that it is indeed a solution. Mathematically speaking, NP is the set of languages for which there are polynomial length proofs that a string is in the language (although there are not necessarily short proofs that a string is *not* in the language). NP-complete problems are a subset of these NP problems which have the property that if any of these NP-complete problems is solvable by an efficient algorithm, then all NP problems are solvable by an efficient algorithm.

There are essentially two lines of argument for why P should be different from NP. The first, which in my opinion is not terribly convincing, is that nobody has yet found a polynomial-time algorithm for solving NP-complete problems. While such an algorithm would generate a complete upheaval of our understanding of computational complexity, similar revolutions have occurred, albiet infrequently, in other branches of mathematics and science. The second argument is barely more rigorous than the first. It relates NP completeness to the difficulty of finding mathematical proofs. If, for instance, a quadratic algorithm was discovered for solving an NP complete problem, then a mathematician could use this algorithm to mechanically check whether a conjectured theorem had a proof of length n using computation time of cn^2 steps for some constant c. Now, let us assume that the primes are in some sense quasi-randomly distributed, as is believed by many mathematicians (although many other quasi-randomly distributed objects could be used in this argument as well). It then seems that it should be very difficult to check the truth of a statement such as

There are 17 primes in arithmetic progression between integers a and b.

without testing a large fraction of the numbers between a and b for primality; here the relative sizes of a and b should be chosen so that the probability of the above statement is roughly $\frac{1}{2}$. On the other hand, if you are given 11 numbers, testing to see if these are indeed primes in arithmetic progression can be done in time polynomial in the length of b. This problem is in the class NP, which means that it can be efficiently translated into a 3SAT problem—a Boolean formula in conjuctive normal form with 3 variables per clause (this translation is essentially the proof of the NP-compleness result). However, this problem appears quite hard, and it is very likely not NP-complete (meaning the reverse translation cannot be done). If any NP-complete problem could be solved in polynomial time, then problems such as the above could be solved in polynomial time. Intuitively, it seems as though it would be very difficult to prove the non-existence of such an arithmetic progression of primes, especially if you believe the distribution of prime numbers is quasi-random. Thus, this is some intuitive evidence towards the conjecture that $P \neq NP$.

Could the use of a quantum computer help solve such problems in NP? In this new question, we now have lost the mathematical intuition that proofs can be much harder to discover than to check. The symmetry between checking and discovering the proof is now gone: we are allowed a quantum computer to discover these proofs, but only permitted a digital computer to check them. Although the argument is not as convincing, it still does not seem likely that quantum computers can solve NP complete problems in less than exponential time. There is more evidence in this direction, in that there is a proof that a quantum computer cannot search a space of size N in less than $O(\sqrt{N})$ time.[10] This result shows that a quantum algorithm for solving NP-complete problems in sub-exponential time will have to use the structure of these problems, and this result can also be used to find an oracle with respect to which NP is not contained in BQP.

If quantum computers cannot indeed solve NP complete problems, then where should we be looking for problems to speed up using quantum algorithms? The obvious place to look is in problems neither known to be in P or to be NP-complete. There are only a few problems in this class. Those handful of these which appear to be related to periodicity, and thus possibly susceptible to attack using quantum Fourier transforms, have received substantial study from the quantum algorithms community. These include the two problems of graph isomorphism and of finding a short vector in a geometrical lattice. The problem of graph isomorphism is: given two graphs, is there a permutation of the nodes which renders them identical? The problem of finding a short vector in a lattice is: given a lattice in d dimensions—i.e., the integer combinations of a set of d independent basis vectors—is it possible to efficiently find a vector that is not much longer than the shortest vector in this lattice? This problem becomes hard for large d. Finding a vector with length within a constant factor of the length of the shortest vector is NP-hard, while the best classical polynomial-time algorithms known can only find a vector having length within a factor that is exponential in the dimension d. While neither of these can be solved efficiently by a quantum algorithm yet, the study of the lattice problem from a quantum point of view has led to a purely classical result that puts this problem (with certain parameters) in the complexity class $NP \cap co\text{-}NP$.[11]

If we moderate our goals somewhat, and look also for quantum algorithms that speed problems up by a polynomial factor, then we have not only all the problems in P to consider, but also the NP-complete problems. Grover's algorithm can be applied to speed up the algorithms for many of these problems by a quadratic factor, and it is conceivable that some of them can be sped up by a larger factor. In my paper,[3] I suggested looking at trying to speed up the solutions of problems in P by quantum algorithms, and I still believe this is a good source of research problems.

3. PROGRESS IN QUANTUM ALGORITHMS

Despite the general lack of progress that appears to have been made on quantum algorithms, there have been a number of results which I consider to represent incremental progress which may eventually lead to new quantum algorithms. In my talks, I generally classify known quantum algorithms into three classes: those using periodicity finding, those using variants of Grover search, and those using quantum computers to simulate quantum mechanics. There has been progress in all three areas of this classification, and a couple of new algorithmic techniques have been proposed which appear promising, although they have so far not resulted in any breakthroughs in the discovery of new algorithms. I will now describe some of this progress; I will not attempt to be comprehensive, but merely to give pointers to some papers which I think show the potential for substantial progress.

We first treat the progress in quantum algorithms that use the Fourier transform, the tool that let us perform periodicity finding. It did not take long after the papers[11,12] to realize that a natural generalization of the factoring and discrete logarithm algorithms was to the abelian hidden subgroup problem: the problem of finding a subgroup of an abelian group which is hidden in the values of a function. Fourier transforms on abelian groups could be used to find periodicity and solve this problem in much the same way that the Fourier transform on the cyclic group was used to factor and find discrete logarithms (see, e.g., Ref. 13). Hallgren[14] has recently shown that the Fourier transform can also be used to find the periodicity of functions with irrational periods, and that this is useful in solving certain number theory problems such as finding solutions to Pell's equation and finding class groups of number fields. There are other properties of the Fourier transform which can be used for purposes other than finding periodicity. For instance, shifts of a periodic function transform nicely under the Fourier transform, and this fact can be used to solve certain hidden shift problem.[15] The Fourier transform can also be defined over non-abelian groups. It is not known how to compute this efficiently for all these groups, but it can be computed for some of them, such as the symmetric group[16] and the dihedral group.[17] Kuperberg has recently been able to give an algorithm solving the hidden subgroup problem over the dihedral group in subexponential time by working directly with the representations returned by the Fourier transform.[18] These results indicate at least that the Fourier transform can be used in ways that are more versatile and powerful than those previously considered.

The second class of quantum algorithms I discuss are the generalization of Grover's algorithm for searching a set of N things in time $O(\sqrt{N})$.[2] Many of these are covered in the survey.[19] The most important is probably that of amplitude amplification,[8,19] which lets one amplify the probability of success of a quantum algorithm which has only a small probability of success with efficiency quadratically better than would be possible classically. Recently, there have been a number of algorithms discovered that combine the techniques of Grover's algorithm with quantum walks to perform certain tasks faster than one can do classically.[22–24] It was first shown that quantum random walks could be used to solve some problems faster than classical algorithms could in;[25] these problems, however, appeared fairly artificial. Ambainis combined these with random walk techniques to give a near-optimal time quantum algorithm for testing whether two elements in a database are distinct.[22] These techniques have also been used to show that certain graphs with locality can be searched quickly by a quantum computer, where the computer program has the option either going from a vertex to a neighboring vertex or testing that vertex to see whether it is the "goal" vertex.[24] A survey of these results appears in.[23]

The third class of quantum algorithms are those simulating quantum mechanics. There are two recent papers showing that the simulation of quantum mechanical processes can be used to solve certain classical problems faster than it is known how to do classically. One of these[26] uses the fact that a certain observable in topological quantum field theories has its expectation value equal to the value of the Jones polynomial evaluated at certain points. While the variance of this observable is too large for a quantum computer to compute the value of this Jones polynomial exactly in polynomial time, it can be approximated by a quantum computer much more efficiently than it is known how to do using a classical computer. Another paper along these lines shows how to approximate zeros of certain finite field zeta functions by a quantum computer.[27] Inspired by the spectral approach to the Riemann hypothesis, which attempts to relate the zeros of the Riemann zeta function to the eigenvalues of a (currently unknown) chaotic quantum system, van Dam shows that the zeros of certain finite field zeta functions are given by the eigenvalues of a quantum circuit, and that this fact can be used to approximate them more efficiently on a quantum computer than it is known how to do classically.

Finally, I want to mention adiabatic quantum computation. This heuristic attempts to find the ground state of a Hamiltonian by tracking its evolution by a quantum computer as the Hamiltonian is evolved from one whose ground state is known to one whose ground state is the desired result of the quantum computation.(28) The adiabatic theorem says that this approach is efficient if there exists a spectral gap of size at least reciprocal polynomial for all the intermediate Hamiltonians in the evolution. Although this approach has not yet been shown to yield an algorithm for an interesting problem, it does appear to me to have promise. It has been shown recently that all polynomial time quantum computations can be translated so they can be solved by this adiabatic method.(29)

REFERENCES

1. P. W. Shor, "Polynomial Time Algorithms for Prime factorization and Discrete logarithms on a Quantum Computer,"
2. L. K. Grover, "Quantum Mechanics Helps in Searching for a Needle in a Haystack", *Phys. Rev. Lett.* **78**, 325–328 (1997).
3. P. W. Shor, "Why haven't More Quantum Algorithms been Found?" *J. ACM* **50**, 87–90 (2003); *Siam J. Comput.* **26**, 1484–1509 (1997).
4. S. Cook, "The Complexity of Theorem Proving Procedures," in *Proc. of the 3rd Annual ACM Symposium on Theory of Computing* (ACM Press, New York 1971), pp. 151–158.
5. R. Karp, "Reducibility Among Combinatorial Problems," in (R. Miller, and J. Thatcher), *Complexity of Computer Computations*, (Plenum, NY, 1972), pp. 85–103.
6. L. A. Levin, *Problems of Information Transmission* "Universal Search Problems," **9**(3), 265–266 (1973) [Russian].
7. M. R. Garey and D. S. Johnson, *Computers and Intractability: A Guide to the Theory of NP-Completeness*, (W. H. Freeman and Company, 1979).
8. S. Cook, "The P versus NP Problem," at http://www.claymath.org/millennium/.
9. M. Sipser, "The History and Status of the P Versus NP Question," in *Proc. 24th ACM Symposium on the Theory of Computing*, 1992, pp. 603–619.
10. C. H. Bennett, E. Bernstein, G. Brassard, and U. Vazirani, "Strengths and Weakness of Quantum Computing", *SIAM J. Comput.* **26**, 1510–1523 (1997).
11. D. Aharonov and O. Regev, "Lattice Problems in NP ∩ co-NP," manuscript in preparation, available at www.tau.ac.il/ odedr/.
12. D. R. Simon, "On the Power of Quantum Computation *Siam", J. Comput.* **26**, 1474–1483 (1997).
13. M. Mosca and A. Ekert,."The Hidden Subgroup Problem and Eigenvalue Estimation on a Quantum Computer," in *Proc. of the 1st NASA International Conference on Quantum Computing and Quantum Communication*, Palm Springs, USA, *Lecture Notes in Computer Science*, **1509** (1999); arXiv: quant-ph/9903071.
14. S. Hallgren, "Polynomial-time Algorithms for Pell's Equation and the Principal Ideal Problem," in *Proc. 34th Annual ACM Symposium on Theory of Computing*, (ACM Press, 2002), pp. 653–658.
15. W. van Dam, S. Hallgren, and L. Ip, "Quantum Algorithms for Some Hidden Shift Problems," pp. 489–498. *Proc. ACM-SIAM Symposium on Discrete Algorithms*, 2003

16. R. Beals, "Quantum computation of Fourier Transforms over Symmetric Groups," in *Proc. 29th Annual ACM Symposium on Theory of Computing*, 1997, pp. 48–53.
17. M. Ettinger and E. Höyer, "On Quantum Algorithms for Non-commutative Hidden Subgroups," arXiv: quant-ph/9807029.
18. G. Kuperberg, "A Subexponential-time Quantum Algorithm for the Dihedral Hidden Subgroup Problem," arXiv: quant-ph/0302112.
19. L. K. Grover and A. M. Sengupta, "From Coupled Pendulums to Quantum Search," in R. K. Brylinski and G. Chen, Eds, *Mathematics of Quantum Computation*, (Chapman&Hall/CRC, Boca Raton, FL), pp. 119–134.
20. G. Brassard, P. Höyer, M. Mosca, and A. Tapp, "Quantum Amplitude Amplification and Estimation," *AMS Conterporary Math Series* **305**, 53–74 *Quantum Computation and Information, Amer. Math. Soc.* (2002).
21. L. K. Grover, "Quantum Computers can Search Rapdily by Using Almost any Transformation," *Phys. Rev. Lett.* **80**, 4329–4332 (1998). "Needle in a Haystack", *Phys. Rev. Lett.* **78**, 325–328 (1997).
22. A. Ambainis, "Quantum Walk Algorithm for Element Distinctness," quant-ph/0311001.
23. A. Ambainis, "Quantum Walks and their Algorithmic Applications," quant-ph/0403120.
24. A. Ambainis, J. Kempe, and A. Rivosh, "Coins Make Quantum Walks Faster," quant-ph/0402107.
25. A. M. Childs, R. E. Cleve, E. Deotto, E.Farhi, S. Gutmann, and D. A. Spielman, "Exponential Algorithmic Speedup by Quantum Walk," pp.59–68. *Proc. 35th ACM Symposium on Theory of Computing*, (ACM Press, 2003).
26. M. Bordewich, M. Freedman, L. Lovász, and D. Welsh, "Approximate Counting and Quantum Computation," available at http://research.microsoft.com/research/theory/freedman/.
27. W. van Dam, "Quantum Computing and Zeroes of Zeta Functions," arXiv: quant-ph/0405081.
28. E. Farhi, J. Goldstone, S. Gutman, and M. Sipser, "Quantum Computation by Adiabatic Evolution," arXiv: quant-ph/0001106.
29. D. Aharonov, W. van Dam, J. Kempe, Z. Landau, S. Lloyd, and O. Regev, "Adiabatic Quantum Computation is Equivalent to Standard Quantum Computation," arXiv: quant-ph/0405098.

NMR Quantum Information Processing

Chandrasekhar Ramanathan,[1] Nicolas Boulant,[1] Zhiying Chen,[1] David G. Cory,[1] Isaac Chuang,[2] and Matthias Steffen[2]

Nuclear magnetic resonance (NMR) has provided a valuable experimental test-bed for quantum information processing (QIP). Here, we briefly review the use of nuclear spins as qubits, and discuss the current status of NMR-QIP. Advances in the techniques available for control are described along with the various implementations of quantum algorithms and quantum simulations that have been performed using NMR. The recent application of NMR control techniques to other quantum computing systems are reviewed before concluding with a description of the efforts currently underway to transition to solid state NMR systems that hold promise for scalable architectures.

KEY WORDS: quantum control; quantum simulation; quantum algorithms; solid-state quantum information processing.

PACS: 03.67.–a, 03.67.Lx.

1. INTRODUCTION

Nuclear spins feature prominently in most condensed matter proposals for quantum computing,(1–10) either directly being used as computational or storage qubits, or being important sources of decoherence. Fortunately, the coherent control of nuclear spins has a long and successful history driven in large part by the development of nuclear magnetic resonance (NMR) techniques in biology, chemistry, physics, and medicine.(11,12) The central feature of NMR that makes it amenable to quantum information processing (QIP) experiments is that, in general, the spin degrees of freedom are separable from the other degrees of freedom in the systems studied, both

[1]Department of Nuclear Engineering, Massachusetts Institute of Technology, Cambridge, MA 02139, USA. E-mail: sekhar@mit.edu

[2]MIT Center for Bits and Atoms & Department of Physics, Massachusetts Institute of Technology, Cambridge, MA 02139, USA.

in the liquid and solid state. We can therefore describe the Hamiltonian of the spin system quite accurately; there is an extensive literature on methods to control nuclear spins, and the hardware to implement such control is quite precise.

This readily accessible control of nuclear spins has led to liquid state NMR being used as a testbed for QIP, as well as to preliminary studies of potentially scalable approaches to QIP based on extensions of solid state NMR. The liquid state NMR QIP testbed, although it is not scalable, has permitted studies of control and QIP in Hilbert spaces larger than are presently available with other modalities, and has helped to provide concrete examples of QIP. Here we review what has been learnt in these initial studies and how they can be extended to the solid state where scalable implementations of QIP appear to be possible.

The DiVincenzo criteria(13) for quantum computation provide a natural starting place to understand why NMR is such a good testbed for QIP, and in particular, for implementing quantum algorithms using liquid state NMR techniques. These criteria concern (1) the qubits, (2) the initial state preparation, (3) the coherence times, (4) the logic gates, and (5) the readout mechanism.

1.1. Quantum Bits

Protons and neutrons are elementary particles which carry spin-1/2, meaning that in a magnetic field $\bar{B}$, they have energy $-\bar{\mu} \cdot \bar{B}$, where the magnetic moment $\bar{\mu} = \mu_N \bar{I}$ ($\bar{I}$ is the spin operator) is quantized into two energy states, $|\downarrow\rangle$ and $|\uparrow\rangle$. These states have an energy scale determined by the nuclear Bohr magneton $\mu_N = eh/2m_N \approx 5.1 \times 10^{-27}$ A/m^2 (Table 1). Since spin is inherently a *discrete* quantum property which exists inside a finite Hilbert space, spin-1/2 systems are excellent quantum bits.

Nuclear spins used in NMR QIP are typically the spin-1/2 nuclei of ^{1}H, ^{13}C, ^{19}F, ^{15}N, ^{31}P, or ^{29}Si atoms, but higher order spins such as spin-3/2 and 5/2 have also been experimentally investigated. In liquid-state NMR, these atoms are parts of molecules dissolved in a solvent, such

Table 1. Atoms with spin-1/2 nuclei typically used in NMR, and their energy scales, expressed as a resonance frequency. [frequencies are given for $|\bar{B}| \sim 11.74$ T.]

^{1}H	^{19}F	^{31}P	^{13}C	^{29}Si	^{15}N
500 MHz	470 MHz	202 MHz	125 MHz	99 MHz	50 MHz

that the system is typically extremely well approximated as being $O(10^{18})$ independent molecules. Each molecule is an N-spin system, with N magnetically distinct nuclei. Typically, this molecule sits in a strong, static magnetic field, B_0, oriented along the $\hat{z}$-axis, such that the N spins precess about $\hat{z}$. The spins interact with each other indirectly via inter-atomic electrons sharing a Fermi contact interaction with two (or more) nuclei. The connectivity of the chemical bonds thus determines which nuclei interact.

Since the energy scale of the interactions is weak compared with typical values of $\mu_N|\bar{B}|$, qubits in molecules can be independently manipulated, and provide a natural tensor product Hilbert space structure. This structure is essential for quantum computing, and in particular, system scalability.

1.2. Initial State Preparation

The energy scale of a nuclear spin in typical magnetic fields is much smaller than that of room temperature fluctuations. As seen in Table 1, at 11.74 T the proton has an NMR resonance frequency of 500 MHz, whereas room temperature thermal fluctuations are $k_B T \approx 25$ meV ≈ 6 THz, about 10^4 times larger. As the Boltzmann distribution governs the thermal equilibrium state of the spins

$$\rho = \frac{\exp\left[-\frac{\mu_N|\bar{B}|}{k_B T}\right]}{\mathcal{Z}}, \tag{1}$$

where $\mathcal{Z}$ is the partition function normalization factor, $\rho \approx \mathbf{1}$ for $k_B T \gg \mu_N|\bar{B}|$. Thus, the room temperature thermal equilibrium state is a very highly random distribution, with spins being in their $|\downarrow\rangle$ and $|\uparrow\rangle$ states with nearly equal probability.

Such a highly mixed state is not ordinarily suitable for quantum computation, which ideally works with a system initialized to a fiducial state such as $|00\cdots0\rangle$. It was the discovery of a set of procedures to circumvent this limitation, which made NMR quantum information processing feasible and interesting.[14,15] The essential observation is that a *computational* procedure can be applied to ρ, such that the only observed signal comes from the net excess population in the $|00\cdots0\rangle$ state of the thermal ensemble. One class of such techniques averages away the signal from all other states. This averaging can be performed sequentially in time using sequences of pulses which symmetrically permute undesired states, spatially using magnetic field gradients which prepare spins differently in different parts of a single sample, or by selecting a special subset of spins depending on the logical state ($|0\rangle$ or $|1\rangle$) of the unselected spins. These techniques, known as temporal, spatial, and logical labeling, do not scale well, and only create a signal strength which decreases exponentially with the number of qubits realized.[16]

Another class of techniques is based on efficient compression,[17] and in contrast, the signal strength obtained is constant with increasing number of qubits realized. Indeed, only $O(\text{poly}(n))$ space and time resources are needed to initialize $O(n)$ spins using this method, which has now been experimentally demonstrated,[18] but there is a constant overhead factor which prevents it from being practical until n is large, or the initial temperature of the spins can be brought lower by several orders of magnitude.

1.3. Coherence Times

Nuclear spins couple very weakly with the external world, primarily due to their small magnetic moment, and the weakness of long-range magnetic forces. Thus, typical nuclei in liquid-state molecules may have a T_1 timescale for energy relaxation of between 1 and 30 seconds, and a T_2 timescale for phase randomization of between 0.1 and 10 seconds. Decoherence may occur due to the presence of quadrupolar nuclei such as ^{35}Cl and ^{2}D, chemical shift anisotropies, fluctuating dipolar interactions, and other higher order effects. Though the coherence lifetimes are long, the number of gates that can be implemented is limited by the relatively weak strength of the qubit–qubit couplings (typically a few hundred Hertz at most).

1.4. Logic Gates

In order to perform arbitrary quantum computations only a finite set of logic gates is required, similar to arbitrary classical computations. One such set consists of arbitrary single qubit rotations and the two-qubit controlled-NOT gate. We describe how each of these is implemented.

The Hamiltonian describing a 2-spin system in an external field $B_0\hat{z}$ is (setting $\hbar = 1$)

$$\mathcal{H} = \omega_A I_{zA} + \omega_B I_{zB} + \mathcal{H}_{A,B}\,. \qquad (2)$$

Here, I_z is the spin angular momentum operator in the $\hat{z}$ direction, and $\omega_i = -\gamma_i(1-\alpha_i)B_0$, where γ_i is the gyromagnetic ratio for spin i, which depends on the nuclear species, α_i is the shielding constant, and $\mathcal{H}_{A,B}$ is the spin–spin coupling. The shielding constant depends on the local chemical environment of the nuclei, which shields the magnetic field, resulting in a shift in frequency by an amount known as the chemical shift such that spins of the same type (e.g. protons) can have different resonance frequencies. That spins have different resonance frequencies is an important requirement because it permits frequency-dependent addressing of single qubits.

Spins are manipulated by applying a much smaller radio-frequency (RF) field, B_1, in the $\hat{x}$–$\hat{y}$ plane to excite the spins at their resonance frequencies ω_i. In the rotating frame, to good approximation, the spin evolves under an effective field $\bar{B} = B_1 \cos(\phi)\hat{x} + B_1 \sin(\phi)\hat{y}$ (where ϕ is the RF phase). The rotation angle and axis (in the transverse plane) can be controlled by varying ϕ, the magnitude of B_1 and the duration of the RF. Since it is possible to generate arbitrary rotations about the $\hat{z}$-axis using combinations of rotations about the $\hat{x}$- and $\hat{y}$-axis, it is possible to implement arbitrary single-qubit rotations using RF pulses.

Two-qubit gates, such as the controlled-NOT gate require spin–spin interactions. These occur through two dominant mechanisms; direct dipolar coupling, and indirect through-bond interactions. The dipolar coupling between two spins is described by an interaction Hamiltonian of the form

$$\mathcal{H}^D_{A,B} = \frac{\gamma_A \gamma_B}{r^3}\left(\bar{I}_A \cdot \bar{I}_B - 3(\bar{I}_A \cdot \hat{n})(\bar{I}_B \cdot \hat{n})\right), \tag{3}$$

where $\hat{n}$ is the unit vector in the direction joining the two nuclei, and $\bar{I}$ is the magnetic moment vector. While dipolar interactions are rapidly averaged away in a liquid, they play a significant role in liquid crystal[19,20] and solid state NMR QIP experiments. Through-bond electronic interactions are an indirect interaction, also known simply as the scalar coupling, and take on the form

$$\mathcal{H}^J_{A,B} = 2\pi J \bar{I}_A \cdot \bar{I}_B = 2\pi J I_{zA} I_{zB} + \pi J \left(I_{A+} I_{B-} + I_{A-} I_{B+}\right), \tag{4}$$

where J is the scalar coupling constant. This interaction is often resolved in liquids. For heteronuclear species (such that the matrix element of the $I_{A+}I_{B-} + I_{A-}I_{B+}$ term is small, when $2\pi J \ll \omega_A - \omega_B$), the scalar coupling reduces to

$$\mathcal{H}^J_{A,B} \approx 2\pi J I_{zA} I_{zB}. \tag{5}$$

Multiple-qubit interactions, such as the controlled-NOT (CNOT) operation, may be performed by inserting waiting periods between pulses so that J-coupled evolution can occur. For the J-coupled two-spin system, a CNOT can be implemented as a controlled phase shift preceded and followed by rotations, given by the sequence $C_{AB} = R_{yA}(270 = -90) R_{zB} R_{zA}(-90) R_{zAB}(180) R_{yA}(90)$. This is shown schematically in Fig. 1.

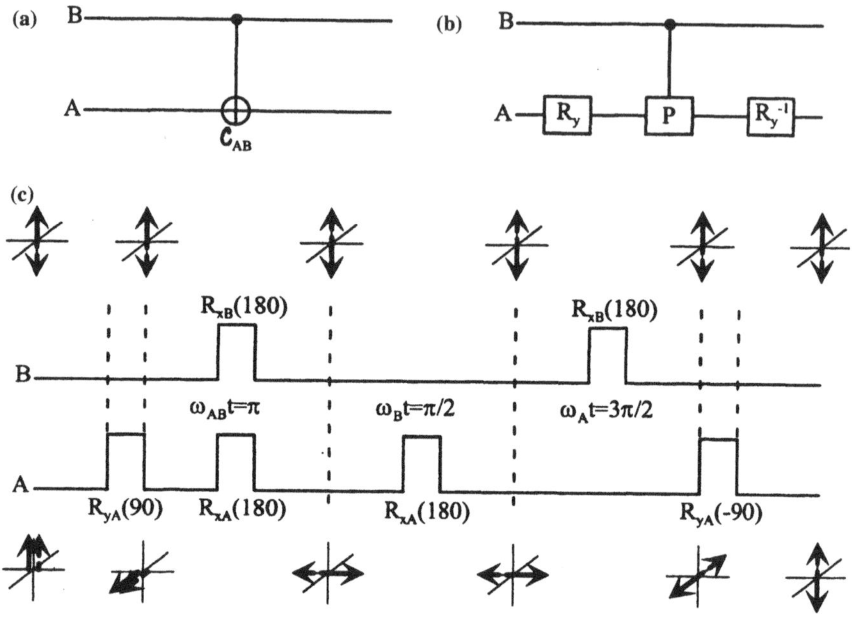

Fig. 1. (a) A controlled-NOT gate acting on two qubits, (b) the controlled-NOT gates implemented by a controlled phase shift gate (specified by a unitary matrix with diagonal elements $\{1, 1, 1, -1\}$) preceded and followed by $\pi/2$ rotations, and (c) the pulse sequence and spin orientations corresponding to the components in (b). Note that, unlike a conventional NMR selective population transfer sequence, extra refocusing is required to preserve the Bell basis exchange symmetry between A and B. The $\hat{z}$ rotations are implemented via $\hat{x}$ and $\hat{y}$-rotations, which are not explicitly shown.

In summary, one and two-qubit gates are implemented by applying a sequence of RF pulses interlaced with waiting periods. In this sense, it is perhaps interesting to note that the sequence of elementary instructions (pulses and delay times) are the machine language of the NMR quantum information processor.

1.5. Readout Mechanism

Readout of the quantum state in NMR QIP is not the usual ideal projective von Neumann measurement. Instead, the system is continually read out by the weak coupling of the magnetic dipole moments to an external pickup coil, across the ends of which a voltage is produced by Faraday induction. This coil is usually the same coil as that used to produce the strong RF pulses which control the spins, and thus it only detects magnetization in the $\hat{x}$–$\hat{y}$ plane. The induced voltage, known as the free induction decay, may be expressed as

$$V(t) = -2V_0 \mathrm{tr}\left[e^{-i\mathcal{H}t}\rho e^{i\mathcal{H}t}(iI_x^k + I_y^k)\right], \tag{6}$$

where $\mathcal{H}$ is the Hamiltonian for the spin system, I_x^k and I_y^k operate only on the kth spin, and V_0 is a constant factor dependent on coil geometry, quality factor, and maximum magnetic flux from the sample volume.

The Fourier transform of $V(t)$ is the NMR spectrum as shown in Fig. 2 for example. When properly calibrated, the NMR spectrum immediately reveals the logical state of qubits which are either $|0\rangle$ or $|1\rangle$. Specifically, for example, if the initial state of a two-spin ^{1}H–^{13}C system is described by a diagonal density matrix,

$$\rho = \begin{bmatrix} a & 0 & 0 & 0 \\ 0 & b & 0 & 0 \\ 0 & 0 & c & 0 \\ 0 & 0 & 0 & d \end{bmatrix} \tag{7}$$

(where the states are 00, 01, 10, and 11, with proton on left and carbon on right and a, b, c, and d denote the occupation probabilities) then after a $R_x(\pi/2)$ readout pulse, the integrals of the two proton peaks (in the proton frequency spectrum) are given by $a - c$ and $b - d$, and the integrals of the two carbon peaks are given by $a - b$ and $c - d$. Both the proton and carbon spectra contain two peaks because of the J-coupling interaction during the measurement period.

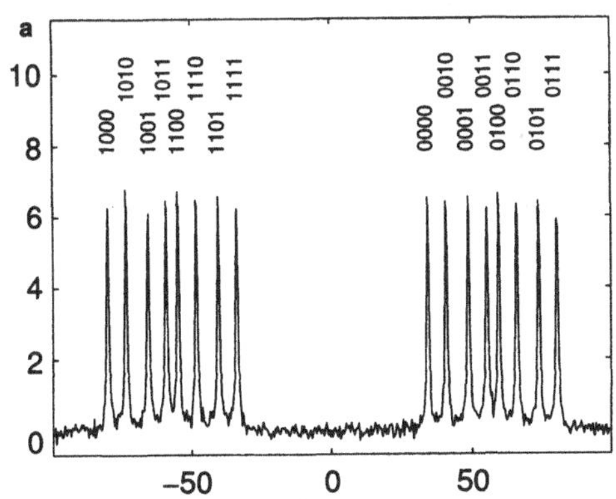

Fig. 2. Thermal equilibrium spectrum of a 5 spin molecule. Each peak corresponds to a certain logical state of the remaining four spins, which is indicated by the binary numbers. The real part of the spectrum is shown, in arbitrary units. Frequencies are given with respect to $\omega_i/2\pi$, in Hertz.(21)

One important issue in the readout of QIP results from NMR arises because the system is an *ensemble*, rather than a single N-spin molecule. The problem is that the output of a typical quantum algorithm is a random number, whose distribution gives information which allows the problem to be solved. However, the average value of the random variable would give no relevant information, and this would be the output if the quantum algorithm were executed without modification on an NMR quantum computer.

This problem may be resolved[15] by appending an additional computational step to the quantum algorithm to eliminate or reduce the randomness of its output. For example, the output of Shor's algorithm is a random rational number c/r, from which classical post-processing is usually employed to determine a number r, which is the period of the modular exponentiation function under examination. However, the post-processing can just as well be performed on the quantum computer itself, such that r is determined on each molecule separately. From r, the desired prime factors can also be found, and tested; only when successful does a molecule announce an answer, so that the ensemble average reveals the factors. Similar modifications can be made to enable proper functioning of all known exponentially fast quantum algorithms[22–24] on an NMR quantum computer.

2. QUANTUM CONTROL

Implementing an algorithm on a quantum computer requires performing both unitary transformations and measurements. Errors in the control and the presence of noise can severely compromise the accuracy with which a unitary transform can be implemented. Quantum control techniques are used to maximize the accuracy with which such operations can be performed, given some model for the system's dynamics. NMR has provided valuable insight into the design of schemes to control quantum systems, as the task of applying pulse sequences to perform operations that are selective, and also robust against experimental imperfections, has been the subject of extensive studies.[25,26] A single, isolated quantum system will evolve unitarily under the Hamiltonian of the system. In an ensemble measurement (whether in space or time), an isolated system can also appear to undergo non-unitary dynamics, called incoherent evolution, due to a distribution of fields over the ensemble.[27,28] An open quantum system, interacting with an environment, will decohere if these interactions are not controlled. The study of quantum control using NMR can therefore be separated into three different subsections:

- *Coherent control*: how can one design RF control schemes to implement the correct unitary dynamics for a single, isolated quantum system?
- *Incoherent noise*: how can coherent control be extended to an ensemble, given that the system Hamiltonian will vary across the ensemble?
- *Decoherent noise*: how can one achieve the desired control, when coupling to the environment is taken into account?

2.1. Coherent Control

The density matrix of a closed system evolves according to the Liouville–Von Neumann equation of motion:

$$\frac{d\rho}{dt} = -i\,[\mathcal{H}_{\text{int}} + \mathcal{H}_{\text{ext}}, \rho], \tag{8}$$

where $\mathcal{H}_{\text{int}}$ is the internal Hamiltonian of the system of qubits, and $\mathcal{H}_{\text{ext}}$ the externally applied control fields. More specifically, extending Eqs. (2) and (4), the internal Hamiltonian for a system of N spin-1/2 nuclei in a large external magnetic field B_0 is

$$\mathcal{H}_{\text{int}} = \sum_{k=1}^{N} -\gamma_k(1-\alpha_k)B_0(r)I_z^k + 2\pi\sum_{j>k}^{N}\sum_{k=1}^{N} J_{kj}\bar{I}^k\cdot\bar{I}^j, \tag{9}$$

where $-\gamma(1-\alpha_k)B_0(\mathrm{r})$ is the chemical shift of the kth spin. The corresponding experimentally controlled RF Hamiltonian is

$$\mathcal{H}_{\text{ext}} = \sum_{k=1}^{N} -\gamma_k f(r)B_{\text{RF}}(t)e^{-i\phi(t)I_z^k}I_x^k e^{i\phi(t)I_z^k}, \tag{10}$$

where the time-dependent functions $B_{\text{RF}}(t)$ and $\phi(t)$ specify the applied RF control field, while $f(r)$ reflects the distribution of RF field strengths over the sample. The spatial variation of the static and RF magnetic fields leads to incoherence.[27] We will return to this in the next section.

If the total Hamiltonian is time-independent, possibly through transformation into a suitable interaction frame, the equation of motion can be integrated easily and yields a unitary evolution of the density matrix $\rho(t) = U(t)\rho(0)U^{\dagger}(t)$. Given an internal Hamiltonian and some control resources, how can we implement a given propagator or prepare a given state? In QIP, it is necessary to implement the correct propagator, which requires designing gates that perform the desired operation regardless of the input state. We will therefore focus our discussion on NMR control techniques that are universal, i.e. whose performance is essentially independent of the input state, although sequences whose performances are state dependent can also be useful for initialization purposes. Average Hamiltonian theory is a powerful tool for coherent control that was initially developed for NMR.(29) Waugh and Harberlen have provided a theoretical framework to implement a desired effective Hamiltonian evolution of a spin system over some period of time. Such a tool fits well into the context of QIP since it aims to implement the correct propagator over the system Hilbert space while refocusing undesired qubit–qubit interactions. The basic idea is to apply a cyclic train of pulses $P=\{P_j\}_{j=1}^{M}$ with $\prod P_j = \mathbf{1}$ to the system which, in its simplest form, are assumed to be infinitely short and equally spaced by $\Delta t > 0$. The net controlled evolution over the period $T = M\Delta t$ can then be expressed as

$$e^{-i\bar{H}T} = \prod_{k=0}^{M} e^{-iH_k\Delta t}, \tag{11}$$

where the "toggling-frame" Hamiltonians $H_k = U_k^{\dagger} H_k U_k$ are expressed in terms of the composite pulses $U_k = \prod_{j=1}^{k} P_j, k = 1, \ldots, M, U_0 = \mathbf{1}$. Any average Hamiltonian (up to a scalar multiple) can be implemented in NMR systems of distinguishable spins.(30) This work has been extended to correct for some experimental imperfections or uncertainties primarily using symmetry arguments. Composite pulses have also been used to design robust control sequences as they can be designed to be self-compensating for small experimental errors.(25,26,31–36)

Strong modulation of the spin system currently represents the state of the art in performing selective, controlled operations in large Hilbert spaces (up to 10 qubits). Strong modulation of the spins permits accurate selective rotations while refocusing the internal Hamiltonian during the RF irradiation of the spins.(37) Figure 3 shows the dependence of the fidelities achievable on the available control resources, using this technique.(27) These are simulation results, assuming only unitary dynamics. The fidelities are seen to improve both with improved distinguishability of the spins (higher field strengths), and stronger modulation (increased RF power). The drawback of this technique is that it is not scalable, as the time necessary to find the numerical solutions grows exponentially with the number of qubits (see(38) for a review of other techniques). An alternative, scalable approach that relies on optimization over single and pairs of nuclei has been described,(39) though it does not appear to perform as well.

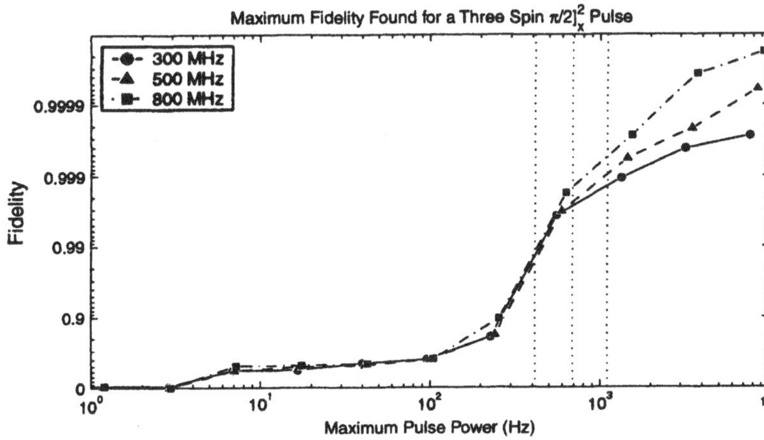

Fig. 3. Maximum fidelity achievable versus external magnetic field strengths and maximum radio-frequency power available for strongly modulating pulses. The frequencies shown are the proton Larmor frequencies at different magnetic fields (300 MHz ≡ 7 T, 500 MHz ≡ 11.7 T, 800 MHz ≡ 18.8 T).

Finally, there is a growing body of work in the area of optimal control theory for quantum systems,(40–42) which has also used NMR as an experimental testbed. It is foreseeable that it might be possible to combine the ideas presented in these studies with strong modulation and pulse-shaping techniques to design optimal control sequences, given some knowledge about the system decoherence and the control parameters.

2.2. Incoherent Noise

Incoherent noise arises from a spatial or temporal distribution of experimental parameters in an ensemble measurement.(27) It is manifested in NMR in the spatial dependence of the Hamiltonian $\mathcal{H}$. The density matrix at a given location in the sample still obeys the Liouville–Von Neumann equation where the internal and external Hamiltonian are now spatially dependent. Since it is the spatially averaged density matrix that is measured, the apparent evolution of the ensemble system is non-unitary and yields the following operator sum representation(43) of the superoperator

$$\rho(t) = \sum_i p_i U_i(t) \rho(0) U_i^{\dagger}(t), \tag{12}$$

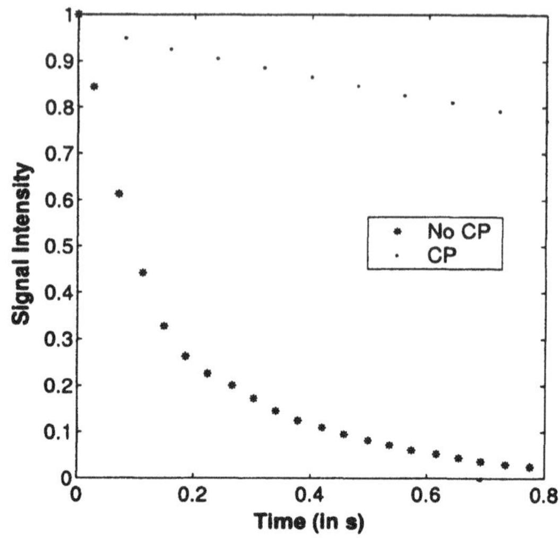

Fig. 4. Decay of the transverse magnetization of a nuclear spin in a liquid state sample with large B_0 field inhomogeneities both with and without a Carr–Purcell (CP) sequence.

where U_i is a unitary operator and p_i represents the fraction of spins that experience a given U_i ($\sum_i p_i = 1$). This incoherent evolution can be counteracted using a different set of techniques than those used to deal with the decoherent errors to be discussed in the next section.

Hahn's pioneering work showed that inhomogeneities in the Hamiltonian could be refocused during an experiment if an external control Hamiltonian orthogonal to the inhomogeneous Hamiltonian was available[(44)] (see Fig. 4). In the case of an inhomogeneous but static Hamiltonian, the correlation time of the noise is infinite. This work was extended by Carr and Purcell to counteract long, but not infinite, correlation time noise fluctuations due to molecular diffusion.[(45)] Composite pulse sequences have also been used to counteract the effects of incoherent processes,[(31–36,46,47)] but often assume specific input states, or still need to be proven effective over the full Hilbert space of multi-qubit systems. Spin decoupling fits into a similar framework, and provides a means of modulating the system in order to average out unwanted interactions with the environment.[(48)] It has inspired coherent approaches to other control problems for error correction purposes.[(49–51)]

The use of strongly modulating pulses has been extended to incorporate incoherent effects, considering local unitary dynamics over the ensemble.[(27)] *A priori* knowledge of the inhomogeneity of the external Hamiltonian was used to find robust pulse sequences yielding a higher fidelity operation over the ensemble. This knowledge was easily obtained by spectroscopic NMR techniques. Though this work focused on counteracting the main source of incoherent errors in an NMR experiment, i.e. RF inhomogeneity, it could easily be extended to compensate for experimental uncertainties such as the phase noise of the RF irradiation or static B_0 field inhomogeneities.

2.3. Decoherent Noise

If the coupling between the system and the environment is weak enough and the correlation time of the noise is short, the evolution of the system is Markovian and obeys the following master equation:

$$\frac{d|\rho(r)\rangle}{dt} = -\left(i\hat{H}(r) + \Gamma\right)|\rho(r)\rangle, \tag{13}$$

where Γ is the Liouville space relaxation superoperator.(12) This equation yields a non-unitary evolution of the density matrix so that pure states can evolve into mixed states. To understand and test models of decoherence, methods based on quantum process tomography (QPT) were developed to measure Γ,(52,53) so that the dynamics of the system could be simulated more accurately. The full model of the system including coherent, incoherent and decoherent dynamics, has been tested extensively with a three qubit QPT of the Quantum Fourier Transform superoperator.(54) When knowledge about the noise operators is available, quantum error correction (QEC) schemes can in principle be designed to allow quantum computing in the presence of imperfect control. NMR has primarily been used to test the ideas of quantum error correction (QEC)(55–57) and of fault-tolerant quantum computations.(58) Experiments were carried out to investigate different QEC scenarios,(59–64) in addition to encoded operations acting on logical qubits.(65) Schemes to implement logical encoded quantum operations while remaining in a protected subspace have also been investigated(61) for a simple system made of two physical qubits and are still being studied for larger systems.

3. QUANTUM ALGORITHMS

Many quantum algorithms have now been implemented using liquid-state NMR techniques. The first quantum algorithms implemented with NMR were Grover's algorithm(66,67) and the Deutsch–Jozsa algorithm(68,69) for two qubits. The quantum counting algorithm was implemented soon afterwards using two qubits.[70] A variety of implementations of Grover's algorithm and the Deutsch–Jozsa algorithm have subsequently been performed. The two-qubit Grover search was re-implemented using a subsystem of a three qubit system,[71] demonstrating state preparation using logical labeling. A three qubit Grover search has been implemented, in which 28 Grover iterations were performed, involving 280 two-qubit gates.[72] A three-qubit Deutsch–Jozsa algorithm using transition selective pulses,[73] another more advanced version using SWAP gates to avoid small couplings,[74] and yet another implementation without SWAP gates.[75] A subset of a five-qubit Deutsch–Jozsa algorithm has also been implemented.[76]

The implementation of quantum algorithms reached a new level with the full implementation of a Shor-type algorithm using five qubits.[21] This work involved the use of exponentiated permutations, combined with the quantum Fourier transform, which had been previously been implemented.[77]

The most complex quantum algorithm realized to-date is the demonstration of Shor's algorithm using liquid-state NMR QIP methods: In this work,[78] a seven-qubit molecular system was used to factor the number 15 into its prime factors. This molecule, shown in Fig. 5, was specially chemically synthesized to give resolvable fluorine spectra, in which the two ^{13}C nuclei, and the five ^{19}F nuclei, could each be addressed independently because of the spread of their resonant frequencies. The NMR spectra of this molecule are quite remarkable; for example, the thermal spectrum of ^{19}F spin number 1 shows 64 lines, corresponding to the random states of the other 6 spins (Fig. 6). Several hundred pulses were applied, with a wide variety of phases, and shapes, at seven different frequencies, in this demonstration of the factoring algorithm (Fig. 7). A comparison of the experimental results with numerical simulations suggests that decoherence was the major source of error in the experiment rather than errors in the unitary control, which is remarkable considering the number of pulses applied.

Most recently, among NMR implementations of quantum algorithms, has been the realization of a three-qubit adiabatic quantum optimization algorithm.[79] In this work, the ^{1}H, ^{13}C, and ^{19}F nuclei in molecules of bromofluoromethane were used as qubits, and the solution to a combinatorial problem, MAXCUT, was obtained, using an optimization algorithm proposed by Farhi and Goldstone[80] and Hogg.[81] This algorithm is notable because it is fundamentally different in nature from Shor-type quantum algorithms, and may obtain useful speedups for a wide variety of optimization problems.

i	$\omega_i/2\pi$	$T_{1,i}$	$T_{2,i}$	J_{7i}	J_{6i}	J_{5i}	J_{4i}	J_{3i}	J_{2i}
1	−22052.0	5.0	1.3	−114.3	25.2	6.6	14.5	−221	37.7
2	−4918.7	10.0	1.7	79.9	2.1	12.9	54.1	−5.7	
3	489.5	13.7	1.8	12.5	3.9	18.6	−3.9		
4	15186.6	2.8	1.8	41.6	19.4	59.5			
5	25088.3	3.0	2.5	1.0	−13.5				
6	4244.3	31.6	2.0	68.9					
7	−4519.1	45.4	2.0						

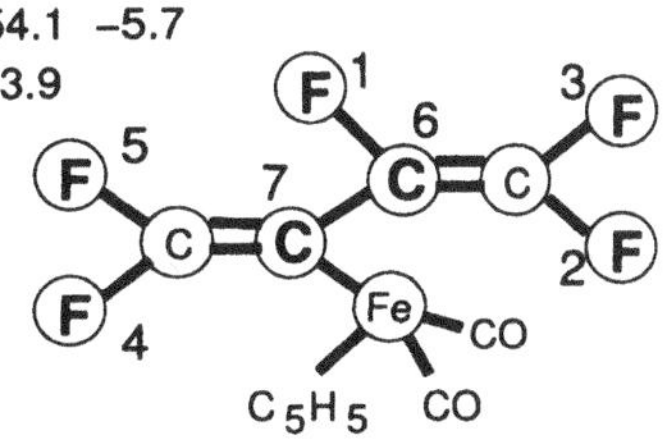

Fig. 5. The seven spin molecule used in the quantum factoring NMR experiment, showing its J-coupling constants, T_1 and T_2 relaxation times (in seconds), and chemical shifts (in Hertz) at 11.74 T.

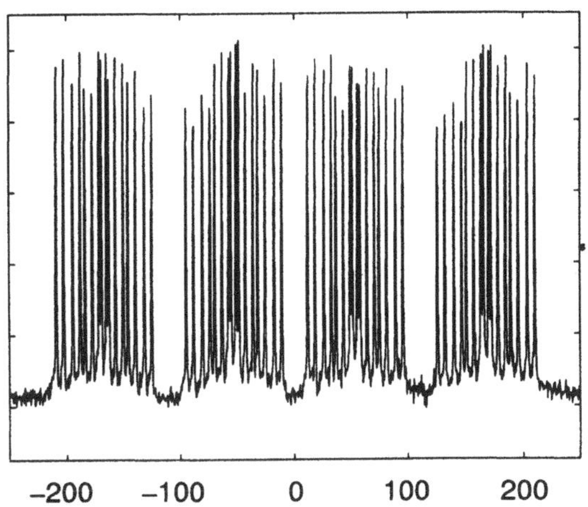

Fig. 6. Experimentally measured thermal equilibrium spectra of the NMR spectrum of fluorine atom number 1, in the molecule of Fig. 5. The real part of the spectrum is shown, in arbitrary units. Frequencies are given with respect to $\omega_i/2\pi$, in Hertz.

3.1. Other Quantum Protocols

The Greenberger–Horne–Zeilinger state and its derivative entangled states of three particles have been studied as well. First, an effective-pure GHZ state was prepared,[82] and later a similar experiment was done with seven spins.[39] The claim of having created entangled states was later refuted based on the fact that spins at room temperature are too mixed to be entangled.[83] GHZ correlations have since been further studied on mixed states.[84]

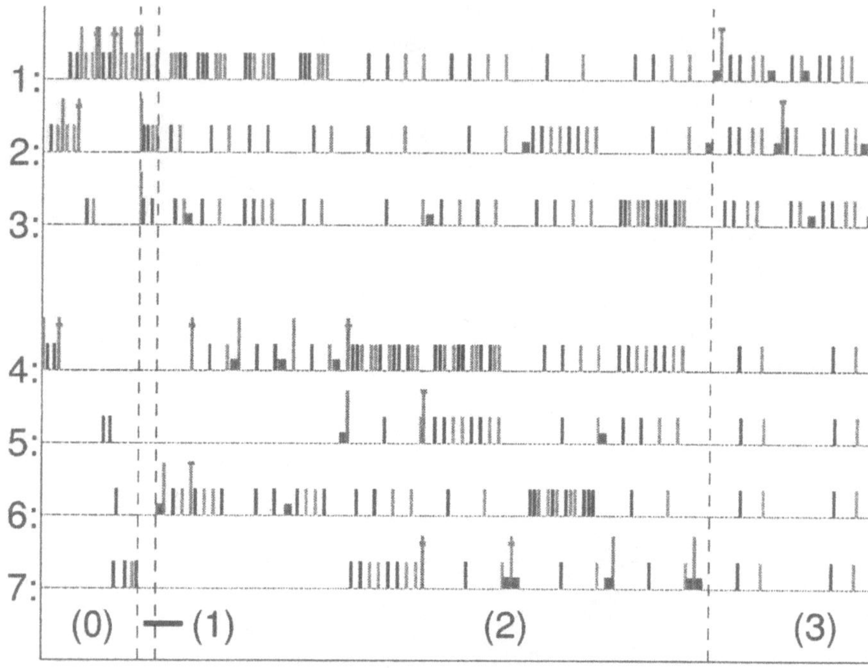

Fig. 7. Pulse sequence for implementing Shor's algorithm to factor $N = 15$ (for case $a = 7$), using seven qubits. The four steps 0 through 3 correspond to different steps in Shor's algorithm. The tall lines represent 90° pulses selectively acting on one of the seven qubits (horizontal lines) about positive $\hat{x}$ (no cross), negative $\hat{x}$ (lower cross) and positive $\hat{y}$ (top cross). Note how single 90° pulses correspond to Hadamard gates and pairs of such pulses separated by delay times correspond to two-qubit gates. The smaller lines denote 180° selective pulses used for refocusing, about positive (darker shade) and negative $\hat{x}$ (lighter shade). Rotations about $\hat{z}$ are denoted by even smaller and thicker lines and were implemented with frame-rotations. Time delays are not drawn to scale. The vertical dashed black lines visually separate the steps of the algorithm; step (0) shows one of the 36 temporal averaging sequences.

A quantum teleportation protocol was implemented using three qubits.(85) Superdense coding has been realized,(86) and an approximate quantum cloning experiment has been implemented (an unknown quantum state cannot be perfectly copied; it can only be approximately cloned).(87) The quantum Baker's map has also been implemented.(88)

Several experiments have been performed in an attempt to increase the thermal polarization of nuclear spins in liquid solution, as this poses a significant challenge for scaling NMR quantum computers to many qubits. An algorithm approach implementing the basic building block of the Schulman–Vazirani cooling scheme has been demonstrated.(18) High initial polarization of the proton and carbon in a chloroform molecule have been obtained by transfer from optically pumped rubidium, through hyperpolarized xenon, and a two-qubit Grover search implemented on this non-thermally polarized system.[89] In a different approach[90] *para*-hydrogen was transformed into a suitable molecule leading to a polarization of 10% which is much larger than the thermal polarization of $O(10^{-4})$. A quantum algorithm was subsequently performed on this molecule. Most recently, a two-spin system was initialized to an effective purity of 0.916 by chemically synthesizing a two-spin molecule using highly polarized *para*-hydrogen.[91]

4. QUANTUM SIMULATION

In 1982, Feynman recognized that a quantum system could efficiently be simulated by a computer based on the principle of quantum mechanics rather than classical mechanics.[92] This is perhaps one of the most important short term applications of QIP. An efficient quantum simulator will also enable new approaches to the study of multibody dynamics and provide a testbed for understanding decoherence.

A general scheme of simulating one system by another is expressed in Fig. 8. The goal is to simulate the evolution of a quantum system S using a physical system P. The physical system is related to the simulated system via an invertible map ϕ , which creates the correspondence of states and propagators between the two systems. In particular, the propagator U in the system S is mapped to $V = \phi U \phi^{-1}$. After the evolution of the physical system from state p to p_T, the inverse map brings it back to the final state $s(T)$ of the simulated system.

The first explicit experimental NMR realization of such a scheme was the simulation of a truncated quantum harmonic oscillator (QHO).[93] The

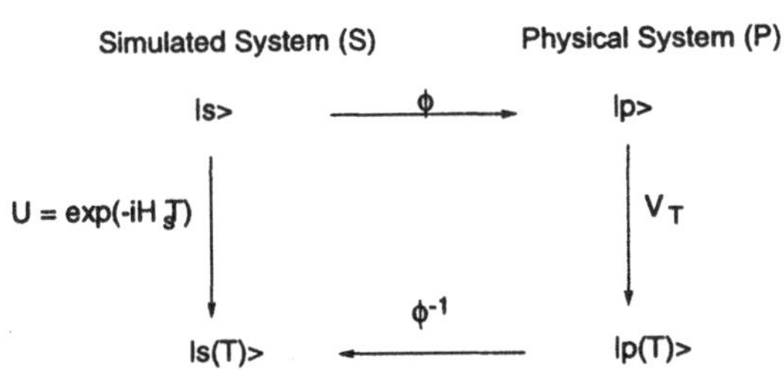

Fig. 8. Correspondence between the simulated and physical system. The initial state s evolves to $s(T)$ under the propagator U. This process is related to the evolution of state p in the physical system by an invertible map ϕ.

states of the truncated QHO were mapped onto a two-qubit system as follows

$$\begin{aligned} |n=0\rangle &\leftrightarrow |0\rangle|0\rangle \equiv |00\rangle \\ |n=1\rangle &\leftrightarrow |0\rangle|1\rangle \equiv |01\rangle \\ |n=2\rangle &\leftrightarrow |1\rangle|0\rangle \equiv |10\rangle \\ |n=3\rangle &\leftrightarrow |1\rangle|1\rangle \equiv |11\rangle . \end{aligned} \tag{14}$$

The propagator of the truncated QHO

$$U = \exp\left\{-i\left(\frac{1}{2}|0\rangle\langle 0| + \frac{3}{2}|1\rangle\langle 1| + \frac{5}{2}|2\rangle\langle 2| + \frac{7}{2}|3\rangle\langle 3|\right)\Omega T\right\} \tag{15}$$

(Ω is the oscillator frequency) was mapped onto the following propagator of a two-spin system

$$V_T = \exp\left\{i\left(2I_z^2\left(1+I_z^1\right)-2\right)\Omega T\right\}. \tag{16}$$

Implementing this propagator on the 2-spin system simulates the truncated QHO as shown in Fig. 9.

Quantum simulation however is not restricted to unitary dyamics. It is sometimes possible to engineer the noise in a system to control the decoherence behavior and simulate non-unitary dynamics of the system.[94]

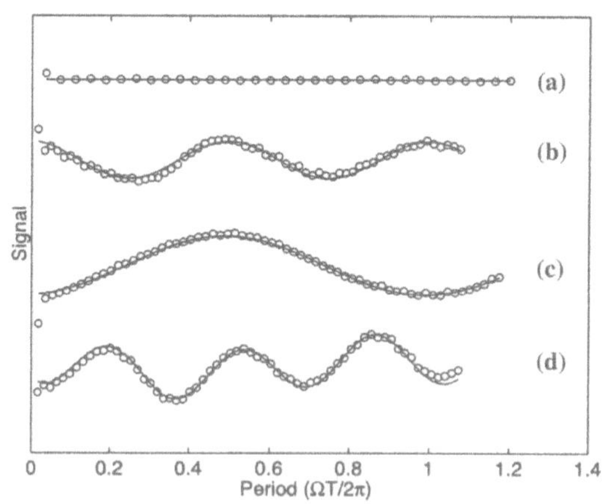

Fig. 9. NMR signals demonstrate a quantum simulation of truncated harmonic oscillator. The solid lines are fits to theoretical expectations. Evolution of the different initial states are shown: (a) evolution of $|0\rangle$ with no oscillation (b) evolution of $|0\rangle + i|2\rangle$, showing 2Ω oscillations (c) evolution of $|0\rangle + |1\rangle + |2\rangle + |3\rangle$, showing Ω oscillation and (d) 3Ω oscillations.

Simple models of decoherence have been shown using a controlled quantum environment in order to gain further understanding about decoherence mechanisms. In one model,[95] the environment is taken to be a large number of spins coupled to a single system spin so that the total Hamiltonian can be expressed as

$$\mathcal{H}=\omega_1 I_z^1+\sum_{k=2}^{N}\omega_k I_z^k+2\pi\sum_{k=2}^{N}J_{1k}I_z^1 I_z^k \tag{17}$$

corresponding to the system, the environment, and the coupling between the system and the environment, respectively (the couplings within the environment were omitted here for simplicity). Note that the form is identical to the weak coupling Hamiltonian of a liquid state NMR sample presented in the previous sections. However, the number of spins in a typical QIP NMR molecule is small, which makes the decoherence arising from the few "system–environment" couplings rather ineffective, as the recurrence time due to a small environment is relatively short. This can be circumvented by using a second "classical" environment which interacts with a smaller quantum environment (see Fig. 10 for an illustration of the model).[96]

In this model, following the evolution of the system and the small quantum environment, a random phase kick was applied to the quantum environment. This has the effect of scrambling the system phase informa-

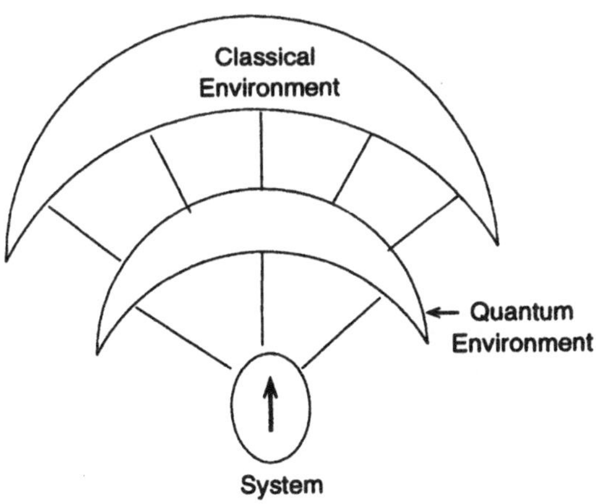

Fig. 10. Basic model for the system, local quantum and classical environments.

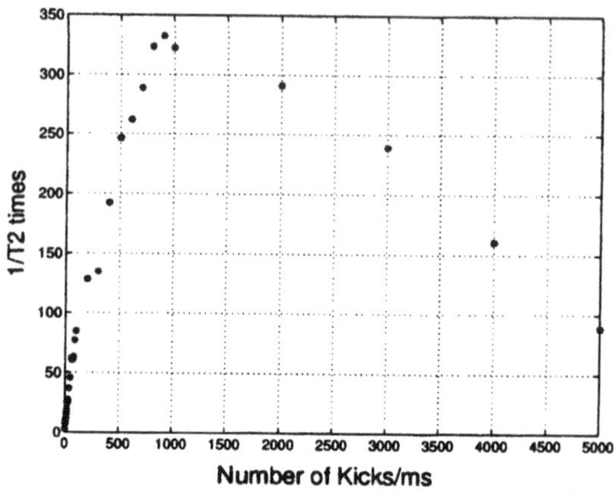

Fig. 11. Simulation showing the dependence of the decay rate on the kick rate, and the onset of the decoupling limit. Beyond 900 kicks/ms the decay rate decreases.

tion stored in the environment during the coupling interaction and therefore emulates the loss of memory. When the kick angles are averaged over small angles, the decay induced by the kicks is exponential and the rate is linear in the number of the kicks.[96] As the kick angles are completely randomized over the interval from 0 to 2π, a Zeno type effect is observed. Figure 11 shows the dependence of the decay rate on the kick frequency: the decay rate initially increases to reach a maximum and then decreases, thereby illustrating the motional narrowing[11] or decoupling[48] limit. This NMR-inspired model thus provides an implementation of controlled decoherence yielding both non-exponential and exponential decays (with some control over the decay rates), and can be extended to investigate other noise processes.

A type-II quantum computer is a hybrid classical/quantum device that can potentially solve a class of classical computational problems.[97,98] It is essentially an array of small quantum information processors sharing information through classical channels. Such a lattice of parallel quantum information processors can be mapped onto a liquid state NMR system by mapping the lattice sites of the quantum computer onto spatial positions in the nuclear spin ensembles. The implementation of a type-II quantum computer using NMR techniques has been demonstrated in solving the diffusion equation.[99] The experimental results show good agreement with both the analytical solutions and numerical NMR simulations. The spatial separation of the different lattice sites in the ensemble allows one to address all the lattice sites simultaneously using frequency selective RF pulses and a magnetic field gradient. This yields a significant savings in time compared to schemes where the sites have to be addressed individually.

5. CONTRIBUTION TO OTHER QC SYSTEM

NMR QIP studies have contributed significantly to enabling quantum computation with other physical systems. Fundamentally, this has been because of the exquisite level of control achievable in NMR, which remains unrivaled. Several of these contributions are briefly summarized below; a complete discussion is available in the literature.[38]

5.1. Composite Pulses: Trapped Ions

The use of *composite pulses* has been an important contribution of NMR to QIP. A single, imperfect pulse is replaced by a sequence of pulses which accomplishes the same operation with less error. Historically, in the art of NMR, such sequences were first invented to compensate for apparatus imperfections, such as frequency offsets and pulse amplitude miscalibrations. For example, the machine may perform rotations

$$\tilde{R}_{\hat{n}}(\theta) = \exp\left[-i(1+\epsilon)\hat{n}\cdot\vec{I}\right], \tag{18}$$

where ϵ is an unknown, systematic pulse amplitude error. Ideally, ϵ is zero, but in practice, it may vary geometrically across a sample, or slowly, with time. Using average gate fidelity as an error metric, this pulse can be shown to have error which grows quadratically with ϵ. In comparison, consider the sequence

$$BB1_\theta = \tilde{R}_\phi(\pi)\tilde{R}_{3\phi}(3\pi)\tilde{R}_\phi(\pi)\tilde{R}_x(\theta), \tag{19}$$

where $\tilde{R}_\phi(\cdot)$ denotes a rotation about the axis $[\cos\phi, \sin\phi, 0]$, and the choice $\phi = \cos^{-1}(-\theta/4\pi)$ is made. This sequence, introduced by Wimperis,[100] gives average gate fidelity error $\sim 21\pi^6\epsilon^6/16384$, which is much better than the $O(\epsilon^2)$ for the single pulse, even for relatively large values of ϵ. Generalizations and extensions of this technique can help correct not just systematic single qubit gate errors, but also coupled gate errors.[25,26]

NMR composite pulses have also recently been successfully employed in quantum computation with trapped ions. In an experiment with a single trapped ^{40}Ca ion, a sequence of $O(10)$ laser pulses was performed using a variety of phases, to implement a proper swap operation between the internal atomic state and the motional state of the ion, and a controlled-phase gate. These steps allowed the full two-qubit Deutsch–Jozsa algorithm to be implemented.[101] Composite pulses have also been used in superconducting qubits demonstrating robustness against detuning in a quantronium circuit.[102]

5.2. Shaped Pulses: Superconducting Qubits

NMR also widely employs shaped pulses to achieve desired control excitations of the spins. Typically, this shaping is performed in the amplitude and phase domain. One goal of this method, for example, is to achieve narrow excitation bandwidths. To first order, the excited bandwidth is the Fourier transform of the temporal width of the pulse. However, because of the non-linear *Bloch* response of the spins to the RF excitation, the first order approximation rapidly breaks down for more than small tip angles.(103) Thus, in order to achieve sharp excitation bandwidths for different tip angles, or uniform excitation of the spins over a certain frequency range, a panoply of pulse shapes, such as gaussians, hermite-gaussians,(104) and fancifully named ones, including BURP and RE-BURP(105) have been designed.

These NMR techniques are applicable to precise control of quantum systems other than NMR. Numerical optimization can be used to sculpt pulse shapes to provide desired unitary transformations,(37) and shaped pulses may be useful for controlling Josephson junction phase qubits.(106)

6. TRANSITION TO SILID STATE NMR

While the liquid state studies have allowed us to explore open system dynamics and to develop means and metrics for obtaining control in small quantum systems, these studies have generally been limited to less than 10 qubits. Though the decoherence times are long (on the order of seconds), the strength of the spin–spin coupling (used to implement two-qubit gates) is small (about 100 Hz), limiting the number of operations that can be performed. In addition, at room temperature the density matrix characterizing the spin system is highly mixed and it is necessary to use pseudo-pure states.(14,15) As the room temperature polarization of the sample is very small (<1 part in 10^5) , the exponential loss in signal as the size of the spin system grows limits the number of qubits that can be observed. Another limitation to scaling liquid state NMR techniques is the use of chemistry for frequency-dependent addressing. As the number of qubits increases, the number of transitions that need to be individually addressed grows as well. These transitions all lie within a fixed (chemistry dependent) bandwidth, making it progressively harder to address a single transition without disturbing any other. While techniques such as algorithmic cooling(17,107) and cellular automata(108) schemes have been proposed to overcome some of these limitations, their experimental feasibility has not been demonstrated to date.

Solid state NMR approaches allow us to obtain control over a much larger Hilbert space, and hold great promise for the study of many body dynamics and quantum simulations. The most important spin–spin interaction is the through-space dipolar coupling, which is on the order of tens of kilohertz in typical dielectric crystals, so that it should be possible to implement a large number of operations (perhaps 10^4) before the spins decohere. Moreover, in the solid state, the spins can be highly polarized by techniques such as polarization transfer from electronic spins.(11) The increased polarization allows an exploration of systems with a larger number of qubits, and also allows preparation of the system close to a pure state. While traditional solid state NMR techniques rely on chemistry for addressing, it is possible to introduce spatial addressing of the spins using extremely strong magnetic field gradients that produce distinguishable Larmor precession frequencies on the atomic scale, or via an auxiliary quantum system such as an electron spin, a quantum dot or even a superconducting qubit that is coupled to the nuclear spin system. For instance, entanglement between an electron and nuclear spin in an ensemble has recently been demonstrated.(109)

A variety of architectures have been proposed for solid state NMR quantum computing, a few of which are enumerated below :

1. Cory *et al.* proposed an ensemble solid state NMR quantum computer, using a large number of *n*-qubit quantum processor molecules embedded in a lattice.(5) The processors are sufficiently far apart that they only interact very weakly with each other. The bulk lattice is a deuterated version of the QIP molecule, with no other spins species present. Paramagnetic impurities in the lattice are used to dynamically polarize the deuterium spins, and this polarization can then be transferred to the QIP molecules using polarization transfer techniques. The addressing is based on the chemistry of the processor molecule.
2. Yamamoto and coworkers have proposed another ensemble NMR processor, using an isotopically engineered silicon substrate containing ^{29}Si spin chains in an ^{28}Si or ^{30}Si lattice.(8,9) The ^{29}Si has spin 1/2 while ^{28}Si and ^{30}Si have spin 0. A microfabricated (dysprosium) ferromagnet is used to produce extremely strong magnetic field gradients to create a variation of the nuclear spin Larmor frequencies at the atomic scale. Detection is performed using magnetic resonance force-microscopy.
3. The use of an N–V defect center in diamond coupled to a cluster of ^{13}C spins as a quantum processor has been proposed by Wrachtrup *et al.*.(6) The hypefine coupling between the electron

spin in the defect and the carbon nuclei allows these carbon nuclei to be addressed individually. Using a combination of optically detected electron nuclear double resonance and single molecule spectroscopy techniques, they suggest that it should be possible to both prepare pure spin states of the system as well as directly detect the result of a computation by performing a single spin measurement.

4. Suter and Lim propose an architecture based on endohedral fullerenes, encapsulating either a phosphorus or nitrogen atom, positioned on a silicon surface.[10] The C_{60} cages form atom traps, and the decoherence time of the phosphorus or nitrogen nuclear spin is consequently very long. Switched magnetic field gradients that can produce observable electron Larmor frequency shifts on nanometer length scales, in combination with frequency selective RF pulses are used to address the spins. Each site represents two physical qubits, the electron spin of the fullerene and the nuclear spin of the trapped atom, and are used to create a single logical qubit. Two qubit gates between different fullerene molecules are mediated by the electron–electron dipolar coupling.

While simple gates have been demonstrated in solid state NMR,[110] the fidelity of these gates is low, and significant experimental challenges remain. The decoherence mechanism in the solid state is primarily due to the indistinguishability of the chemically equivalent nuclear spins. The Hamiltonian of a homonuclear spin system is[11]

$$\mathcal{H} = \omega \sum_i I_z^i + \sum_{i<j} d_{ij} \left(3 I_z^i I_z^j - \bar{I}^i \cdot \bar{I}^j \right) \tag{20}$$

where the first term corresponds to the Zeeman energy and second term is the truncated dipolar Hamiltonian in strong magnetic fields. While the total dipolar Hamiltonian commutes with the total Zeeman Hamiltonian, the Zeeman and dipolar terms do not commute on a term by term basis. The phase memory of the spins is scrambled as they undergo energy-conserving spin flips with other spins in the system. Control of the dipolar interaction is therefore an essential element of any NMR solid state proposal. An important example of the precision of control is the ability to effectively suppress all internal Hamiltonians and preserve quantum information. Figure 12(a) shows the measured free induction decays for a crystal of calcium fluoride at two different crystal orientations, while Fig. 12(b) shows the signal while refocusing the dipolar interaction. It is therefore possible to experimentally extend the coherence times of the ^{19}F spins in a single crystal of calcium fluoride from 100 μs (no modulation),

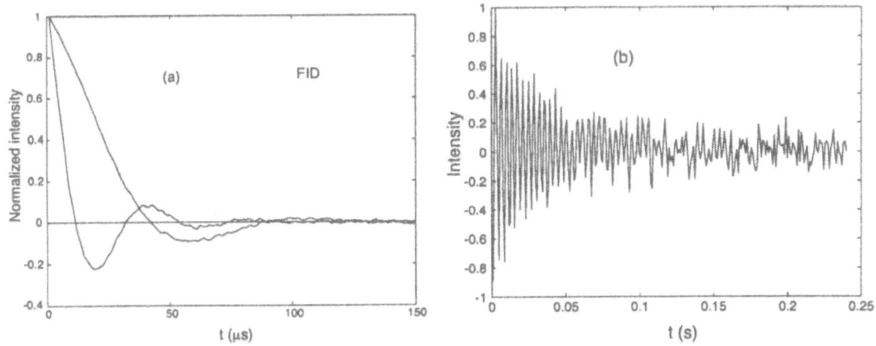

Fig. 12. (a) Free induction decay of a single crystal of calcium fluoride, at two different crystal orientations (b) Decay of the ^{19}F signal while the dipolar coupling between the nuclear spins is decoupled. (The sequence was modified to show an oscillation, so the decoupling is not optimal here.)

to 2 ms using standard NMR techniques, and finally to 500 ms using recently developed methods.[111] This represents an increase by approximately 4 orders of magnitude. In the limit of perfect coherent control of the dipolar couplings, it should be possible to significantly further extend the coherence time of the spins. In addition to improving coherent control, such studies provide insight into the next important contribution to decoherence.

The decay of the observed FID in Fig. 12(a) is due to the mutual dipolar couplings of the spins. These couplings produce correlated many spin states that are not directly observable using standard NMR techniques. However, using multiple quantum encoding techniques,[112] it is possible to directly measure the growth of the spin system under the dipolar coupling. The truncated dipolar Hamiltonian shown above is a zero quantum Hamiltonian when examined in the basis of the quantizing Zeeman Hamiltonian as expected. However, we can requantize the system in another basis (such as the x-basis for example) via a similarity transformation, and explore the growth of multiple quantum coherences in this new basis. The dipolar Hamiltonian in the x-basis is

$$\begin{aligned}\mathcal{H} = &-\frac{1}{2}\sum_{i<j} d_{ij}\left\{2I_x^i I_x^j - \frac{1}{2}\left(I_{x+}^i I_{x-}^j + I_{x-}^i I_{x+}^j\right)\right\}\\ &-\frac{3}{4}\sum_{i<j} d_{ij}\left(I_{x+}^i I_{x+}^j + I_{x-}^i I_{x-}^j\right)\end{aligned} \tag{21}$$

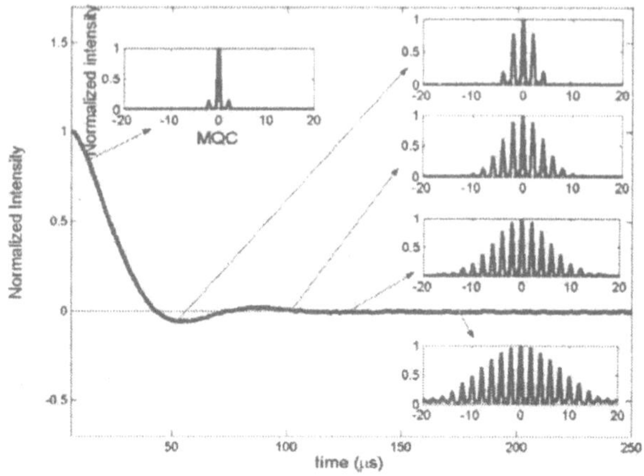

Fig. 13. Multiple quantum encoding, combined with evolution reversal sequences allow us to follow the growth of the correlated spin states during the course of a free induction decay. It is seen that even when the macroscopic signal appears to have decayed away, the spin system remains highly coherent, and states involving up to 20 correlated spins are observed.

and is thus seen to contain both zero and double quantum terms. It is possible to directly observe the growth of these x-basis coherences. Figure 13 shows the results of this experiment, illustrating the growth in the number of the correlated spins from 1 to about 20 in the first 150 μs following the application of a $\pi/2$ pulse.

In an early demonstration of the capability to explore many body dynamics in spin systems, solid state NMR techniques have been used to directly measure the rate of spin diffusion of Zeeman and dipolar energy in a single crystal of calcium fluoride.(113,114) As seen from the Hamiltonian above, these are both constants of the motion and are independently conserved. Spin diffusion is a coherent process caused by the mutual spin flips induced by the dipolar coupling between spins, that appears diffusive in the long-time, long-wavelength limit.(115) It is estimated that up to 10^{18} spins are involved at the long timescales explored in these experiments. It was found that while the measured diffusion coefficients for Zeeman order were in good agreement with theoretical predictions, the diffusion of dipolar order was observed to be significantly faster than previously predicted. While these experiments were performed in highly mixed thermal states, future experiments planned at low temperatures, and high polarizations should enable a more complete exploration of the large Hilbert space dynamics.

7. CONCLUSIONS

NMR implementations of QIP have thus yielded a wealth of information by providing experimental realizations of a number of proposed schemes. This in turn has guided our understanding of the relevant issues involved in scaling these testbed systems up in size. The methodologies developed are relevant across most of the physical platforms that have been proposed for QIP, and manifestations of this "cross-fertilization" are beginning to appear in the literature.

Solid state NMR holds great promise for scalable QIP architectures. The efforts currently underway to characterize and control large spin systems are essential to determining how the methodologies of control and system decoherence scale as a function of the system Hilbert space size.

8. ACKNOWLEDGMENTS

This work was supported by funds from ARDA/ARO, DARPA, the NSF and the Air Force Office of Sponsored Research.

REFERENCES

1. B. E. Kane, *Nature* **393**, 133 (1998).
2. G. Burkard, H.-A. Engel, and D. Loss, *Fortschr. Phys.* **48**, 965 (2000).
3. A. Imamoglu, *Fortschr. Phys.* **48**, 987 (2000).
4. V. Privman, I. D. Vagnet, and G. Kventsel, *Phys. Lett. A* **239**, 146 (1998).
5. D. G. Cory, R. Laflamme, E. Knill, L. Viola, T. F. Havel, N. Boulant, G. Boutis, E. Fortunato, S. Lloyd, R. Martinez, C. Negrevergne, M. Pravia, Y. Sharf, G. Teklemariam, Y. S. Weinstein, and W. H. Zurek, *Fortschr. Phys.* **48**, 875 (2000).
6. J. Wrachtrup, S. Y. Kilin, and A. P. Nizovtsev, *Opt. Spectrosc+* **91**, 429 (2001).
7. G. P. Berman, G. W. Brown, M. E. Hawley, and V. I. Tsifrinovich, *Phys. Rev. Lett.* **87**, 097902 (2001).
8. T. D. Ladd, J. R. Goldman, F. Yamaguchi, and Y. Yamamoto, *Phys. Rev. Lett.* **89**, 017901 (2002).
9. E. Abe, K. M. Itoh, T. D. Ladd, J. R. Goldman, F. Yamaguchi, and Y. Yamamoto, *J. Superconductivity* **16**, 175 (2003).
10. D. Suter and K. Lim, *Phys. Rev. A* **65**, 052309 (2002).
11. C. P. Slichter, *Principles of Magnetic Resonance,* 3rd Ed. (Springer-Verlag, Berlin, 1990).
12. R. R. Ernst, G. Bodenhausen, and A. Wokaun, *Principles of Nuclear Magnetic Resonance in One and Two Dimensions* (Oxford University Press, Oxford, 1990).
13. D. P. DiVincenzo, *Fortschr. Physik* **48**, 771 (2000).
14. D. G. Cory, A. F. Fahmy, and T. F. Havel, *Proc. Natl. Acad. Sci. USA* **94**, 1634 (1997).
15. N. Gershenfeld and I. L. Chuang, *Science* **275**, 350 (1997).
16. W. Warren, *Science* **277**, 1688 (1997).

17. L. J. Schulman and U. Vazirani, arXive eprint *quant-ph/9804060* (1998).
18. D. Chang, L. Vandersypen, and M. Steffen, *Chem. Phys. Lett.* **338**, 337 (2001).
19. C. Yannoni, M. Sherwood, L. Vandersypen, M. Kubinec, D. Miller, and I. Chuang, *Appl. Phys. Lett.* **75**, 3563 (1999).
20. M. Marjanska, I. Chuang, and M. Kubinec, *J. Chem. Phys.* **112**, 5095 (2000).
21. L. Vandersypen, M. Steffen, G. Breyta, C. Yannoni, R. Cleve, and I. L. Chuang, *Phys. Rev. Lett.* **85**, 5452 (2000).
22. D. Deutsch and R. Jozsa, *Proc. R. Soc. Lond. A* **439**, 553 (1992).
23. D. Simon, in *booktitle Proc. 35th Annual Symposium on Foundations of Computer Science* (IEEE Computer Society Press, Los Alamitos, CA, 1994), pp. 116–123.
24. A. Y. Kitaev, LANL E-print *quant-ph/9511026* (1995).
25. H. K. Cummins and J. Jones, *New. J. Phys.* **2**, 6.1–6.12 (2000).
26. J. Jones, *arXive E-print quant-ph/0301019* (2003).
27. M. A. Pravia, N. Boulant, J. Emerson, E. Fortunato, T. F. Havel, and D. G. Cory, *J. Chem. Phys.* **119**, 9993 (2003).
28. N. Boulant, S. Furuta, J. Emerson, T. F. Havel and D. G. Cory, arXive E-print quant-ph/0312116.
29. U. Haeberlen and J. S. Waugh, *Phys. Rev.* **175**, 453 (1968).
30. S. J. Glaser, T. Schulte-Herbruggen, M. Sieveking, O. Schedletzky, N. C. Nielsen, O. W. Sorensen, and C. Griisigner, *Science* **280**, 421 (1998).
31. M. Levitt, *Prog. Nucl. Magn. Reson. Spectrosc.* **18**, 61 (1986).
32. A. Shaka and R. Freeman, *J. Magn. Reson.* **55**, 487 (1983).
33. J. Baum, R. Tycko, and A. Pines, *Phys. Rev. A* **32**, 3435 (1985).
34. M. S. Silver, R. I. Joseph, and D. I. Hoult, *Phys. Rev. A* **31**, 2753 (1985).
35. H. Cummins, G. Llewellyn, and J. Jones, *Phys. Rev. A* **67**, 042308 (2003).
36. R. Tycko, A. Pines, and J. Guckenherimer, *J. Chem. Phys.* **83**, 2775 (1985).
37. E. M. Fortunato, M. A. Pravia, N. Boulant, G. Teklemariam, T. F. Havel, and D. G. Cory, *J. Chem. Phys.* **116**, 7599 (2002).
38. L. Vandersypen and I. L. Chuang, *Rev. Mod. Phys.* Vol. 76 (2004) arXive E-print quant-ph/0404064.
39. E. Knill, R. Laflamme, R. Martinez, and C.-H. Tseng, *Nature* **404**, 368 (2000).
40. D. Stefanatos, N. Khaneja, and S. J. Glaser, *Phys. Rev. A* **69**, 022319 (2004).
41. N. Khaneja, S. J. Glaser, and R. Brockett, *Phys. Rev. A* **65**, 032301 (2002).
42. N. Khaneja, R. Brockett, and S. J. Glaser, *Phys. Rev. A* **63**, 032308 (2001).
43. K. Kraus, *Ann. Phys.* **64**, 311 (1971).
44. E. L. Hahn, *Phys. Rev.* **80**, 580 (1950).
45. H. Y. Carr and E. M. Purcell, *Phys. Rev.* **93**, 749 (1954).
46. R. Tycko, H. M. Cho, E. Schneider, and A. Pines, *J. Magn. Reson. A* **61**, 90 (1980).
47. R. Tycko, *Phys. Rev. Lett.* **51**, 775 (1983).
48. J. S. Waugh, *J. Mag. Res.* **50**, 30 (1982).
49. E. Knill, R. Laflamme, and L. Viola, *Phys. Rev. Lett.* **84**, 2525 (2000).
50. L. Viola, E. Knill, and S. Lloyd, *Phys. Rev. Lett.* **82**, 2417 (1999).
51. L. Viola and S. Lloyd, *Phys. Rev. A.* **58**, 2733 (1998).
52. A. M. Childs, I. L. Chuang, and D. W. Leung, *Phys. Rev. A* **64**, 012314 (2001).
53. N. Boulant, T. F. Havel, M. A. Pravia, and D. G. Cory, *Phys. Rev. A* **67**, 042322 (2003).
54. Y. Weinstein, T. F. Havel, J. Emerson, N. Boulant, M. Saraceno, S. Lloyd, and D. G. Cory, *J. Chem. Phys.*, in press.
55. P. Shor, *Phys. Rev. A* **52**, 2493 (1995).
56. A. M. Steane, *Phys. Rev. Lett.* **77**, 793 (1996).
57. J. Preskill, *Proc. Roy. Soc. London A* **454**, 385 (1998).

58. M. A. Nielsen and I. L. Chuang, *Quantum Computation and Quantum Information* (Cambridge University Press, Cambridge, 2000).
59. L. Viola, E. M. Fortunato, M. A. Pravia, E. Knill, R. Laflamme, and D. G. Cory, *Science* **293**, 2059 (2001).
60. D. G. Cory, M. D. Price, W. Maas, E. Knill, R. Laflamme, W. H. Zurek, T. F. Havel, and S. S. Somaroo, *Phys. Rev. Lett.* **81**, 2152 (1998).
61. E. M. Fortunato, L. Viola, J. Hodges, G. Teklemariam, and D. G. Cory, *New J. Phys.* **4**, 5.1-5.20 (2002).
62. N. Boulant, M. A. Pravia, E. M. Fortunato, T. F. Havel, and D. G. Cory, *QIP* **1**, 135 (2002).
63. E. Knill, R. Laflamme, R. Martinez, and C. Negrevergne, *Phys. Rev. Lett.* **86**, 5811 (2001).
64. D. Leung, L. Vandersypen, X. Zhou, M. Sherwood, C. Yannoni, M. Kubinec, and I. L. Chuang, *Phys. Rev. A* **60**, 1924 (1999).
65. J. E. Ollerenshaw, D. A. Lidar, and L. E. Kay, *Phys. Rev. Lett.* **91**, 217904 (2003).
66. I. L. Chuang, N. Gershenfeld, and M. Kubinec, *Phys. Rev. Lett.* **18**, 3408 (1998).
67. J. A. Jones, M. Mosca, and R. H. Hansen, *Nature* **393**, 344 (1998), quant-ph/9805069.
68. I. L. Chuang, L. M. K. Vandersypen, X. L. Zhou, D. W. Leung, and S. Lloyd, *Nature* **393**, 143 (1998).
69. J. A. Jones and M. Mosca, *J. Chem. Phys.* **109**, 1648 (1998), *arXive E-print quant-ph/9801027*.
70. J. Jones and M. Mosca, *Phys. Rev. Lett.* **83**, 1050 (1999).
71. L. M. K. Vandersypen, C. S. Yannoni, M. H. Sherwood, and I. L. Chuang, *Phys. Rev. Lett.* **83**, 3085 (1999).
72. L. Vandersypen, M. Steffen, M. H. Sherwood, C. Yannoni, G. Breyta, and I. L. Chuang, *Appl. Phys. Lett.* **76**, 646 (2000).
73. N. Linden, H. Barjat, and R. Freeman, *Chem. Phys. Lett* **296**, 61 (1998).
74. D. Collins, N. Linden, and S. Popescu, *arXive eprint quant-ph/0005102* (2000).
75. H. Kim, J.-S. Lee, and S. Lee, *Phys. Rev. A* **62**, 022312 (2000).
76. R. Marx, A. Fahmy, J. Myers, W. Bermel, and S. Glaser, *Phys. Rev. A* **62**, 123310 (2000).
77. Y. Weinstein, M. Pravia, E. Fortunato, S. Lloyd, and D. Cory, *Phys. Rev. Lett.* **86**, 1889 (2001).
78. L. Vandersypen, M. Steffen, G. Breyta, C. Yannoni, R. Cleve, and I. L. Chuang, *Nature* **414**, 883 (2001).
79. M. Steffen, W. van Dam, T. Hogg, G. Breyta, and I. Chuang, *Phys. Rev. Lett.* 90, 067903 (2003).
80. E. Farhi and J. Goldstone, *Science* **292**, 472 (2001).
81. T. Hogg, *Phys. Rev. A* **61**, 052311 (2000).
82. R. Laflamme, E. Knill, W. H. Zurek, P. Catasti, and S. V. S. Mariappan, *Phil. Trans. Roy. Soc. Lond. A* **356**, 1941 (1998), *arXive E-print quant-ph/9709025*.
83. S. L. Braunstein, C. M. Caves, R. Jozsa, N. Linden, S. Popescu, and R. Schack, *Phys. Rev. Lett.* **83**, 1054 (1999).
84. R. Nelson, D. Cory, and S. Lloyd, *Phys. Rev. A* **61**, 022106(5) (2000).
85. M. A. Nielsen, E. Knill, and R. Laflamme, *Nature* **396**, 52 (1998).
86. X. Fang, X. Zhu, M. Feng, X. Mao, and F. Du, *Phys. Rev. A* **61**, 022307(5) (2000).
87. H. K. Cummins, G. Llewellyn, and J. Jones, *arXive eprint quant-ph/0208092* (2002).
88. Y. Weinstein, S. Lloyd, J. Emerson, and D. Cory, *Phys. Rev. Lett.* **89**, 157902 (2002).
89. A. Verhulst, O. Liivak, H. Vieth, C. Yannoni, and I. Chuang, *Appl. Phys. Lett.* **79**, 2480 (2001).
90. P. Hübler, J. Bargon, and S. Glaser, *J. Chem. Phys.* **113**, 2056 (2000).

91. J. A. Jones, *arXive E-print quant-ph/0312014*.
92. R. Feynman, *Int. J. Theor. Phys.* **21**, 467 (1982).
93. S. Somaroo, C. H. Tseng, T. F. Havel, R. Laflamme, and D. G. Cory, *Phys. Rev. Lett.* **82**, 5381 (1999).
94. D. Bacon, A. M. Childs, I. L. Chuang, J. Kempe, D.W. Leung, and X. L. Zhou, *Phys. Rev. A* **64**, 062302 (2001).
95. W. H. Zurek, *Phys. Rev. D* **26**, 1862 (1982).
96. G. Teklemariam, E. M. Fortunato, C. C. Lopez, J. Emerson, J. P. Paz, T. F. Havel, and D. G. Cory, *Phys. Rev. A* **67**, 062316 (2003).
97. J. Yepez, *Int. J. Mod. Phys.* **C9**, 1587 (1998).
98. J. Yepez, *Int. J. Mod. Phys.* **C12**, 1285 (2001).
99. M. A. Pravia, Z. Chen, J. Yepez, and D. G. Cory, *QIP* **2**, 97 (2003).
100. S. Wimperis, *J. Magn. Reson. B* **109**, 221 (1994).
101. S. Gulde, M. Riebe, G. Lancaster, C. Becher, J. Eschner, H. Häffner, F. Schmidt-Kaler, I. Chuang, and R. Blatt, *Nature* **421**, 48 (2003).
102. E. Collin, G. Ithier, A. Aasime, P. Joyez, D. Vion, and D. Esteve, *LANL E-print cond-mat/0404507*.
103. J. Pauly, P. L. Roux, and D. Nishimura, *IEEE Trans. Med. Imag.* **10**, 53 (1991).
104. W. Warren, *J. Chem. Phys.* **81**, 5437 (1984).
105. M. Green and R. Freeman, *J. Magn. Reson.* **93**, 93 (1991).
106. M. Steffen, J. Martinis, and I. L. Chuang, *Phys. Rev. B* **68**, 224518 (2003).
107. P. O. Boykin, T. Mor, V. Roychowdhury, F. Vatan, and R. Vrijen, *Proc. Natl. Acad. Sci.* USA **99**, 3388 (2002).
108. S. Lloyd, *Science* **261**, 1569 (1993).
109. M. Mehring, J. Mende, and W. Scherer, *Phys. Rev. Lett.* **90**, 153001 (2003).
110. G. M. Leskowitz, N. Ghaderi, R. A. Olsen, and L. J. Mueller, *J. Chem. Phys.* **119**, 1643 (2003).
111. G. S. Boutis, P. Cappellaro, H. Cho, C. Ramanathan, and D. G. Cory, *J. Magn. Reson.* **161**, 132 (2003).
112. C. Ramanathan, H. Cho, P. Cappellaro, G. S. Boutis, and D. G. Cory, *Chem. Phys. Lett.* **369**, 311 (2003).
113. W. Zhang and D. G. Cory, *Phys. Rev. Lett.* **80**, 1324 (1998).
114. G. S. Boutis, D. Greenbaum, H. Cho, D. G. Cory, and C. Ramanathan, *Phys. Rev. Lett.* **92**, 137201 (2004).
115. N. Bloembergen, *Physica (Amsterdam)* **15**, 386 (1949).

Quantum Computing with Trapped Ion Hyperfine Qubits

B. B. Blinov,[1,3] D. Leibfried,[2] C. Monroe,[1] and D. J. Wineland[2]

We discuss the basic aspects of quantum information processing with trapped ions, including the principles of ion trapping, preparation and detection of hyperfine qubits, single-qubit operations and multi-qubit entanglement protocols. Recent experimental advances and future research directions are outlined.

KEY WORDS: Ion trapping; hyperfine qubits; quantum gates; qubit detection; qubit initialization; entanglement.

PACS: 03.67.Lx, 32.80.Pj, 32.80.Qk, 42.50.Vk.

1. OVERVIEW

Trapped atomic ions were first proposed as a viable quantum computing candidate by Cirac and Zoller in 1995.(1) Almost 10 years later, the trapped ion system remains one of the few candidates that satisfies the general requirements for a quantum computer as outlined by DiVincenzo(2): (i) a scalable system of well-defined qubits, (ii) a method to reliably initialize the quantum system, (iii) long coherence times, (iv) existence of universal gates, and (v) an efficient measurement scheme. Most of these requirements have been demonstrated experimentally with trapped ions, and there exist straightforward (albeit technically difficult) paths to solving the remaining problems.

Experimental approaches in ion trap quantum computing can be divided by the type of qubit, in terms of the qubit level energy splitting,

[1]FOCUS Center and Department of Physics, University of Michigan Ann Arbor, MI 48109, USA; e-mail: bblinov@umich.edu

[2]Time and Frequency Division, National Institute of Standards and Technology Boulder, CO 80303, USA.

[3]To whom correspondence should be addressed.

and the couplings required to drive quantum logic gates between the qubit states. The two primary types of trapped ion qubit architectures are **optical qubits** derived from a ground state and an excited metastable state separated by an optical frequency, and **hyperfine qubits** derived from electronic ground-state hyperfine (HF) levels separated by a microwave frequency. In this paper, we highlight the latter case of HF qubits. For a review of optical qubits we refer to the paper by R. Blatt included in this volume. More general accounts and various reviews are given in Refs. 3–16. Moreover, the Advanced Research and Development Activity (ARDA) has posted and updated a "roadmap" to highlight accomplishments and problems with the various possible implementations of quantum information processing including those based on ion traps; various references can be found at this site (http://qist.lanl.gov). Many experimental groups worldwide have addressed various aspects of trapped-ion quantum information processing; these include groups at University of Aarhus, IBM-Almaden, NIST-Boulder, University of Hamburg, McMaster University, University of Innsbruck, Los Alamos, University of Michigan, Max Planck Inst.-Garching, Oxford University, and NPL-Teddington.

2. ION TRAPS

Ion traps come in various forms[17]; for brevity, we restrict our discussion to the linear RF (Paul) trap shown schematically in Fig. 1. Linear Paul traps with particular application to quantum information processing are discussed by various groups; see, for example, Refs. 3–16 and 18–27. The linear trap is essentially a quadrupole mass filter plugged on the axis by superimposing a static electric potential well. In the x, y plane of the figure, ions are bound by a ponderomotive pseudopotential

$$U_{x,y}(\mathbf{r}) = \frac{q^2}{2m\Omega_T^2}\langle E^2(\mathbf{r})\rangle \simeq \frac{q^2 V_0^2}{4m\Omega_T^2 R^4}(x^2+y^2), \qquad (1)$$

where q is the ion's charge, m its mass, E is the RF electric field (resulting from a potential $V_0\cos(\Omega_T t)$ applied to the dark electrodes of Fig. 1), $\mathbf{r}$ is the radial distance from the trap axis, and R is the distance between the trap axis and the nearest electrode surface. The oscillation frequency of an ion in this pseudopotential is given by

$$\omega_{x,y} \simeq \frac{qV_0}{\sqrt{2}\Omega_T m R^2}, \qquad (2)$$

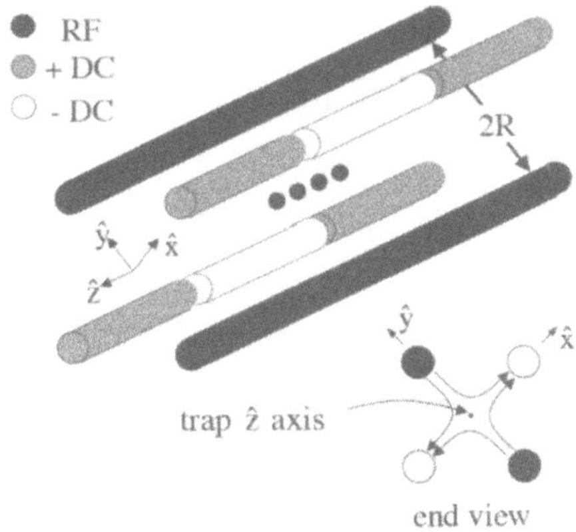

Fig. 1. Electrode configuration for a linear RF (Paul) trap. A common RF potential $V_0 \cos(\Omega_T t)$ is applied to the dark electrodes; the other electrodes are held at RF ground through capacitors (not shown) connected to ground. The lower-right portion of the figure shows the x, y electric fields from the applied RF potential at an instant when the RF potential is positive relative to ground. A static electric potential well is created (for positive ions) along the z-axis by applying a positive potential to the outer segments (grey) relative to the center segments (white).

where we assume the pseudopotential approximation ($\omega_{x,y} \ll \Omega_T$) and assume that $\omega_z \ll \omega_{x,y}$ so that the static radial forces are much smaller than the pseudopotential forces. From the form of Eq. (1), we see that the ions seek the region of minimum $|E(\mathbf{r})|$. As an example, for some typical experiments using $^9Be^+$ ions,[28] $q = 1\,e$, $V_0 = 500\,V$, $\Omega_T/2\pi \simeq 100$ MHz, and $R = 200\,\mu$m, so that $\omega_{x,y}/2\pi \simeq 24$ MHz.

For the purposes of quantum computing, to a good approximation, we can view the linear trap as providing a three-dimensional harmonic well for ion qubits, where the strength of the well in two directions (x and y in Fig. 1) is much stronger than in the third direction (z). When a small number of ions is trapped and cooled, each ion seeks the bottom of the trap well, but the mutual Coulomb repulsion between ions results in an equilibrium configuration in the form of a linear array, like beads on a string. To give an idea of array size, two ions in such a trap are spaced by $2^{1/3}s$, and three ions are spaced by $(5/4)^{1/3}s$ where $s \equiv q^2/(4\pi\epsilon_0 m\omega_z{}^2)^{1/3}$. Expressed equivalently, for singly charged ions, the spacing parameter in micrometers is $s(\mu\text{m}) = 15.2(M(u)\nu_z^2(\text{MHz}))^{-1/3}$, where the ion's mass is expressed in a.m.u. and the axial z frequency in MHz. For ν_z = 5 MHz, two $^9Be^+$ ions are separated by $3.15\,\mu$m.

Although simple gate operations among a few ion qubits have been demonstrated, a viable quantum computer architecture must accomodate very large number of qubits. As the number of ions in a trap increases, several difficulties are encountered. For example, the addition of each ion adds three vibrational modes. It soon becomes nearly impossible to spectrally select the desired vibrational mode unless the speed of operations is slowed to undesirable levels.[4,29] Furthermore, since error correction will most likely be incorporated into any large processor, it will be desirable to reset (or possibly measure) ancilla qubits without disturbing the coherence of logical qubits. Since ion qubits are typically reset by means of state-dependent laser scattering, the scattered light from ancilla qubits held in a common trap may disturb the coherence of the logical qubits.

For these and other reasons, it appears that a scalable ion-trap system must incorporate arrays of interconnected traps, each holding a small number of ions. The information carriers between traps might be photons,[19,30–32] or ions that are moved between traps in the array. In the latter case, a "head" ion held in a movable trap could carry the information by moving from site-to-site as in the proposal of Ref. 22. Similarly, as suggested in Refs. 4 and 26, qubit ions themselves could be shuttled around in an array of interconnected traps. In this scheme, the idea is to move ions between nodes in the array by applying time-dependent potentials to "control" electrode segments. To perform logic operations between selected ions, these ions are transferred into an "accumulator" trap for the gate operation. Before the gate operation is performed, it may be necessary to sympathetically re-cool the qubit ions with "refrigerant" ions.[4,26,33–35] Subsequently, these ions are moved to memory locations or other accumulators. This strategy always maintains a relatively small number of motional modes that must be considered and minimizes the problems of ion-laser-beam addressing using focused laser beams. Such arrays also enable highly parallel processing and ancilla qubit readout in a separate trapping region so that the logical ions are shielded from the scattered laser light.

Most gate schemes for trapped ions have a speed that is limited to, or proportional to, the oscillation frequency of the ions in the trap. From Eq. 2, we therefore want to maximize V_0/R^2. As R becomes smaller it is more difficult to control the relative dimensions of the electrode structures. Refs. 3–16 and 18–27 discuss some approaches to making small traps with accurate dimensions.

3. TRAPPED ION HYPERFINE QUBITS

Ions can be confined for days in an ultra-high vacuum with minimal perturbations to their internal atomic structure, making particular internal states ideal for representing quantum bits. Electric field perturbations are small, because localized ions experience a vanishing time-averaged electric field. Although magnetic field perturbations to internal structure can be important, the coherence between two internal levels can be made magnetic field-insensitive (to lowest order) by operating near an extremum of the energy separation between the two levels with respect to the magnetic field. Qubit coherence in such atomic ground states has been observed for times exceeding 10 min in the context of trapped ion frequency standards.[36,37]

Qubits stored in metastable levels separated by optical frequencies [1,38] enjoy the simplicity of single-photon optical transitions, provided the radiative decay rate is sufficiently slow (some weakly allowed optical transitions in atomic ions have lifetimes $\gg$1 s) However, phase-stable narrow-linewidth lasers are required in order to realize the full benefit of the long decay times.[39,40] In addition, Stark shifts from coupling to non-resonant allowed transitions become important for these longer qubit lifetimes since the laser intensity must be high for appreciable transition rates.[41]

Ground state hyperfine levels, or states of nuclear vs. electronic spin, typically separated by microwave frequencies, have extremely long radiative lifetimes. Trapped ion HF levels are arguably the most attractive choice for qubit states, and form the thesis of this paper. Figure 2 displays the lowest energy levels of the $^{111}Cd^+$ ion, for concreteness (nuclear spin $I = 1/2$). We will be interested primarily in two electronic states, the $^2S_{1/2}(F = 1, m_F = 0)$ and $^2S_{1/2}(F = 0, m_F = 0)$ hyperfine ground states (denoted by $|\downarrow\rangle$ and $|\uparrow\rangle$ respectively), separated by frequency ω_{HF} ($\omega_{HF}/2\pi \simeq 14.53$ GHz for $^{111}Cd^+$). These long-lived spin states will form the basis for a quantum bit. Other candidate ions with similar HF structure (non-zero nuclear spin) include $^9Be^+$, $^{25}Mg^+$, $^{43}Ca^+$, $^{87}Sr^+$, $^{137}Ba^+$, $^{173}Yb^+$, and $^{199}Hg^+$.

3.1. Qubit Initialization and Detection

Standard optical pumping techniques allow the HF qubits to be initialized into either $|\downarrow\rangle$ or $|\uparrow\rangle$ states. Subsequent detection of the spin states can be accomplished using the technique of quantum jumps.[42] For example, in $^{111}Cd^+$, a circularly polarized laser beam resonant with the $^2S_{1/2} -^2 P_{3/2}$ transition near $\lambda \simeq 214.5$ nm (Fig. 2) scatters many photons if the atom is in the $|\downarrow\rangle$ spin state (a "cycling" transition), but essentially no photons when the atom is in the $|\uparrow\rangle$ spin state. Even if a modest number of the scattered photons are detected, the efficiency of discrimination between these two states approaches 100%. In general, the HF qubit detection efficiency with such cycling transition is given by

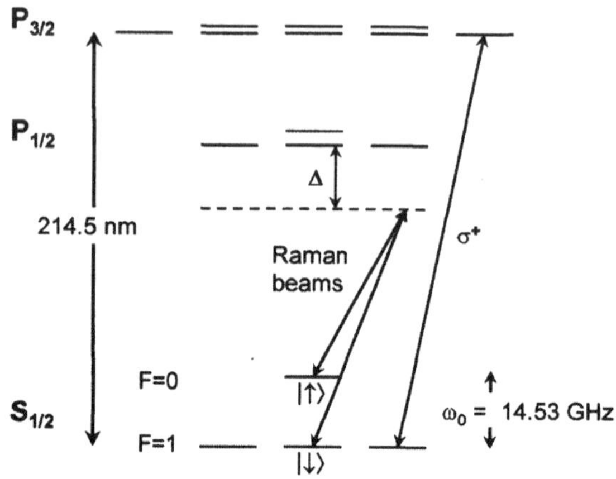

Fig. 2. Electronic (internal) energy levels (not to scale) of the $^{111}Cd^{+}$ion. The $^2S_{1/2}(F=1, m_F=0)$ and $^2S_{1/2}(F=0, m_F=0)$ hyperfine ground states (denoted by $|{\downarrow}\rangle$ and $|{\uparrow}\rangle$ respectively), separated in frequency by $\omega_{HF}/2\pi \simeq 14.53$ GHz, and magnetic field insensitive to first order, form the basis of a quantum bit. Detection of the internal HF state is accomplished by illuminating the ion with a σ^+-polarized "detection" beam near $\lambda_{Cd} \simeq 214.5$ nm and observing the fluorescence from the cycling transitions between $|{\downarrow}\rangle$ and the $^2P_{3/2}|(F=2, m_F=2)\rangle$ state. The excited P state has radiative linewidth $\gamma_e/2\pi \simeq 47$ MHz. Also drawn are a pair of σ^+-polarized Raman beams that are used for quantum logic gates.

$1-(M/\epsilon_{\text{phot}})(\gamma_e/\omega_{\text{HF}})^2$, where M includes appropriate atomic branching ratios and is of order unity, and ϵ_{phot} is the *photon* detection efficiency of the ion fluorescence. In Fig. 3, the number of photons scattered in 0.2 ms by a single $^{111}Cd^{+}$ion (net quantum-efficiency $\epsilon_{\text{phot}} \sim 10^{-3}$) is plotted for the "bright" ($|{\downarrow}\rangle$) and the "dark" ($|{\uparrow}\rangle$) states. By placing the discriminator between two and three detected photons, a HF qubit detection efficiency of 99.7% is realized.

3.2. HF Qubit Rotations: Single Qubit Gates

Single-qubit rotations of HF states can be accomplished by either applying microwave radiation tuned to the energy splitting between the two levels ω_{HF}, or by driving stimulated Raman transitions (SRT) with two laser fields that are properly detuned from an excited state and differ in frequency by ω_{HF}.

When microwaves are used, it is necessary to efficiently couple radiation with ~cm wavelengths into a sub-millimeter size ion trap. Low gain microwave horns (with small opening angles) provide reasonable means of generating a strong beam of microwaves. Rabi frequencies approaching 100 kHz have been achieved with modest microwave powers (<1 W) applied

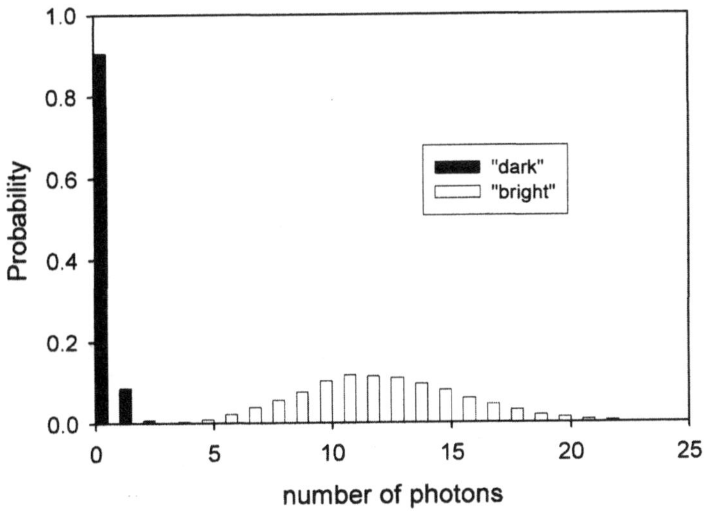

Fig. 3. Detection histograms of a single trapped $^{111}Cd^+$ ion. The white bars correspond to the distribution of the number of fluorescence photons detected by a CCD camera for a $^{111}Cd^+$ion prepared initially in the $|\downarrow\rangle$ state upon application of a σ^+-polarized detection laser for 0.2 ms. The black bars correspond to the photon distribution for the ion initially prepared in the $|\uparrow\rangle$ state under the same conditions. The very small overlap between the two distributions corresponds to a detection efficiency of >99.7%.

through a horn within 10 cm of the ion. Microwave qubit rotations can easily be made very clean by using stable RF sources, and are useful for joint rotations of all qubits. However, individual addressing of trapped ions with microwaves is difficult, unless magnetically sensitive qubit states are employed and substantial magnetic field gradients are applied.(44)

In the case of $^2S_{1/2}$ HF qubit rotations with SRT,(5) two co-propagating laser fields are applied to the ion, each with a detuning $\Delta \gg \gamma_e$ from an excited $^2P_{1/2}$ or $^2P_{3/2}$ state, denoted by $|e\rangle$ with radiative linewidth γ_e. The difference frequency of the two fields is set to the HF qubit resonance, resulting in an effective field that coherently rotates the HF qubit similar to the microwave case, except because the SRT laser beams can be focussed, individual ions can be addressed. The SRT Rabi frequency is given by $\Omega_{\text{SRT}} = g_1 g_2^*/\Delta$, where g_i are the resonant Rabi frequencies for the two laser beams respectively driving transitions to the excited state $|e\rangle$. The probability of spontaneous emission from off-resonant excitation to the state $|e\rangle$ during a SRT π-pulse decreases with increasing Δ as $P_{se} \simeq \gamma_e/\Delta$.

However, Δ cannot be increased indefinitely. In order for SRT to effectivly couple HF states and effectively flip the state of electron +

nuclear spin, the excited state $|e\rangle$ must be populated for a sufficient time so that the spin-orbit interaction allows the spin to flip. In the case of $S_{1/2}$ hyperfine ground state qubits coupled to excited atomic $P_{1/2}$ or $^2P_{3/2}$ states, this time-scale is set by the inverse fine-structure splitting $1/\Delta_{\rm FS}$, and we find that the probability of spontaneous emission in a given qubit-dependent operation is roughly $P_{se} \simeq \gamma_e/\Delta_{\rm FS}$, which can be as low as 10^{-5} in candidate ion species.(13)

3.3. Interactions Between HF Qubits: Entangling Qubit Gates

Trapped ion qubits can be controllably coupled through their mutual Coulomb interaction. While the internal HF qubit states are essentially unaffected by the Coulomb interaction directly, external control fields can generate an effective coupling between qubits that relies on a qubit state-dependent force. This external control field can thus entangle trapped ion qubits through the "data bus" represented by the Coulomb interaction.

For a qubit stored in atomic $S_{1/2}$ HF ground states $|\downarrow\rangle$ and $|\uparrow\rangle$, the coupling to an ion's position $\hat{\mathbf{x}}$ proceeds via a dipole coupling of one (or both) qubit states to an excited atomic $P_{1/2}$ or $P_{3/2}$ state $|e\rangle$ having the form

$$H_I = -(\hat{\mu}_{\uparrow,\mathbf{e}} + \hat{\mu}_{\downarrow,\mathbf{e}}) \cdot \mathbf{E}(\hat{\mathbf{x}}). \tag{3}$$

Here, $\hat{\mu}_{\mathrm{S},\mathbf{e}}$ is the electric dipole operator between qubit state $|S\rangle$ and $|e\rangle$, and $\mathbf{E}(\hat{\mathbf{x}})$ is the electric field of the laser as a function of the position of the ion $\hat{\mathbf{x}}$. This interaction can be sequentially or simultaneously applied to different trapped ion qubits in order to generate entanglement. While the ion position is thus used to entangle trapped ion qubits, successful gate schemes rely on the quantum state of position not becoming ultimately entangled with the qubit states following the gate. Below we discuss two of the most common methods for coupling the qubit and position of trapped ions: motion-sensitive stimulated Raman transitions, and qubit state-dependent forces.

3.3.1. Motion-sensitive stimulated Raman transitions

Optical stimulated Raman transitions are not only useful for simple HF qubit rotation operations as discussed above, but can be critical for coupling qubit states to the external motional state of the ions. In this case, the two Raman beams are directed onto the ion(s) with a non-zero wavevector difference $\delta\mathbf{k}$ along the relevant direction of motion to be coupled. For a single ion, the resulting coupling under the rotating wave approximation[4] is

$$H = \eta\Omega_{\mathrm{SRT}}(\sigma_+ e^{i\eta(a+a^\dagger)} + \sigma_- e^{i\eta(a+a^\dagger)}), \tag{4}$$

where $\eta = \delta\mathbf{k}\cdot\mathbf{x_0}$ is the Lamb–Dicke parameter associated with the coordinate x and $x_0 = (\hbar/2m\omega)^{1/2}$ is the 0-point spread of the ion wavepacket. The raising and lowering operators for the qubit (motion) are given by σ_+ and σ_- ($a^\dagger$ and a). The Lamb–Dicke parameter η sets the scale for the coupling between the light field and the position of the ion, related to the gradient of the light field experienced by the ion.

We provide two specific examples of SRT couplings from Eq. 4 that can be exploited for entangling the quantum state of a single ion with a collective mode of motion. Subsequently, similar operations can be applied to other ions sharing the motion, resulting in a net entangling quantum gate between ion qubits. In the original Cirac-Zoller proposal,[1] the difference frequency between the two stimulated Raman fields is tuned to a "motional sideband" at frequency $\omega_{\mathrm{HF}} \pm k\omega$, where k is an integer describing the sideband order. Raman sideband operations coherently rotate the HF qubit state while simultaneously affecting the quantum state of motion. The resulting coupling for the first lower sideband ($k = -1$) takes the form of the classic Jaynes–Cummings Hamiltonian:

$$H_{-1} = \eta\Omega_{\mathrm{SRT}}(\sigma^+ a + \sigma^- a\dagger). \tag{5}$$

The above expression assumes that the ion is confined to within the Lamb–Dicke limit, although this is not essential. For a single trapped ion initially prepared in the vibrational ground state ($|0\rangle$), this coupling results in the mapping of an arbitrary qubit state $(\alpha|\downarrow\rangle + \beta|\uparrow\rangle)|0\rangle$ to $|\downarrow\rangle(\alpha|0\rangle + \beta|1\rangle)$. This interaction is the basis for the Cirac-Zoller[1] and Mølmer-Sørensen[46,47] quantum logic gate schemes.

When a pair of non-copropagating Raman beams are tuned to the carrier transition, multi-qubit entangling gates can also be realized. In this case, we find that qubit transitions are driven without accompanying motional state transitions, although the qubit Rabi frequency acquires a dependence upon the motional state of the form

$$\Omega_{n,n} = \eta\Omega_{\mathrm{SRT}} e^{-\eta^2/2} L_n(\eta^2), \tag{6}$$

where $L_n(X)$ is the Laguerre polynomial of order n. For the lowest three values of n, we have $L_0(\eta^2) = 1$, $L_1(\eta^2) = 1 - \eta^2$, and $L_2 = 1 - 2\eta^2 + \eta^4/2$. This motion-dependent qubit rotation can be used to construct quantum logic gates[51,52]

3.3.2. Spin-dependent optical forces

An alternative method for coupling hyperfine qubits with the ion motion is to use laser beams to generate a dipole force that depends upon the state of the qubit $|S\rangle$ through atomic selection rules and appropriate polarization of the light. As an example, we consider the case of an optical coupling between $S_{1/2}$ and excited $P_{1/2}$ states, with a nuclear spin $I = 1/2$. In this case, there are four ground states and four excited states, as depicted in Fig. 2. If the $S_{1/2}$ qubit states are $|\uparrow\rangle$ and $|\downarrow\rangle$ states, then a σ^+-polarized laser beam will couple the $|\uparrow\rangle$ to the excited state, with $|\downarrow\rangle$ decoupled.

Such a "spin-dependent force" can take many forms. For instance, the ion can be placed in an intensity gradient of a laser beams through focusing or through application of a standing-wave.[22] Alternatively, off-resonant laser beams with a difference frequency near the trap frequency ω can be applied to the ion (a "walking wave" field), resulting in a resonant (qubit state-dependent) displacement of the motional state in phase space.[53,56,57] Spin-dependent optical forces underly the "push" gate of Ref. 22 and the geometric phase gate discussed below[53]. Finally, we note the possibility of applying pairs of counterpropagating light pulses of duration $\tau \ll 1/\gamma_e$, that resonantly drive transitions from one qubit state to the excited P state and back down, accompanied by a $2\hbar k$ impulse from the recoil of the absorption.[55] This is the basis for the fast gate scheme proposed by Garcia-Rippol *et al.*.[54]

3.3.3. Comparison of couplings

The above two methods can be considered as formally equivalent, both involving a qubit state-dependent interaction with the ion coordinate that can be subsequently coupled to another ion through the Coulomb force. The original Cirac-Zoller coupling[1] requires the preparation of the quantum state of ion motion to the $|n = 0\rangle$ ground state, whereas most of the other couplings require preparation of the motion to within the Lamb-Dicke regime, where $\eta^2(\langle n\rangle + 1) \ll 1$, where $\langle n\rangle$ is tha average number of (thermal) vibrational quanta in the ion motion.

HF qubit gates based on SRT couplings allow the creation of entangling quantum gates that change the qubit state, such as the CNOT gate, whereas spin-dependent optical forces generally provide gates that do not change the qubit state, such as the phase gate. Therefore, spin-dependent optical force gates can provide better isolation from errors associated with qubit rotations such as the residual coupling to spectator atomic levels. In particular, this imples that the gate speed for this method can be higher than that of SRT-based gates.[54] However, spin-dependent optical forces appear to be applicable only to magnetically sensitive HF qubit states, thus the qubits in this case may be more susceptable to fluctuating magnetic fields. Moreover, HF qubit rotations must ultimately be applied in any case for universal quantum logic, so it is likely that both methods will be important in future work. As mentioned in the case of single qubit rotations based on stimulated Raman transitions, a common fundamental source of error in all coupling schemes is spontaneous emission from the excited state $|e\rangle$ (lifetime $1/\gamma_e$) during the laser-induced coupling. We find that the probability of spontaneous emission per entangling gate is roughly $P_{se} \simeq \gamma_e/\eta\Delta_{FS}$, which can be as low as$10^{-4}$ [13] for certain ion candidates.

4. GATE SCHEMES AND DEMONSTRATIONS

The basic elements of the original Cirac-Zoller gate,[1] a CNOT gate between the motion and the internal state of a single trapped ion was implemented in 1995.[45] The full Cirac-Zoller gate on two ions was implemented with about 70% fidelity in 2003.[38] In 1999, Sørensen and Mølmer[46,47] and also Solano *et al.*[48] suggested an alternative gate scheme. Compared to the original Cirac and Zoller gate,[1] the latter proposal has the advantages that (i) laser-beam focusing (for individual ion addressing) is not required, (ii) it can be carried out in one step, (iii) it does not require use of an additional internal state, and (iv) it does not require precise control of the motional state (as long as the Lamb–Dicke limit is satisfied). Based on this approach, the NIST group realized a universal gate between two spin qubits[49,50] that was also used to demonstrate a particular four-qubit gate.[49]

A variation of the original Cirac-Zoller gate was demonstrated in 2002, relying on Raman carrier operations as discussed above. Here, the Lamb–Dicke parameter η was set so that the carrier operations depended upon the motional state in a particular way[51,52], resulting in a π-phase gate with a single pulse of light. Compared to the previously realized CNOT gate between motion and spin,[45] this gate has the advantages that (i) it requires one step instead of three, (ii) it does not require an auxiliary internal state, and (iii) it is immune to Stark shifts caused by off-resonant sideband couplings.[4]

In 2003, the NIST group demonstrated a π-phase gate between two trapped ion qubits[53] based on a qubit state-dependent force; with this gate a Bell state with 97% fidelity was generated. The gate realized the transformations: $|\downarrow\rangle|\downarrow\rangle \rightarrow |\downarrow\rangle|\downarrow\rangle$, $|\downarrow\rangle|\uparrow\rangle \rightarrow e^{i\pi/2}|\downarrow\rangle|\uparrow\rangle$, $|\uparrow\rangle|\downarrow\rangle \rightarrow e^{i\pi/2}|\uparrow\rangle|\downarrow\rangle$, and $|\uparrow\rangle|\uparrow\rangle \rightarrow |\uparrow\rangle|\uparrow\rangle$. Combined with single bit rotations, this operation can yield either a π-phase gate or the CNOT operation.

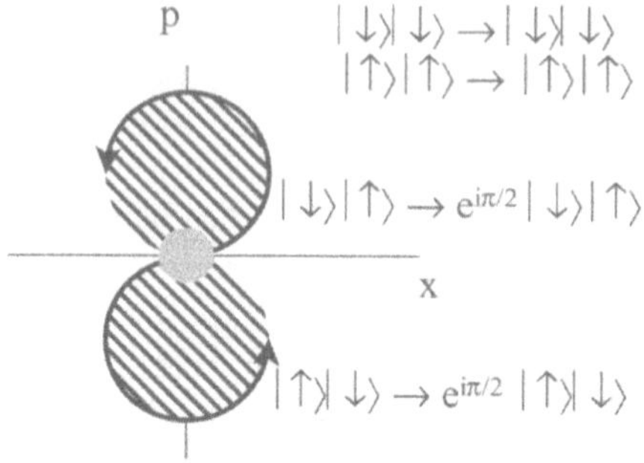

Fig. 4. Schematic representation of the displacements of the axial stretch-mode amplitude in phase space for the four basis states of the two spin qubits. The detuning and amplitude of the displacements are chosen to give a $\pi/2$ phase shift on the $|\downarrow\rangle|\uparrow\rangle$ and $|\uparrow\rangle|\downarrow\rangle$ states while the $|\downarrow\rangle|\downarrow\rangle$ and $|\uparrow\rangle|\uparrow\rangle$ states are unaffected because the optical dipole forces for these states do not couple to the stretch mode.

The gate relies in part on properties of motional states as they are displaced in phase space. For a closed trajectory in phase space, the overall quantum state acquires an phase shift that depends on the area enclosed by the path. The second element required for the gate is to make the path area be spin-dependent. This is accomplished by making the displacement in phase space with a spin-dependent optical dipole force as was done in previous experiments.[56,57]

To implement this gate on two ions, the Raman transition beams were separated in frequency by $\sqrt{3}\omega_z + \delta$, where $\sqrt{3}\omega_z$ is the stretch-mode frequency for two ions and δ is a small detuning (below). The separation of the ions was adjusted to be an integer multiple of $2\pi/\Delta k$ so that the optical-dipole force (from the "walking" standing wave) on each ion was in the same direction if the ions were in the same spin state but, due to the choice of laser polarizations, in opposite directions if the spin states were different. This had the effect that the application of the laser beams to the $|\downarrow\rangle|\uparrow\rangle$ and $|\uparrow\rangle|\downarrow\rangle$ states caused excitation on the stretch mode but the $|\downarrow\rangle|\downarrow\rangle$ and $|\uparrow\rangle|\uparrow\rangle$ states were unaffected. The detuning δ and duration of the displacement pulses were chosen to make one complete (circular) path in phase space with an area that gave a phase shift of $\pi/2$ on the $|\downarrow\rangle|\uparrow\rangle$ and $|\uparrow\rangle|\downarrow\rangle$ states. This is shown schematically in Fig. 4.

5. CONCLUSION

The trapped ion system is arguably one of the most attractive candidates for large-scale quantum computing. Here, we have concentrated on the use of atomic hyperfine ground states as qubits, the most stable quantum bit known. With a rich variety of schemes for generating entangling quantum logic gates between HF qubits based on externally applied laser fields, it appears that the scale up procedure is now limited by the fabrication of more complex trap arrays and the precise control of laser fields to produce high fidelity gates.

REFERENCES

1. J. I. Cirac and P. Zoller, *Phys. Rev. Lett.* **74**, 4091 (1995).
2. D. P. DiVincenzo, in *Scalable Quantum Computers*, edited by S. L. Braunstein, H. K. Lo, and P. Kok (Wiley-VCH, Berlin, 2001), pp. 1–13.
3. A. Steane, *Appl. Phys. B* **64**, 623 (1997).
4. D.J. Wineland, C.R. Monroe, W.M. Itano, D. Leibfried, B.E. King, and D.M. Meekhof, *J. Res. Nat. Inst. Stand. Tech.* **103**, 259 (1998).
5. D.J. Wineland, C.R. Monroe, W.M. Itano, B.E. King, D. Leibfried, D.M. Meekhof, C.J. Myatt, and C.S. Wood, *Fortschritte der Physik* **46**, 363 (1998).
6. D. F. V. James, in *Scalable Quantum Computers*, edited by S. L. Braunstein, H. K. Lo, and P. Kok (Wiley-VCH, Berlin, 2001), pp. 53–68.
7. G. J. Milburn, S. Schneider, and D. F. V. James, in *Scalable Quantum Computers*, edited by S. L. Braunstein, H. K. Lo, and P. Kok (Wiley-VCH, Berlin, 2001), pp. 31–40.
8. J. F. Poyatos, J. I. Cirac, and P. Zoller, in *Scalable Quantum Computers*, edited by S. L. Braunstein, H. K. Lo, and P. Kok (Wiley-VCH, Berlin, 2001), pp. 15–30.
9. A. Sørensen and K. Mølmer, in *Scalable Quantum Computers*, edited by S. L. Braunstein, H. K. Lo, and P. Kok (Wiley-VCH, Berlin, 2001), pp. 41–52.
10. A. M. Steane and D. M. Lucas, in *Scalable Quantum Computers*, edited by S. L. Braunstein, H. K. Lo, and P. Kok (Wiley-VCH, Berlin, 2001), pp. 69–88.
11. C. Monroe, *Nature* **416**, 238 (2002).
12. D. J. Wineland, in *Experimental Quantum Computation and Information, Proc. Int. School of Physics "Enrico Fermi"*, edited by F. D. Martini and C. Monroe (IOS Press, Amsterdam, 2002), Vol. 148, pp. 165–196.
13. D.J. Wineland, M. Barrett, J. Britton, J. Chiaverini, B.L. DeMarco, W.M. Itano, B.M. Jelenkovic, C. Langer, D. Leibfried, V. Meyer, T. Rosenband, and T. Schaetz, *Phil. Trans. R. Soc. Lond. A* **361**, 1349 (2003).
14. S. Gulde, H. Häffner, M. Riebe, G. Lancaster, A. Mundt, A. Kreuter, C. Russo, C. Becher, J. Eschner, F. Schmidt-Kaler, I.L. Chuang, and R. Blatt, in *Proceedings of the XVIII International Conference on Atomic Physics*, edited by H. R. Sadeghpour, E. J. Heller, and D. E. Pritchard (World Scientific, Singapore, 2003), pp. 293–302.
15. D.J. Wineland, D. Leibfried, B.L. DeMarco, V. Meyer, M.A. Rowe, A. Ben Kish, M. Barrett, J. Britton, J. Hughes, W.M. Itano, B.M. Jelenkovic, C. Langer, D. Lucas and T. Rosenband, in *Proceedings of the XVIII International Conference on Atomic Physics*, edited by H. R. Sadeghpour, E. J. Heller, and D. E. Pritchard (World Scientific, Singapore, 2003), pp. 263–272.
16. C. Wunderlich and C. Balzer, quant-ph/0305129 (2003).
17. P. K. Ghosh, *Ion Traps* (Clarendon Press, Oxford, 1995).
18. C. A. Schrama, E. Peik, W. W. Smith, and H. Walther, *Opt. Commun.* **101**, 32 (1993).
19. R. G. DeVoe, *Phys. Rev. A* **58**, 910 (1998).

20. H.C, Nägerl, W. Bechter, J, Eschner, F. Schmidt-Kaler, and R. Blatt , *Appl. Phys. B* **66**, 603 (1998).
21. P. A. Barton, C. J. S. Donald, D. M. Lucas, D. A. Stevens, A. M. Steane, and D. N. Stacey, *Phys. Rev. A* **62**, 032503 (2000).
22. J. I. Cirac and P. Zoller, *Nature* **404**, 579 (2000).
23. G.R. Guthöhrlein, M. Keller, K. Hayasaka, W. Lange, H. Walther, *Nature* **414**, 49 (2001).
24. L. Hornekær, N. Kjærgaard, A. M. Thommesen, and M. Drewsen, *Phys. Rev. Lett.* **86**, 1994 (2001).
25. D. J. Berkeland, *Rev. Sci. Instrum.* **73**, 2856 (2002).
26. D. Kielpinski, C. Monroe, and D. J. Wineland, *Nature* **417**, 709 (2002).
27. M. Madsen, D. Stick, W. K. Hensinger, J. A. Rabchuk, and C. Monroe, quant-ph 0401047 (2004).
28. M.A. Rowe, A. Ben Kish, B.L. DeMarco, D. Leibfried, V. Meyer, J. Britton, J. Hughes, W.M. Itano, B.M. Jelenkovic, C. Langer, T. Rosenband, and D.J. Wineland, *Quant. Inform. Comp.* **2**, 257 (2002).
29. A. Steane, C. F. Roos, D. Stevens, A. Mundt, D. Leibfried, F. Schmidt-Kaler, and R. Blatt, *Phys. Rev. A* **62**, 042305 (2000).
30. J. I. Cirac, P. Zoller, H. J. Kimble, and H. Mabuchi, *Phys. Rev. Lett.* **78**, 3221 (1997).
31. T. Pellizzari, *Phys. Rev. Lett.* **79**, 5242 (1997).
32. L.-M. Duan, B. B. Blinov, D. L. Moehring, and C. Monroe, quant-ph 0401032 (2004).
33. H. Rohde, S. T. Gulde, C. F. Roos, P. A. Barton, D. Leibfried, J. Eschner, F. Schmidt-Kaler, R. Blatt, *J. Opt. B: Quantum Semiclass. Opt.* **3**, S34 (2001).
34. B. B. Blinov, L. Deslauriers, P. Lee, M. J. Madsen, R. Miller, and C. Monroe, *Phys. Rev. A* **65**, 040304 (2002).
35. M. Barrett, B.L. DeMarco, T. Schaetz, D. Leibfried, J. Britton, J. Chiaverini, W.M. Itano, B.M. Jelenkovic, J.D. Jost, C. Langer, T. Rosenband, and D.J. Wineland, *Phys. Rev. A* **68**, 042302 (2003).
36. P. T. H. Fisk, *Rep. Prog. Phys.* **60**, 761 (1997).
37. In *Proceedings of the 6th Symposium on Frequency Standards and Metrology*, edited by P. Gill (World Scientific, Singapore, 2002).
38. F. Schmidt-Kaler, H. Häffner, M. Riebe, S. Gulde, G. P. T. Lancaster, T. Deuschle, C. Becher, C. F. Roos, J. Eschner and R. Blatt, *Nature* **422**, 408 (2003).
39. R. J. Rafac, B. C. Young, J. A. Beall, W. M. Itano, D. J. Wineland, and J. C. Bergquist, *Phys. Rev. Lett.* **85**, 2462 (2000).
40. F. Schmidt-Kaler, S. Gulde, M. Riebe, T. Deuschle, A. Kreuter, G. Lancaster, C. Becher, J. Eschner, H. Häffner, and R. Blatt, *J. Phys. B: At. Mol. Opt. Phys.* **36**, 623 (2003).
41. H. Häffner, S. Gulde, M. Riebe, G. Lancaster, C. Becher, J. Eschner, F. Schmidt-Kaler, and R. Blatt, *Phys. Rev. Lett.* **90**, 143602 (2003).
42. R. Blatt and P. Zoller, *Eur. J. Phys.* **9**, 250 (1988).
43. D.J. Wineland, J.J. Bollinger, W.M. Itano, F.L. Moore, and D.J. Heinzen, *Phys. Rev. A* **46**, R6797 (1992).
44. F. Mintert and C. Wunderlich, *Phys. Rev. Lett.* **87**, 257904 (2001).
45. C.R. Monroe, D.M. Meekhof, B.E. King, W.M. Itano, and D.J. Wineland, *Phys. Rev. Lett.* **75**, 4714 (1995).
46. A. Sørensen and K. Mølmer, *Phys. Rev. Lett.* **82**, 1971 (1999).
47. A. Sørensen and K. Mølmer, *Phys. Rev. A* **62**, 02231 (2000).
48. E. Solano, R. L. de Matos Filho, and N. Zagury, *Phys. Rev. A* **59**, 2539 (1999).
49. C.A. Sackett, D. Kielpinski, B.E. King, C. Langer, V. Meyer, C.J. Myatt, M.A. Rowe, Q.A. Turchette, W.M. Itano, D.J. Wineland, and C.R. Monroe, *Nature* **404**, 256 (2000).

50. D. Kielpinski, V. Meyer, M.A. Rowe, C.A. Sackett, W.M. Itano, C.R. Monroe, and D.J. Wineland, *Science* **291**, 1013 (2001).
51. B. DeMarco, A. Ben-Kish, D. Leibfried, V. Meyer, M. Rowe, B.M. Jelenkovic, W.M. Itano, J. Britton, C. Langer, T. Rosenband, D.J. Wineland, quant-ph/0208180 (2002).
52. C.R. Monroe, D. Leibfried, B.E. King, D.M. Meekhof, W.M. Itano, and D.J. Wineland, *Phys. Rev. A* **55**, R2489 (1997).
53. D. Leibfried, B.L. DeMarco, V. Meyer, D. Lucas, M. Barrett, J. Britton, W.M. Itano, B.M. Jelenkovic, C. Langer, T. Rosenband, and D.J. Wineland, *Nature* **422**, 412 (2003).
54. J. J. Garcia-Ripoll, J. I. Cirac, and P. Zoller, *Phys. Rev. Lett.* **91**, 157901 (2003).
55. J. F. Poyatos, J. I. Circa, R. Blatt, and P. Zoller, *Phys. Rev. A* **54**, 1532 (1996).
56. C. Monroe, D. M. Meekhof, B. E. King, and D. J. Wineland, *Science* **272**, 1131 (1996).
57. C.J. Myatt, B.E. King, Q.A. Turchette, C.A. Sackett, D. Kielpinski, W.M. Itano, C.R. Monroe, and D.J. Wineland, *Nature* **403**, 269 (2000).
58. X. Wang, A. Sørensen, and K. Mølmer, *Phys. Rev. Lett.* **86**, 3907 (2001).

Ion Trap Quantum Computing with Ca^+ Ions

R. Blatt,[1,2] H. Häffner,[1] C. F. Roos,[1] C. Becher,[1,3] and F. Schmidt-Kaler[1]

The scheme of an ion trap quantum computer is described and the implementation of quantum gate operations with trapped Ca^+ ions is discussed. Quantum information processing with Ca^+ ions is exemplified with several recent experiments investigating entanglement of ions.

KEY WORDS: Quantum computation; trapped ions; ion trap quantum computer; Bell states; entanglement; controlled-NOT gate; state tomography; laser cooling.

PACS: 03.67.Lx; 03.67.Mn; 32.80.Pj.

1. INTRODUCTION

Quantum information processing was proposed and considered first by Feynman and Deutsch.[(1,2)] The requirements for a quantum processor are nowadays known as the DiVincenzo criteria.[(3)] Storing and processing quantum information requires: (i) scalable physical systems with well-defined qubits; which (ii) can be initialized; and have (iii) long lived quantum states in order to ensure long coherence times during the computational process. The necessity to coherently manipulate the stored quantum information requires: (iv) a set of universal gate operations between the qubits which must be implemented using controllable interactions of the quantum systems; and finally, to determine reliably the outcome of a quantum computation (v) an efficient measurement procedure. In recent years, a large variety of physical systems have been proposed and investigated for their use in

[1]Institut für Experimentalphysik, Universität Innsbruck, Technikerstrasse 25, A-6020 Innsbruck, Austria.
[2]Institut für Quantenoptik und Quanteninformation, Österreichische Akademie der Wissenschaften, Technikerstrasse 25, A-6020 Innsbruck, Austria.
[3]To whom correspondence should be addressed. E-mail: christoph.becher@uibk.ac.at

quantum information processing and are considered in other articles of this issue.

In this paper, quantum information processing is discussed using trapped Ca^+ ions where the qubit is encoded in long-lived (ground and metastable) electronic states. A possible different approach for encoding qubits uses two hyperfine levels of hydrogen-like ions, e.g., Be^+ ions (see article by Blinov *et al.*(4) in this volume) or Cd^+ ions.(5)

This paper is organized as follows: after a brief introduction to the concept of the ion trap quantum computer in Sec. 2, some crucial details of the Ca^+-based approach are outlined in Sec. 3. Coherent manipulation of the ions is briefly described in Sec. 4 and the basic two-ion gate operation is reviewed in Sec. 5. The preparation of two-qubit entangled states is summarized in Sec. 6 and future developments of a Ca^+-based ion trap computer are outlined in Sec. 7.

2. CONCEPT OF THE ION TRAP QUANTUM COMPUTER

Strings of trapped ions were proposed in 1995 for quantum computation by Ignacio Cirac and Peter Zoller.(6) With such a system, all requirements for a quantum information processor(3) can be met. Using strings of trapped ions in a linear Paul trap, qubits can be realized employing either metastable excited states, long-lived hyperfine states or corresponding Zeeman sub-states. A set of universal quantum gate operations is then given by: (i) single-qubit rotations (which are realized by Rabi oscillations of individual ions); (ii) the controlled-NOT (CNOT) operation between any two qubits. As a first step the entire ion string is cooled to the ground state of its harmonic motion in the ion trap. Since the mutual Coulomb repulsion spatially separates the ions, any induced motion couples to all ions equally. By applying a laser pulse to the controlling ion its internal excited state amplitude is mapped to a single phonon quantum motion of that ion. This phonon, however, is now carried by the entire string, and an operation on the target qubit which depends on whether or not there is motion in the string, allows one to realize the CNOT-gate operation.

Any algorithm can be implemented using a series of such one- and two-qubit operations and therefore this set of instructions constitutes a universal quantum gate.(7) Thus, the realization of these quantum gates allows one to build and operate a quantum computer. Moreover, in principle, this concept provides a scalable approach towards quantum computation and has therefore attracted quite some attention.

During recent years, several other techniques have been proposed to implement gate operations with trapped ions. Sørensen and Mølmer

[8,9] and, with a different formulation, Milburn[10] proposed a scheme for "hot" quantum gates, i.e., their procedures for gate operations do not require ground state cooling of an ion string. Although successfully applied to trapped Be^+ ions,[11] with the trapping parameters currently available, these gate procedures are not easily applicable to Ca^+ ions. Other gates based on ac Stark shifts have been suggested by Jonathan *et al.*[12] and holonomic quantum gates (using geometric phases) have been proposed by Duan *et al.*[13] A different CNOT-gate operation also based on the ac Stark effect which does not require individual addressing and ground state cooling has been realized with trapped Be^+ ions.[14]

3. SPECTROSCOPY IN ION TRAPS

Ions are considered to be trapped in a harmonic potential with frequency ν_z, interacting with the travelling wave of a single mode laser tuned close to a transition that forms an effective two-level system.

Internal state detection of a trapped ion is achieved using the electron shelving technique. For this, one of the internal states of the trapped atom is selectively excited to a third short-lived state thereby scattering many photons on that transition if the coupled internal state was occupied. If, on the other hand, the atom's electron resides in the uncoupled state of the qubit (i.e., the electron is shelved in that state) then no photons are scattered and thus the internal state can be detected with an efficiency of nearly 100%.[15]

Figure 1 shows the relevant levels of the Ca^+ ion which are populated in the experiment. The qubit is implemented using the narrow quadrupole transition at 729 nm, i.e., $|g\rangle = |S_{1/2}\rangle$ and $|e\rangle = |D_{5/2}\rangle$. For optical cooling and state detection, resonance fluorescence on the $S_{1/2}$–$P_{1/2}$ transition is

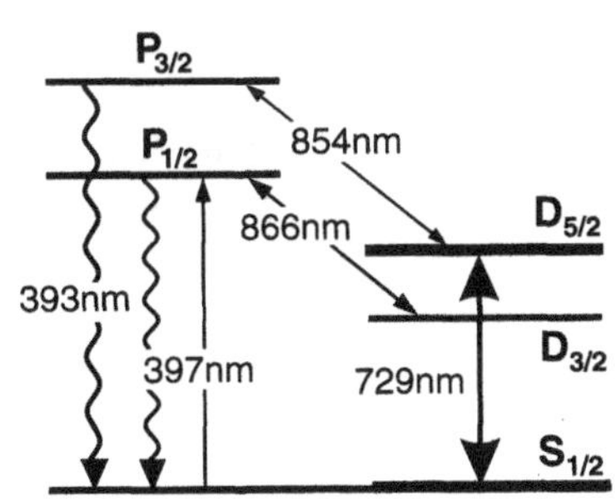

Fig. 1. Level scheme of $^{40}Ca^+$. The qubit is implemented using the narrow quadrupole transition. All states split up into the respective Zeeman sublevels.

scattered by excitation with 397 and 866 nm radiation. The laser at 854 nm is applied to repump the excited state $|e\rangle$, for example after a shelving operation.

3.1. Laser Cooling of Ion Strings

A prerequisite of the Cirac–Zoller (CZ) scheme is that the initial state of the quantum register is prepared in its motional ground state, i.e., we require that the motional mode which carries the coupling between the qubits is initially in the ground state.

Laser cooling of trapped ions is therefore one of the key techniques for an ion trap quantum computer.(15,16) Usually so-called sideband cooling(17,18) is used to cool one mode of an ion string to its motional ground state. This is experimentally achieved using optical pumping schemes involving either Raman transitions(19) or coupled transitions.(18,20) More elaborate cooling schemes using electromagnetic transparency(21,22) or sympathetic cooling(23–25) have been investigated and and can be employed for cooling multiple vibrational modes simultaneously or cooling an ion string by addressing just one ion, respectively.

3.2. Addressing of Individual Ions

The implementation of the CZ CNOT-gate operation requires that individual ions can be addressed in order to rewrite internal information onto the vibrational ("phonon") mode using appropriate transitions. Therefore, the Innsbruck experiments were designed to operate in a regime where the minimum ion distance is on the order of a few μm such that focussing a laser beam at 729 nm is feasible to individually address the single ions.(26) In the current setup, Ca^+ ions are stored with axial trap frequencies of about 1–1.2 MHz and thus the inter-ion distance of two and three ions is approximately 5 μm. The laser beam at 729 nm is focussed to a waist diameter of approximately 2.5 μm such that with the Gaussian beam profile neighboring ions are excited with less than 10^{-3} of the central intensity. Beam steering and individual addressing is achieved using electrooptic beam deflection which allows for fast switching ($\sim 15\,\mu$s) between different ion positions.(27)

4. COHERENT MANIPULATION OF QUANTUM INFORMATION

Quantum information processing requires that individual qubits are coherently manipulated. We realize single-qubit rotations by coherent manipulation of the $S_{1/2}(m=-1/2) \leftrightarrow D_{5/2}(m=-1/2)$ transition in Ca^+. Coupling of two qubits requires the precise control of the motional state of a single ion or a string of ions. Both operations can be performed by applying laser pulses at the carrier (i.e., not changing the vibrational quantum number, $\Delta n_z = 0$) or at one of the sidebands of the S–D transition (i.e., laser detuned by $\pm\nu_z$, thus changing the vibrational quantum number by $\Delta n_z = \pm 1$).

All qubit transitions are described as rotations on a corresponding Bloch sphere and they are written as unitary operations $R(\theta,\phi)$, $R^-(\theta,\phi)$, $R^+(\theta,\phi)$ on the carrier, red sideband and blue sideband, respectively. The parameter θ describes the angle of the rotation and depends on the strength and the duration of the applied pulse. ϕ denotes its phase, i.e., the relative phase between the optical field and the atomic polarization and determines the axis about which the Bloch vector rotates.(27) Typical pulse durations for a π-pulse range from about 1 to several 10 μs for the carrier transition and 50–200 μs on the sideband transition, with the chosen time depending on the desired speed and precision of the operations. Such pulses are the primitives for quantum information processing with trapped ions. By concatenating pulses on the carrier and sidebands, gate operations and, eventually whole quantum algorithms, can be implemented.(27) Even the simplest gate operations require several pulses, therefore it is imperative to control the relative optical phases of these pulses in a very precise manner or, at least, to keep track of them such that the required pulse sequences lead to the desired operations. This requires the precise consideration of all phases introduced by the light shifts of the exciting laser beams.(28)

5. CIRAC–ZOLLER CNOT-GATE OPERATION

For the realization of the CZ CNOT-gate operation, two ions are loaded into the linear trap and, by means of an intensified CCD camera, the fluorescence is monitored separately for each ion.(27) If no information on a particular qubit is needed, the signal of a more sensitive photomultiplier tube is used to infer the overall state population and, thus, the exposure time can be reduced.

As proposed by Cirac and Zoller, the common vibration of an ion string is used to convey the information for a conditional operation (bus-mode).(6) Accordingly, the gate operation can be achieved with a sequence of three steps after the ion string has been prepared in the ground state $|n_b = 0\rangle$ of the bus-mode. First, the quantum information of the control ion is mapped onto this vibrational mode. As a result, the entire string of ions is moving and thus the target ion participates in the common motion. Second, and conditional upon the motional state, the target ion's qubit is inverted. Finally, the state of the bus-mode is mapped back onto the control ion. Note that this gate operation is not restricted to a two-ion crystal since the vibrational bus-mode can be used to interconnect any of the ions in a large crystal, independent of their position.

We realize this gate operation[29] with the following sequence of laser pulses. A blue sideband π-pulse, $R^+(\pi, 0)$, on the control ion transfers its quantum state to the bus-mode. Next, we apply the CNOT-gate operation

$$R_{\mathrm{CNOT}} = R\left(\frac{\pi}{2}, 0\right) R^+\left(\pi, \frac{\pi}{2}\right) R^+\left(\frac{\pi}{\sqrt{2}}, 0\right) R^+\left(\pi, \frac{\pi}{2}\right) R^+\left(\frac{\pi}{\sqrt{2}}, 0\right) R\left(\frac{\pi}{2}, \pi\right) \tag{1}$$

to the target ion. Finally, the bus-mode and the control ion are reset to their initial states by another π-pulse $R^+(\pi, \pi)$ on the blue sideband. The resulting gate fidelity of about 71–78% is well understood in terms of a collection of experimental imperfections.[29] Most important is dephasing due to laser frequency noise and ambient magnetic field fluctuations that cause a Zeeman shift of the qubit levels.[27] As quantum computing might be understood as a multi-particle Ramsey interference experiment, a faster execution of the gate operation would help to overcome this type of dephasing errors. However, a different type of error increases with the gate speed: with higher Rabi frequencies, the off-resonant excitation of the nearby and strong carrier transition becomes increasingly important,[30] even if the corresponding phase shift is compensated. Additional, but minor, errors are due to the addressing imperfection, residual thermal excitation of the bus-mode and spectator modes as well as laser intensity fluctuations.

If the qubits are initialized in the superposition state $|\text{control, target}\rangle = |S+D, S\rangle$, the CNOT operation generates an entangled state $|S, S\rangle + |D, D\rangle$. Using local operations with varying phase then allows the preparation of arbitrary Bell states using the CNOT-gate operation[31] (see Fig. 2).

6. BELL STATE GENERATION AND ENTANGLEMENT STUDIES

Bell states are very important for an investigation of entanglement with the capability to produce them at the push of a button is one of the major advantages of an ion trap quantum computer. However, while conceptually simple and straightforward, Bell states need not be generated

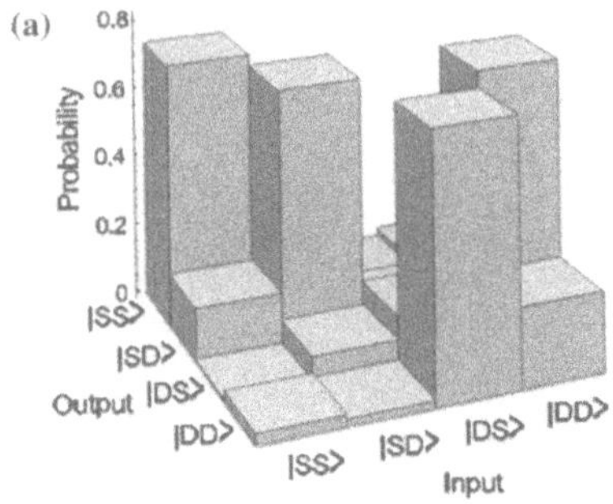

(b)

	$\|SS\rangle$	$\|SD\rangle$	$\|DS\rangle$	$\|DD\rangle$
$\|SS\rangle$	0.74(3)	0.13(3)	0.05(3)	0.08(3)
$\|SD\rangle$	0.15(3)	0.71(5)	0.06(1)	0.08(2)
$\|DS\rangle$	0.01(2)	0.08(3)	0.14(4)	0.77(3)
$\|DD\rangle$	0.03(3)	0.02(1)	0.72(6)	0.22(4)

Fig. 2. Truth table of CZ CNOT-gate operation. The amplitude of the controlling ion (first entry of the state notation) controls the state of the target ion (second entry), i.e., when the controlling ion's amplitude is $|S\rangle$, the target ion's state remains the same, when it is $|D\rangle$, the target's ion state is flipped. (a) graphical representation; (b) numerical results as shown in (a).[29]

using CNOT-gate operations. With trapped ions, there are simpler and more efficient procedures to produce and investigate these states.

Using a string of two ions and the individual addressing capability in the Innsbruck experiment, we create all Bell states by applying laser pulses to ions 1 and 2 on the blue sideband and on the carrier transition. To produce the Bell state $\Psi_{\pm} = 1/\sqrt{2}(|S, D\rangle \pm |D, S\rangle)$ we use the pulse sequence $U_{\Psi_\pm} = R_2^+(\pi, \pm\pi/2) \cdot R_2(\pi, \pi/2) \cdot R_1^+(\pi/2, -\pi/2)$ applied to the $|S, S\rangle$ state. Here, the indices 1 (2) refer to pulses applied to ions 1 and 2, respectively. The first pulse $R_1^+(\pi/2, -\pi/2)$ entangles the motional and the internal degrees of freedom. The next two pulses $R_2^+(\pi, \pm\pi/2) \cdot R_2(\pi, \pi/2)$ map the motional degree of freedom onto the internal state of ion 2. Appending another π-pulse on the carrier transition, $R_2(\pi, 0)$, to the sequence $U_{\Psi_\pm}$ produces the state $\Phi_{\pm}$. This entire pulse sequence takes less than $200\,\mu$s and is much simpler than a full CZ CNOT-gate operation which takes about twice that time and is thus more sensitive to decoherence.

Investigation of the prepared state and a characterization of the achieved entanglement then is obtained by a quantum state analysis using a tomographic procedure. Quantum state tomography allows the estimation of an unknown quantum state that is available in many identical copies. It has been experimentally demonstrated for a variety of physical systems, among them the quantum state of a light mode,[32] the vibrational state of a single ion,[33] and the wave packets of atoms of an atomic beam.[34] Multi-particle states have been investigated in nuclear magnetic resonance experiments[35] as well as in experiments involving entangled photon pairs. Here, we apply this technique to entangled massive particles of a quantum register for an investigation of entanglement and studies of decoherence.

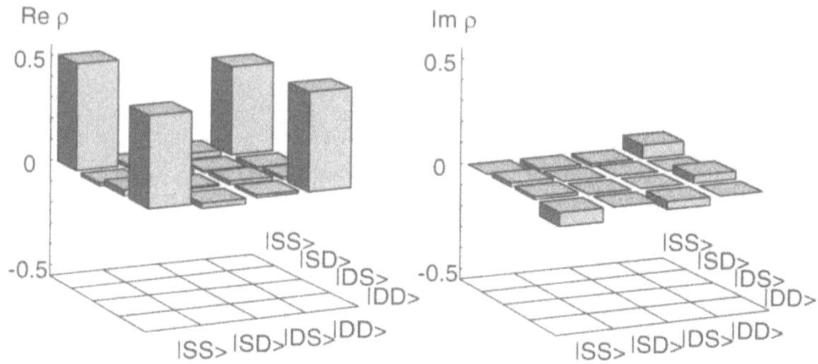

Fig. 3. Real and imaginary part of the density matrix ρ_{Φ_+} that approximates $1/\sqrt{2}(|S, S\rangle + |D, D\rangle)$. The measured fidelity is $F = \langle\Phi_+|\rho_{\Phi_+}|\Phi_+\rangle = 0.91$.

The tomographic method consists of individual single-qubit rotations, followed by a projective measurement. For the analysis of the data, we employ a maximum likelihood estimation of the density matrix following the procedure as suggested in Refs. 36 and 37 and implemented in experiments with pairs of entangled photons.[(38)] As an example, Fig. 3 shows the reconstructed density matrix ρ of one out of four Bell states. To monitor the evolution of these entangled states in time we introduce a waiting interval before performing state tomography. We expect that Bell states of the type $\Psi_\beta = |S, D\rangle + e^{i\beta}|D, S\rangle$ are immune against collective dephasing due to fluctuations of the qubit energy levels or the laser frequency.[(39)] However, a magnetic field gradient that gives rise to different Zeeman shifts on qubits 1 and 2 leads to a deterministic and linear time evolution of the relative phase $e^{i\beta}$ between the $|S, D\rangle$ and the $|D, S\rangle$ component of the $\Psi_\pm$ states. Experimentally, we find that the lifetime of entangled states of this type is indeed no longer limited by the technical constraints (i.e., magnetic field and laser frequency fluctuations) but is only limited by the spontaneous decay from the upper $D_{5/2}$-level (lifetime $\tau_D \simeq 1$ s) of the qubit. Finally, we can specify the entanglement of the four Bell states, using the entanglement of formation,[(40)] and find $E(\Psi_-) = 0.79(4)$, $E(\Psi_+) = 0.75(5)$, $E(\Phi_+) = 0.76(4)$ and $E(\Phi_-) = 0.72(5)$.[(31)]

7. FUTURE DEVELOPMENTS OF THE CA$^+$ ION TRAP QUANTUM COMPUTER

With the availability of one- and two-qubit operations, the individual addressing and the near perfect readout features, a Ca^+-based ion trap quantum computer can be envisioned. Currently, the techniques described above are extended to work with three and more ions which already offer a vast variety of experimental possibilities, ranging from the preparation and investigation of generalized 3-qubit entangled states to an implementation of teleportation and rudimentary error correction protocols.

While detection efficiencies and noise considerations are quite favorable for an optical qubit transition, there are a number of technical limitations. Most of these limitations are not of a fundamental nature, but are rather given by technical shortcomings, such as the sensitivity of the qubit transition with respect to external magnetic fields and spurious laser frequency and intensity fluctuations. The only fundamental limitation is the lifetime of the pertaining qubit states, here in particular that of the $D_{5/2}$ state (1.16 s), which, however, is orders of magnitude larger than typical gate operation times. The limitations discussed above might lead to reconsidering the use of ground state Zeeman and hyperfine splittings for encoding the quantum information. We illustrate here the specific pros and cons considering respective transitions in $^{40}Ca^+$ and $^{43}Ca^+$ ions. Whereas the current experiments work with an optical qubit (i.e. $|0\rangle = |D_{5/2}, m_J = -1/2\rangle$ and $|1\rangle = |S_{1/2}, m_J = -1/2\rangle$, cf. Fig. 1) in the even isotope $^{40}Ca^+$, an alternative implementation would work with the odd isotope $^{43}Ca^+$ (nuclear spin $I = 7/2$) and the hyperfine ground states $|0\rangle = |F = 4, m_F = 0\rangle$ and $|1\rangle = |F = 3, m_F = 0\rangle$ (see Fig. 4). In the latter case, optical manipulation of the qubit would be achieved using Raman transitions.

To a large extent the coherence properties of the qubits depend on the respective sensitivity on external field fluctuations, e.g., magnetic and laser field fluctuations. Therefore, in the optical case, a highly stabilized laser is required for the qubit transition whereas in the case of a Raman transition, both Raman beams can be derived from the same laser source where the required stable relative phase relation can be achieved with only modest technical efforts. The large fine-structure splitting of $\Delta\nu_{FS}$ = 6.7 THz between the $P_{1/2}$ and $P_{3/2}$ states allows a large detuning of the Raman light fields from the P-levels and thus high fidelity gate operations, as spontaneous emission processes are largely suppressed. The fine-structure splitting of $^{43}Ca^+$ can be compared to that of other favorable qubit candidates, e.g. $^{9}Be^+$ with $\Delta\nu_{FS}$ = 0.2 THz and $^{111}Cd^+$ with $\Delta\nu_{FS}$ = 74 THz.

Aside from these more technical constraints, encoding the qubit in the hyperfine ground states ensures that decay from spontaneous emission is completely avoided and thus, very long coherence times (many seconds and even minutes have been demonstrated with trapped Be^+ ions) may be potentially achieved. Furthermore, the qubits will, ideally, depend only in second order on the external magnetic field ($\Delta m_F = 0$ transitions, see Fig. 4). While many of these advantages are available already with Be^+ and Cd^+ ions, the $^{43}Ca^+$ ion offers additionally, a quadrupole

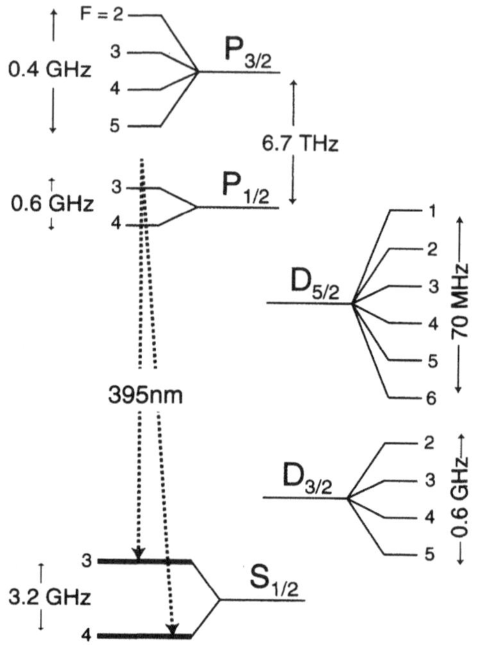

Fig. 4. Level scheme of the $^{43}Ca^+$ isotope. A qubit can be encoded in the hyperfine ground states $|0\rangle = |F=4, m_F=0\rangle$ and $|1\rangle = |F=3, m_F=0\rangle$.

transition that can be advantageously used for shelving and efficient detection without the need for a technically advanced laser source. Therefore, the next generation of a Ca^+-based ion trap quantum computer will ideally combine the advantages of the ground state encoding of the qubit and the optical shelving and detection techniques.

8. SUMMARY AND PERSPECTIVES

On the road towards a scalable quantum processor(41) with ion traps, single-qubit rotations and universal two-qubit operations gate have been realized. With trapped Ca^+ ions, we present an experimental setup which allows one to flexibly control a register of two qubits. With the universal set of quantum gates all unitary operations can be implemented. Therefore, arbitrary two-qubit states can be synthesized with high fidelities and analyzed via state tomography. The currently available experiments demonstrate the operation of a small quantum computer and allow one to develop the basic tools of experimental quantum information processing.

One of the most striking features is that the ion trap quantum information processor is scalable in principle, i.e., adding more qubits is straightforward and at least up to about 10 qubits this should not pose insurmountable technical difficulties. Larger systems will require special architectures such as ion trap arrays,[42] moving ions in structured ion traps[43] or even interconnecting several small ion-trap quantum computers using cavities and photons as a quantum channel.[44,45] While all these techniques require tremendous technical efforts, to the best of our current knowledge there are no principal limitations to scaling up an ion-trap quantum computer.

The current experiments demonstrate that ion trap quantum information processors offer a realistic route towards the realization of large-scale quantum computing and they provide ideal means for the engineering of quantum objects and controlling quantum processes at mesoscopic and macroscopic scales.

ACKNOWLEDGMENTS

We thank Stephan Gulde, Mark Riebe, Gavin Lancaster, and Wolfgang Hänsel for their continuous support in the laboratory. We gratefully acknowledge support by the European Commission (QUEST (HPRN-CT-2000- 00121) and QGATES (IST-2001-38875) networks), by the ARO (DAAD19-03-1-0176), by the Austrian Science Fund (FWF), and by the Institut für Quanteninformation GmbH. H. H. acknowledges support by the Marie-Curie program of the European Commission.

REFERENCES

1. R. P. Feynman, *Int. J. Theor. Phys.* **21**, 467 (1982).
2. D. Deutsch, *Proc. R. Soc. Lond. A* **400**, 97 (1985).
3. D. P. DiVincenzo, *Quantum Information Comput.* **1**, Special 1–6 (2001).
4. B. B. Blinov *et al.*, see article in this issue.
5. P. J. Lee, B. B. Blinov, K. Brickman, L. Deslauriers, M. J. Madsen, R. Miller, D. L. Moehring, D. Stick, and C. Monroe, *Opt. Lett.* **28**, 1582 (2003).
6. J. I. Cirac and P. Zoller, *Phys. Rev. Lett.* **74**, 4091 (1995).
7. D. P. DiVincenzo, *Phys. Rev. A* **51**, 1015 (1995).
8. A. Sørensen and K. Mølmer, *Phys. Rev. Lett.* **82**, 1971 (1999); K. Mølmer and A. Sørensen, *Phys. Rev. Lett.* **82**, 1835 (1999).
9. K. Mølmer and A. Sørensen, *Phys. Rev. A* **62**, 022311 (2000).
10. G. J. Milburn, quant-ph/9908037.
11. C. A. Sackett, D. Kielpinski, B. E. King, C. Langer, V. Meyer, C. J. Myatt, M. Rowe, Q. A. Turchette, W. M. Itano, D. J. Wineland, and C. Monroe, *Nature* **404**, 256 (2000).

12. D. Jonathan, M. B. Plenio, and P. Knight, *Phys. Rev. A* **62**, 042307 (2000); D. Jonathan and M. B. Plenio, *Phys. Rev. Lett.* **87**, 127901 (2001).
13. L. M. Duan, I. Cirac, and P. Zoller, *Science* **292**, 1695 (2001).
14. D. Leibfried, B. DeMarco, V. Meyer, D. Lucas, M. Barrett, J. Britton, W. M. Itano, B. Jelenković, C. Langer, T. Rosenband, and D. J. Wineland, *Nature* **422**, 412 (2003).
15. D. Leibfried, R. Blatt, C. Monroe, and D. Wineland, *Rev. Mod. Phys.* **75**, 281 (2003).
16. J. Eschner, G. Morigi, F. Schmidt-Kaler, and R. Blatt, *J. Opt. Soc. Am. B* **20**, 1003 (2003).
17. F. Diedrich, J. C. Bergquist, W. M. Itano, and D. J. Wineland, *Phys. Rev. Lett.* **62**, 403 (1989).
18. Ch. Roos, Th. Zeiger, H. Rohde, H. C. Nägerl, J. Eschner, D. Leibfried, F. Schmidt-Kaler, and R. Blatt, *Phys. Rev. Lett.* **83**, 4713 (1999).
19. C. Monroe, D. M. Meekhof, B. E. King, S. R. Jefferts, W. M. Itano, D. J. Wineland, and P. Gould, *Phys. Rev. Lett.* **75**, 4011 (1995).
20. I. Marzoli, J. I. Cirac, R. Blatt, and P. Zoller, *Phys. Rev. A* **49**, 2771 (1994).
21. G. Morigi, J. Eschner, and C. H. Keitel, *Phys. Rev. Lett.* **85**, 4458 (2000).
22. C. F. Roos, D. Leibfried, A. Mundt, F. Schmidt-Kaler, J. Eschner, and R. Blatt, *Phys. Rev. Lett.* **85**, 5547 (2000).
23. D. Kielpinski, B. E. King, C. J. Myatt, C. A. Sackett, Q. A. Turchette, W. M. Itano, C. Monroe, and D. J. Wineland, *Phys. Rev. A* **61**, 032310 (2000).
24. H. Rohde, S. T. Gulde, C. F. Roos, P. A. Barton, D. Leibfried, J. Eschner, F. Schmidt-Kaler, and R. Blatt, *J. Opt. B* **3**, S34 (2001).
25. B. B. Blinov, L. Deslauriers, P. Lee, M. J. Madsen, R. Miller, and C. Monroe, *Phys. Rev. A* **65**, 040304(R) (2002).
26. H. C. Nägerl, D. Leibfried, H. Rohde, G. Thalhammer, J. Eschner, F. Schmidt-Kaler, and R. Blatt, *Phys. Rev. A* **60**, 145 (1999).
27. F. Schmidt-Kaler, H. Häffner, S. Gulde, M. Riebe, G. P. T. Lancaster, T. Deuschle, C. Becher, W. Hänsel, J. Eschner, C. F. Roos, and R. Blatt, *Appl. Phys. B* **77**, 789 (2003).
28. F. Schmidt-Kaler, H. Haffner, S. Gulde, M. Riebe, G. Lancaster, J. Eschner, C. Becher, and R. Blatt, *Europhys. Lett.* **65**, 587 (2004).
29. F. Schmidt-Kaler, H. Häffner, M. Riebe, S. Gulde, G. P. T. Lancaster, T. Deuschle, C. Becher, C. F. Roos, J. Eschner, and R. Blatt, *Nature* **422**, 408 (2003).
30. A. Steane, C. F. Roos, D. Stevens, A. Mundt, D. Leibfried, F. Schmidt-Kaler, and R. Blatt, *Phys. Rev. A* **62**, 042305 (2000).
31. C. F. Roos, G. P. T. Lancaster, M. Riebe, H. Häffner, W. Hänsel, S. Gulde, C. Becher, J. Eschner, F. Schmidt-Kaler and R. Blatt, Phys. Rev. Lett. 92, 220402 (2004).
32. D. T. Smithey, M. Beck, M. G. Raymer, and A. Faridani, *Phys. Rev. Lett.* **70**, 1244 (1993).
33. D. Leibfried, D. M. Meekhof, B. E. King, C. Monroe, W. M. Itano, and D. J. Wineland, *Phys. Rev. Lett.* **77**, 4281 (1996).
34. Ch. Kurtsiefer, T. Pfau, and J. Mlynek, *Nature* **386**, 150 (1997).
35. I. L. Chuang, N. Gershenfeld, M. G. Kubinec, and D. Leung, *Proc. R. Soc. London A* **454**, 447 (1998).
36. Z. Hradil, *Phys. Rev. A* **55**, R1561 (1997).
37. K. Banaszek, G. M. D'Ariano, M. G. A. Paris, and M. F. Sacchi, *Phys. Rev. A* **61**, 010304 (1999).
38. D. F. V. James, P. G. Kwiat, W. J. Munro, and A. G. White, *Phys. Rev. A* **64**, 052312 (2001).
39. D. Kielpinski, V. Meyer, M. A. Rowe, C. A. Sackett, W. M. Itano, C. Monroe, and D. J. Wineland, *Science* **291**, 1013 (2001).

40. W. K. Wootters, *Phys. Rev. Lett.* **80**, 2245 (1998).
41. *Quantum Information Science and Technology Roadmapping Project (ARDA)*, available from http://qist.lanl.gov/
42. J. I. Cirac and P. Zoller, *Nature* **404**, 579 (2000).
43. D. Kielpinski, C. R. Monroe, and D. J. Wineland, *Nature* **417**, 709 (2002).
44. J. I. Cirac, P. Zoller, H. J. Kimble, and H. Mabuchi, *Phys. Rev. Lett.* **78**, 3221 (1997).
45. A. B. Mundt, A. Kreuter, C. Becher, D. Leibfried, J. Eschner, F. Schmidt-Kaler, and R. Blatt, *Phys. Rev. Lett.* **89**, 103001 (2002).

Quantum Information Processing in Cavity-QED

S. J. van Enk,[1,2] H. J. Kimble,[1] and H. Mabuchi[1]

We give a brief overview of cavity-QED and its roles in quantum information science. In particular, we discuss setups in optical cavity-QED, where either atoms serve as stationary qubits, or photons serve as flying qubits.

KEY WORDS: Cavity QED; atom trapping & cooling; quantum computing; quantum communication.

PACS: 42.50.Pq; 03.67.Lx; 03.67.Hk; 32.80.Pj.

1. INTRODUCTION

Cavity-QED provides an important paradigm for studying the controlled, coherent coupling of optical and atomic qubits. The central phenomenon of cavity-QED is strong coupling between the internal state of an atom and the state of a single mode of the electromagnetic field. In a high-finesse microcavity, atoms and photons interact much more strongly than they do in free space, and this enhancement can be used to implement quantum logic. For example, with a sufficiently good cavity it becomes possible to observe nonlinear optical effects with just one photon per mode, and this provides a mechanism for controlled phase gates between qubits encoded in the polarizations of single photon wave-packets.(1) Likewise, conditional mappings of the quantum state of 'control' and 'target' atoms trapped inside a cavity should be realizable by virtue of their respective strong couplings to a common cavity eigenmode.(2) The direct atom–photon coupling itself enables quantum logic interactions between qubit representations with very different physical properties, suggesting applications in quantum communication where trapped atoms are

[1]Bell Labs, Lucent Technologies and California Institute of Technology.
[2]To whom correspondence should be addressed. E-mail: vanenk@cs.caltech.edu

used for local storage of quantum information while photons are used for long-distance transmission.[3] In each of these scenarios, the key ingredient is the use of a resonant cavity to enhance the coherent interaction of atoms with a single localized field mode over their dissipative couplings to the electromagnetic continuum (spontaneous emission).

A discussion of the role of cavity-QED in quantum information processing divides naturally into two parts. Below we will first discuss schemes in which a single atom inside a cavity is used to mediate interactions between photonic qubits. We will then turn to proposals in which one or more atoms inside a cavity store qubits in their internal states, and a combination of external laser fields and the quantized cavity mode are used to perform 1-bit and 2-bit operations. Included among the latter are quantum state mapping operations in which an atom's internal state is 'read out' into the state of the intracavity field, which can then be coupled out from the cavity for transmission via free space or an optical fiber. *Flying qubits* (i.e., photon wavepackets) could thus be used for long-distance transmission of quantum information or as intermediaries to perform quantum logic gates between atoms trapped at remote locations, as illustrated in Fig. 1.

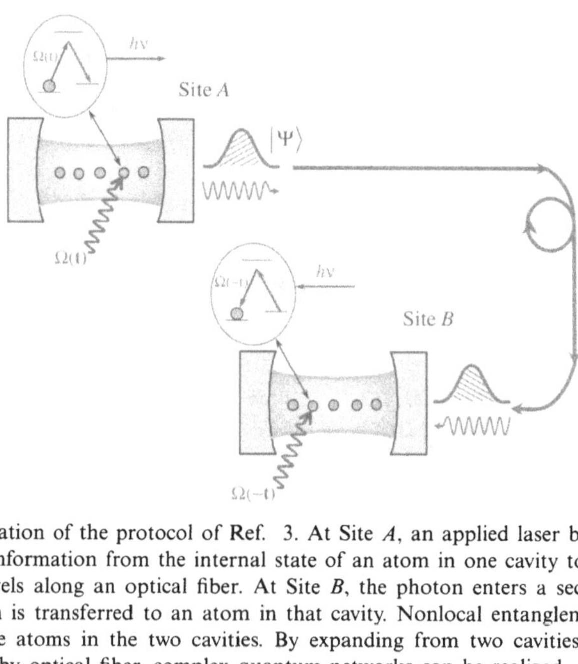

Fig. 1. Illustration of the protocol of Ref. 3. At Site A, an applied laser beam $\Omega(t)$ transfers quantum information from the internal state of an atom in one cavity to a photon state $|\Psi\rangle$, which travels along an optical fiber. At Site B, the photon enters a second cavity, and the information is transferred to an atom in that cavity. Nonlocal entanglement can be created among the atoms in the two cavities. By expanding from two cavities to a larger set interconnected by optical fiber, complex quantum networks can be realized.

From a practical perspective, photonic and atomic qubits each have inherent advantages and disadvantages. Photons are relatively easy to produce and are perfect for the transmission of (quantum) information, but are hard to store. On the other hand, atoms can be trapped for long times with minimal dephasing of ground-state hyperfine or Zeeman coherences, but it is quite difficult to transport them over large distances, although encouraging advances have recently been made in experiments[(4)] that demonstrated coherent transport of ions within a segmented trap. While it is relatively straightforward to couple photons into or out of a microcavity with low insertion loss, it will likely prove challenging to preserve the spatial and spectral purity of optical wave-packets that are cascaded through many cavities in a row. It was a major recent advance to realize a trapping scheme that can confine atoms in a cavity in a state insensitive fashion that does not interfere with or otherwise preclude diverse protocols for quantum information processing.[(5)] Indeed, by now multiple sequential operations per atom have been demonstrated.[(6,7)]

There are two very different parameter regimes in which strong coupling in cavity-QED has been achieved, namely for optical and microwave cavities as appropriate to the wave-length of the electromagnetic field that is resonantly coupled to the cavity. Each of these regimes has its own characteristic advantages and problems. For example, at room temperature a narrow-bandwidth optical cavity contains a completely negligible thermal field, whereas the large number of microwave photons from black-body radiation requires that a cavity operating at frequencies 10–50 GHz must be cooled to well below 1 K to eliminate thermal photons. Microwave photons, therefore, cannot be readily employed for Quantum Communication or distributed Quantum Computation. On the other hand, the dimensions of optical cavities for strong coupling (1–100 μm) are quite small, and so it becomes difficult to trap and to address individually atoms inside an optical cavity.

In microwave cavity-QED experiments,[(8,9)] it is the atoms that move in and out of the cavity, interact with the microwave photons, and that can be entangled and measured. Indeed, atoms have been entangled both with other atoms and with photons in microwave cavity-QED experiments.[(10)] In recent optical cavity-QED experiments, on the other hand, it is light that traverses the cavity and is being measured, after having interacted with a single atom trapped inside the cavity.

Another difference between the two regimes of cavity-QED is the type of atomic transition resonant with the cavity. In the microwave regime, transitions between high-lying circular Rydberg states are used, with a large transition-dipole moment but a small spontaneous emission probability. The disadvantage is that such atoms are difficult to prepare and state measurements are not perfect, the error being in the 10% range. In the optical regime, dipole transitions between ground and electronic excited states are employed, and state detection is efficiently accomplished by way of variations on the 'quantum jump' or 'electron shelving' method as used in ion-trap experiments.

In the remainder of this review we will focus on optical cavity-QED. In order nevertheless to give the reader some pointers to the literature on cavity-QED in general and microwave cavity-QED in particular, we first note there are numerous excellent reviews.(8,9,11) Moreover, there have been reviews of cavity-QED in the context of quantum information processing,(12) and reviews of optical cavity-QED.(13,14)

Before turning to specifics we would like to discuss some lessons learned from research in the area of quantum information processing in cavity-QED. When compared with approaches such as ion trap or ensemble NMR quantum computing, cavity-QED may seem to have made relatively modest progress towards large-scale demonstrations despite considerable effort for a number of years. Of course, it must not be forgotten that all three of these experimental paradigms today build upon decades of prior technical development motivated by goals quite distinct from quantum computing. To a large extent, the different rates of progress have reflected the relative state of advancement in these fields of achieving spatial (and temporal) localization of qubits relative to the structures utilized for implementing quantum logic gates. In optical cavity-QED, techniques have been developed for manipulating atomic motion in ways that do not interfere with the strong atom-cavity coupling that is essential for information processing.(5) The time scale for these advances has been commensurate with previous progress in other areas, such as trapped ions. The "learning ground" for these advances has been largely experiments in bulk Fabry–Perot cavities, which carry a large technical overhead and which clearly are not suited to brute-force scaling of experiments to encompass huge qubit registers. For many years, we have appreciated that the long-term prospects for quantum information processing with cavity QED are tied to the development of integrated hardware platforms that could exploit monolithic optical microcavities, photonic circuits and microfabricated optical or magnetic traps for atom manipulation.(15–17) While it is plausible that such a technical approach could alleviate most difficulties involved in scaling to large systems, it remains to be seen whether the low decoherence and fidelity of control required for quantum information processing can be achieved in such novel apparatus.

Generally speaking it must be kept in mind that scaling from few-qubit demonstrations to complex processors will require not only favorable fabrication methods, but also a validated understanding of decoherence mechanisms and their spatiotemporal correlation properties. Without such an understanding we must check our confidence in architectural strategies for fault tolerance. This is manifestly true for any implementation scheme, but cavity-QED here benefits from a long history of close contact between experiments and modelling via the theory of open quantum systems.[18]

It also seems worth commenting that one obvious approach to solving the qubit localization problem would be to push solid state cavity-QED into the regime of strong coupling, e.g., with rare earth ions or quantum dots taking the place of gas-phase neutral atoms. So far there has not been a convincing demonstration that the requisite combination of coherence time and interaction strength can be achieved by this approach. Detailed quantitative modelling of decoherence effects is likely to be much more difficult than in the case of gas-phase atoms. As the implicit trapping mechanism of embedding qubits in a solid matrix seems inevitably to lead to relatively short coherence times, one feels pushed to investigate optical resonators with extremely low mode volume to maximize the speed of cavity enhanced radiative interactions. It is interesting to note that the very short ($\lesssim 0.1$ ns) gate times that seem reachable with $\lesssim 1$ cubic wavelength optical resonators (such as photonic crystal defect cavities[19]) compare quite favorably with the maximum processing speeds that can be contemplated for implementations based on magnetic dipole couplings. This line of investigation has some potential finally to make good on early observations that quantum logic gates based on the interaction of optical fields with systems of bound charges have a naturally high speed limit.

An important legacy of theoretical research in cavity-QED is a set of novel proposals for exploiting a detailed understanding of decoherence mechanisms in the formulation of efficient schemes for error correction and robust quantum gates. Indeed, such work in cavity-QED has provided some of the earliest illustrations of how general principles such as detection/correction of syndromes[2,20] and gate purification[21] can be set in realistic physical models. Cavity-QED arguably provides the most natural experimental setting in which to study conditional (selective) evolution in quantum systems subject to indirect measurement, as in the experiments in Ref. 22, which is a topic that has become quite central to quantum information theory. In this sense, *cavity-QED highlights the essential physics of quantum error correction.* It is likewise clear that cavity-QED provides a canonical paradigm for investigating how quantum information (and entanglement) can be mapped from intrinsically local representations (the internal states of bound systems of charges) to freely propagating fields, and vice versa. While the above issues are also at the heart of recent work on measurement induced atomic spin-squeezing[23,24] and the

Duan–Lukin–Cirac–Zoller scheme for quantum communication,[25] it does not seem possible for such studies to compete with the extreme degree of ideality (and hence the transparency of modelling) achieved by cavity-QED in the limit of one atom and one photon.

From the perspective of the preceding paragraph it seems clear that experiments in cavity-QED will continue to play a key role in shedding light on the fundamental physics underlying quantum information science, even if it should slip from the vanguard in the race towards construction of full-scale quantum processors. Given its focus on elementary atom–photon interactions, cavity-QED should naturally retain a leading role in the growing fields of quantum communication and distributed quantum information processing, which seek more directly to explore the practical consequences of quantum nonlocality. The ability to generate and distribute entanglement over a quantum network could serve as a valuable adjunct to provide "quantum wiring" for large-scale quantum computers based upon diverse physical mechanisms other than interactions in cavity-QED *per se*. The goals of implementing quantum state synthesis,[26] quantum logic gates,[1] and quantum state mapping[3] have in any case had a profoundly positive effect on the development of cavity-QED as evidenced by the remarkable experimental advances they have motivated.

2. THE BASICS OF CAVITY-QED

The whole concept of cavity-QED is based on the following. If one quantizes the electromagnetic field inside a conductor or between mirrors or inside a dielectric, one first expands the electric and magnetic fields in mode functions that satisfy the Maxwell equations with the appropriate boundary conditions. The energy of one photon ($\hbar\omega$) is given by the integral over the mode volume V of the total (electric and magnetic) energy, which is quadratic in the electric field. The electric field per photon thus scales as $(\hbar\omega/V)^{1/2}$. By making the volume V sufficiently small, one can make the electric field per photon in principle arbitrarily large. It can in fact be so large that just one photon suffices to saturate an optical transition (which in free space typically requires a laser field of a few mW/cm^2).

The possibility to cause nonlinear optical effects at the single-photon level can be very useful for diverse tasks in quantum information science. Indeed, the very first cavity-QED experiment in quantum information used such an effect to perform a nonlinear gate between two different photons.[1] That is, the presence or absence of a photon determines how an atom inside a cavity interacts with a second photon (distinguished from the first photon by polarization or frequency or both). We will return to this means of implementing quantum logic for quantum computing below.

Moreover, the photon inside a cavity is an excitation of a well-defined mode, with well-defined frequency, polarization, propagation direction and transverse mode profile, which makes it itself a good candidate for a qubit. Furthermore, it has been demonstrated that the cavity photon can be coupled out of the cavity and be transformed into a freely propagating single-photon wave-packet, with well-defined spatiotemporal properties.(6)

For now, let us define the most relevant cavity-QED parameters, with reference to Fig. 2. They can all be expressed as rates, from which two dimensionless ratios can be formed: the saturation photon number and the saturation atom number.(13) First, there is the rate g at which the atom couples to the electric field of a single-photon. In general, this quantity depends on the position of the atom. This position dependence can be exploited in various ways, but also may lead to potential problems, as detailed below. The rate κ determines the cavity decay rate, half the rate at which photons escape through the mirrors of the cavity. Finally, the rate Γ describes the spontaneous emission rate of the relevant atomic excited level. In the optical regime, all rates are in the MHz range, with g in current experiments being an order of magnitude larger than both κ and Γ (this is called the strong-coupling regime). In the microwave regime the rates are in the kHz range, with g again being an order of magnitude larger than κ, and Γ much smaller than κ. The critical photon and atom numbers, respectively, are defined as

$$\begin{aligned} n_0 &= \frac{\gamma^2}{2g^2}, \\ N_0 &= \frac{2\gamma\kappa}{g^2}, \end{aligned} \tag{1}$$

where $\gamma = \Gamma/2$. Roughly speaking, n_0 determines the number of photons inside the cavity needed to appreciably affect the atom. Conversely, the number N_0 is the number of atoms needed to appreciably affect the cavity field. In optical cavity-QED, n_0 can be as small as 10^{-4} and $N_0 \approx 10^{-2}$.(13)

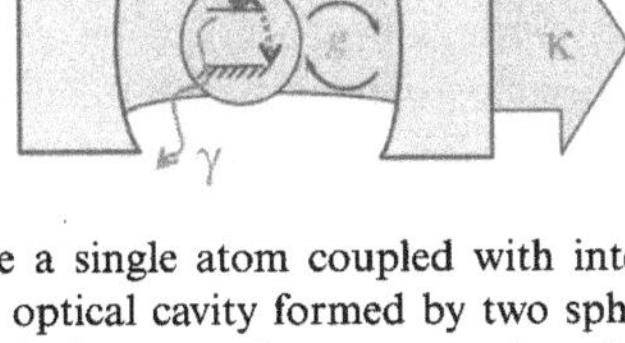

Fig. 2. Cartoon to illustrate a single atom coupled with interaction energy $\hbar g$ to a single mode of a resonator, here an optical cavity formed by two spherical mirrors. Decay from this mode proceeds at rate κ, while the atom decays to modes of the electromagnetic field other than the privileged mode at rate γ.

With these numbers being small, the coherent coupling (g) is much larger than the dissipative couplings (γ, κ).

Another important set of quantities determined by the rates just given are the (generalized) eigenenergies or resonance frequencies $\omega_{\pm}$ (here we consider only the transitions between the ground state and the two eigenstates states with one atom–cavity excitation; hence, there are two transition frequencies) of the combined atom–cavity system. In turn the resonance frequencies $\omega_{\pm}$ determine the response of the system to light of arbitrary frequency ω impinging upon the cavity. The eigenenergies are the roots of the quadratic equation

$$(-\omega_{\pm} + \omega_{C} - i\kappa)(-\omega_{\pm} + \omega_{A} - i\gamma) = g^2 \tag{2}$$

with ω_C and ω_A the bare cavity and atomic resonance frequencies, and with the imaginary part of $\omega_{\pm}$ describing decay, both due to spontaneous emission and cavity decay. In general, the real part of $\omega_{\pm}$ will depend on g and thereby on the position of the atom. In particular, the resonance frequency depends on the presence ($g \neq 0$) or absence ($g = 0$) of the atom. Thus, the transmission or reflection of a probe beam depends on whether there an atom is coupled to the cavity or not. Both effects can be put to good use, as we will discuss later on. Finally, the quality factor Q of a cavity is defined as the ratio between its resonance frequency ω_C and its photon loss rate 2κ.

3. QUANTUM INFORMATION PROCESSING WITH CAVITY-QED

We first discuss setups where the photons act as the qubits performing a quantum computation, with atoms inside cavities used to mediate interactions between photons, and thus to perform quantum logic gates. Subsequently, we discuss the setups where atoms are the qubits and light is used to manipulate the atoms.

3.1. Photons as Flying Qubits

As we have just discussed, the atom–cavity coupling can be made so strong that nonlinear optical effects occur even in the presence of just a single photon, or more precisely, even if the average number of photons inside the cavity is much less than unity. This is possible when the saturation photon number n_0, defined above, is much less than unity. If the atom is near resonance with two *different* cavity modes, then the atom can be used to perform nonlinear gates on two photons from those two modes.

In particular, in Refs. 1 and 27 one finds an experimental demonstration of such a nonlinear effect. The two modes in that experiment differed both in frequency and polarization. Hence, there are two photonic qubits, one for each mode. The states of the qubits, $|0\rangle_k$ and $|1\rangle_k$, correspond to the absence and presence, respectively, of a single photon in mode k. The states undergo phase shifts due to the atom-cavity interaction, but only the state $|1\rangle|1\rangle$ undergoes a *nonlinear* phase shift. In words, the phase shift of one photon is conditioned upon the presence of the other photon. Thus a 2-bit entangling gate is performed, but the nonlinear phase shift in the experiment[1] was only 16°, with larger phase shifts bringing increased levels of dissipation. A new proposal[28] is much more promising in various respects: it is much easier to generate conditional phase shifts of π for photonic qubits, and the gate (i.e., the value of the nonlinear phase shift) is more or less independent of the atomic position. Rather than having the photons enter the cavity, the new proposal relies on the way photons are reflected from a single-sided cavity (i.e., a Fabry–Perot cavity with one high quality input mirror with reflectivity $R_{\text{in}} = 1 - \epsilon_{\text{in}}$ and an output mirror with $R_{\text{out}} = 1 - \epsilon_{\text{out}}$, where $\epsilon_{\text{in}} \sim 10^{-5} - 10^{-6} \gg \epsilon_{\text{out}}$. If the incident light is resonant with the common frequency of the atom-cavity system (i.e., $\omega_0 = \omega_{\text{C}} = \omega_{\text{A}}$), the field will be reflected from the input mirror without appreciable build-up in the cavity, since the atom-cavity system responds at the coupled frequencies $\omega_\pm$ and not at the bare resonances $\omega_{\text{C}}, \omega_{\text{A}}$. In this case, the incident field is reflected with near unity efficiency and with phase shift that we take as a reference of 0. By contrast, if the atom is placed in a state with transition frequency ω_{B} that has large detuning from the bare cavity resonance at ω_{C}, then to all intensive purposes the atom-cavity behaves as two independent systems, with $\omega_\pm \approx \omega_{\text{B,C}}$. The incident field at frequency $\omega_0 = \omega_{\text{C}}$ will then be reflected from the cavity with phase shift π. Hence, the phase shift for a field upon reflection can be conditioned upon the internal state of the atom.

In the new proposal of Ref. [28], the two photonic qubits involved have the same carrier ω_0 frequency but different times of arrival at the cavity. The qubit states now correspond to single-photon states, with horizontal and vertical polarization encoding $|1\rangle$ and $|0\rangle$, respectively. In a simple version of the protocol, vertical polarization is reflected from a polarizing beam splitter, while the horizontally polarized photons are reflected from the cavity. An atom inside the cavity is prepared in an equal superposition of two ground states, one with transition frequency ω_{A} that is resonant with the cavity at ω_{C} and one with frequency ω_{B} that is far from the cavity resonance. A sequence of reflections of two horizontally polarized photons in combination with $\pi/2$ transformations of the atomic ground states then leads to a π phase-shift, whereas all other combinations of two polarized photons lead to a zero phase shift (with the atomic state unaffected at the end of the process). As demonstrated in Ref. 28, the value π for the phase shift is more or less independent of the position of the atom as the only criterion is that g be sufficiently large to shift the atom–cavity resonance frequencies $\omega_{\pm}$ appreciably compared to their linewidths.

Finally, it is important to note in this context that although single photons are not easy to produce, cavity-QED setups are in fact able to generate single photons with high efficiency and with well-defined coherence and spatio-temporal mode profiles, as has been demonstrated in recent experiments.(6) Thus, initial demonstrations of photonic quantum logic using Fabry–Perot cavities seems possible in the near future, with encouraging prospects for scalable quantum computation by employing the protocol of Ref. 28 with lithographically fabricated micro-resonators.(16,17)

3.2. Atoms as Qubits

Although there are several ways of implementing material qubits in cavity-QED, so far we have principally considered Fabry–Perot cavities, consisting of two mirrors of extremely high quality placed at very short distance, with single neutral atoms inside. There are, however, other types of resonators and other types of material qubits that can play the same role as a Fabry–Perot cavity with neutral atoms. In particular one may think of microspheres,(29) microdisks,(30) microtoroids,(31) or photonic bandgap cavities.(19,32) Moreover, one may think of using quantum dots in cavities.(33) Finally, it is certainly possible to build a cavity around an ion trap, as has been demonstrated recently in two different research groups(34,35) (where the finesse of the cavities, though, is not as high as achieved in the experiments with neutral atoms, and strong coupling has not yet been achieved). At the moment quantum information processing experiments in Fabry–Perot cavities with neutral atoms are clearly the more advanced. We thus split this section into two parts, with one subsection that discusses the current advances and obstacles in Fabry–Perot cavity-QED and another that provides references for some novel approaches that may be realized in the future.

3.2.1. *Atoms in Fabry–Perot Cavities*

One important obstacle in the way of quantum computing with cavity-QED arises from the motion of the atoms. Whereas ions can be cooled down to the Lamb–Dicke limit, where the typical excursions of the ions are limited to much less than the (effective) wavelength, for atoms inside a cavity this limit has not yet been reached. In fact, in early experiments atoms traversed the cavity and were not trapped at all. Since then, great advances in trapping and cooling techniques have routinely allowed trapping times of single atoms of several seconds.[5] There are several issues related to the motion of atoms that are of direct interest to quantum information processing. We discuss these one by one.

(1) The motion of atoms inside a cavity can in fact be monitored in real time, by measuring a probe laser beam traversing the cavity. As mentioned above, the resonance frequencies $\omega_{\pm}$ of the atom-cavity system depend on the value of the atom-cavity coupling rate g. The transmission of a probe beam thus depends on g which in turn depends on the atom's position. This dependence can be inverted and together with the atomic equations of motion, one can reconstruct the atom's motion as a function of time. See Refs. 36–40 for experiments that investigate this possibility, with Ref. 38 having actually achieved in the inversion to realize an "atom-cavity microscope." An important goal for continuing research is to use such tracking to actively cool the atomic motion via real-time feedback.[41–43]
(2) The intracavity motion of atoms can certainly be suppressed via adaptations of standard laser cooling techniques, albeit not yet to the Lamb–Dicke limit. As it turns out the cavity field itself may cool the atoms (to about the Doppler limit),[44] and the addition of a far-off resonant trapping beam (FORT) can further control the motion to some degree.[45] A potential problem is that the trapping and cooling laser beams would interfere with the lasers that perform gates on the atoms. However, one important trick is to make the trapping potential independent of the internal state of the atom[5] by choosing an appropriate frequency of the FORT laser beams, thus making the system more robust. Indeed, in recent experiments employed a single atom (that is, one-and-the-same atom) trapped in a FORT to realize a one-atom-laser,[7] and to produce a well-controlled deterministic sequence of single photons.[6]
(3) An important method to mitigate the effects of atomic motion in cavity-QED is by using so-called adiabatic passage schemes.[26] Such schemes rely on the fact that a system starting out in the ground state of a Hamiltonian will stay in a time-dependent ground state of a time-dependent Hamiltonian if the latter is varied sufficiently slowly in time. Thus population can be transferred almost perfectly from one state to another without having to control the Hamiltonian very precisely: what matters is only that it be varied *slowly*. Such

schemes were indeed proposed in this context for transferring quantum information from one cavity to another.(46) Moreover, for operations within one cavity, too, it has been shown that adiabatic passage schemes are far more robust against motional decoherence(47) than other schemes that rely on designing particular laser pulse shapes and accurate timing. Finally, yet another adiabatic passage scheme has been proposed to perform gates in a way that is independent of the atomic position.(48) The latter makes use of the fact that a particular eigenstate of the Hamiltonian depends only on the ratio of two parameters, both of which are functions of the atom's position. If both parameters, one being the coupling of the atom to a probe laser, the other the coupling to the cavity field, depend on position in the same way, the ratio is independent of position. This type of method is, generally speaking, the best way to proceed in quantum information processing. It pays to put a lot of effort into *avoiding* decoherence errors, as the alternatives, error correction and fault-tolerant quantum computing, require enormous amounts of overhead. Adiabatic passage schemes are intrinsically more robust and, one could say, intrinsically fault tolerant.

An issue that has not been addressed yet experimentally is scaling. Although recent experiments did succeed in distinguishing a single trapped atom from 2, or 3, or more trapped atoms, and in monitoring the decay, one by one, of multiple atoms out of the cavity,(49) control of the motion of multiple atoms, and individual addressing of multiple atoms in a cavity, lies still in the future. In fact, at this moment most of our own experimental efforts are aimed more towards implementing complex quantum communication protocols, than directly at scaling up quantum computing schemes using atoms as qubits. However, it is important to note that quantum communication techniques based upon cavity-QED may prove very useful for other implementations of quantum computing, such as the ion-trap quantum computer, to connect qubits separated by relatively large distances.

3.2.2. Other types of cavity-QED

While experiments in optical cavity-QED have so far relied almost exclusively on the traditional Fabry–Perot resonator, a growing menagerie of new monolithic microcavities is becoming available as a result of the explosive development of fabrication technology.(50) The first type of monolithic cavity investigated for cavity-QED was the quartz microsphere,(29) which can be easily fabricated from fiber-quality fused silica preforms using either a gas flame or infrared laser. These whispering-gallery mode resonators can achieve high quality factor by virtue of the high material transparency and the low surface roughness created by the surface tension of molten glass. While high-Q ($Q \simeq 10^9 - 10^{10}$) and low mode volume microspheres have been available in several atomic physics laboratories for a number of years, system integration issues associated with optical coupling and atom delivery have slowed progress. Still, it seems likely that strong coupling will be achieved by combining microspheres with 'atom chip' technology (see Refs. 16, 17 and references therein), in the near future.

Recently some important advances have been made in adapting the basic idea of fused silica whispering-gallery mode resonators, to facilitate their incorporation in chip-like hardware platforms. In particular, new types of high-Q microdisk(30) and microtoroid(31) optical resonators have been demonstrated that can be fabricated by combining lithography with either etching or laser-induced reflow; these methods lead to lithographically positioned microcavities that can be efficiently coupled via tapered optical fibers. The open geometry of whispering-gallery mode resonators presents a number of advantages over the closed Fabry–Perot configuration, in terms of the ease with which cold atoms can be controllably delivered into the optical mode volume.

The fused silica microcavities mentioned above all provide relatively high quality factor ($Q \sim 10^6 - 10^{10}$) in combination with small mode volumes for strong coupling. In contrast, planar photonic crystal defect cavities(32) seem to have limited Q but should provide sufficiently small mode volumes to still make strong coupling possible(19). Indeed, recent experiments have demonstrated that photonic crystal defect cavities with with $N_0 \sim 10^{-4}$ and $n_0 \sim 10^{-8}$ (the numbers defined in Eq. (1)) can be fabricated and probed spectroscopically via robust tapered-fiber optical couplers.(51) If fiber-coupled photonic crystal structures of this type can be successfully combined with atom chip technology for cold atom delivery,(16) the resulting systems could provide an interesting new paradigm for cavity-QED. While some concerns remain about deleterious effects of surface interactions, such research is strongly motivated the promise of extremely high interaction rates ($g \sim 20\,\text{GHz}$) and the expectation of reduced susceptibility to acoustic vibrations.

4. CONCLUSIONS

Cavity-QED has an established tradition of generating experimental and theoretical results that drive our understanding of coherence and decoherence in open quantum systems. In the new era of quantum information science, it has continued to play such a role and will in all likelihood do so for years to come. Basic operations such as quantum state synthesis, quantum logic, and quantum state mapping have all been demonstrated in recent cavity-QED experiments in the optical and/or microwave regimes. Finally, we would like to note that research in cavity-QED has blazed conceptual and methodological trails that will provide crucial guidance for analogous systems in mesoscopic physics.(52)

REFERENCES

1. Q. A. Turchete, C. J. Hood, W. Lange, H. Mabuchi, and H. J. Kimble, *Phys. Rev. Lett.* **75**, 4710 (1995).
2. T. Pellizzari, S. A. Gardiner, J. I. Cirac, and P. Zoller, *Phys. Rev. Lett.* **75**, 3788 (1995).
3. J. I. Cirac, P. Zoller, H. J. Kimble, and H. Mabuchi, *Phys. Rev. Lett.* **78**, 3221 (1997).
4. D. Kielpinski, C. Monroe, and D. J. Wineland, *Nature* **417**, 709 (2002); M. A. Rowe, A. Ben-Kish, B. Demarco, D. Leibfried, V. Meyer, J. Beall, J. Britton, J. Hughes, W. M. Itano, B. Jelenkovic, C. Langer, T. Rosenband, and D. J. Wineland, *Quantum Inform. Comput.* **2**, 257 (2002).
5. J. McKeever, J. R. Buck, A. D. Boozer, A. Kuzmich, H. C. Nagerl, D. M. Stamper-Kurn, and H. J. Kimble, *Phys. Rev. Lett.* **90**, 133602 (2003).
6. J. McKeever, A. Boca, A. D. Boozer, R. Miller, J. R. Buck, A. Kuzmich, and H. J. Kimble, *Science* **303**, 1992 (2004).
7. J. McKeever, A. Boca, A. D. Boozer, J. R. Buck, and H. J. Kimble, *Nature* **425**, 268 (2003).
8. H. Walther, *Adv. Chem. Phys.* **122**, 167 (2002).
9. J. M. Raimond, M. Brune, and S. Haroche, *Rev. Mod. Phys.* **73**, 565 (2001).
10. M. Brune, E. Hagley, J. Dreyer, X. Maitre, A. Maali, C. Wunderlich, J. M. Raimond, and S. Haroche, *Phys. Rev. Lett.* **77**, 4887 (1996); E. Hagley, X. Maitre, G. Nogues, C. Wunderlich, M. Brune, J.-M. Raimond, and S. Haroche, *Phys. Rev. Lett.* **79**, 1 (1997); A. Rauschenbeutel, G. Nogues, S. Osnaghi, P. Bertet, M. Brune, J.-M. Raimond, and S. Haroche, *Science* **288**, 2024 (2000); A. Rauschenbeutel, P. Bertet, S. Osnaghi, G. Nogues, M. Brune, J.M. Raimond, and S. Haroche, *Phys. Rev. A* **64**, 050301 (2001).
11. P. R. Berman (ed.), *Cavity Quantum Electrodynamics* (Academic Press, San Diego, 1994).
12. Ph. Grangier, G. Reymond, and N. Schlosser, *Fortschr. d. Physik* **48**, 859 (2000); H. Mabuchi, and A. C. Doherty, *Science* **298**, 1372 (2002).
13. H. J. Kimble, *Phys. Scripta* **T76**, 127 (1998).
14. J. Ye and T. W. Lynn, *Adv. Atom. Mol. Opt. Phys.* **49**, 1 (2003).
15. H. Mabuchi, M. Armen, B. Lev, M. Loncar, J. Vuckovic, H. J. Kimble, J. Preskill, M. Roukes, and A. Schere, *Quantum Inform. Comput.* **1**, 7 (2001).
16. B. Lev, K. Srinivasan, P. Barclay, O. Painter, and H. Mabuchi, quant-ph/0402093.
17. T. J. Kippenberg, S. M. Spillane, K. J. Vahala, and H. J. Kimble, "Cavity QED with single atoms strongly coupled to micro-toroidal resonators."
18. H. J. Carmichael, *An Open Systems Approach to Quantum Optics* (Springer, Berlin, 1993).
19. J. Vuckovic, M. Loncar, H. Mabuchi, and A. Scherer, *Phys. Rev. E* **65**, 016608 (2001).
20. H. Mabuchi and P. Zoller, *Phys. Rev. Lett.* **76**, 3108 (1996).

21. S. J. van Enk, J. I. Cirac, and P. Zoller, *Phys. Rev. Lett.* **79**, 5178 (1997); see also, *Science* **279**, 205 (1998).
22. G. Rempe, R. J. Thopmson, R. J. Brecha, W. D. Lee, and H. J. Kimble, *Phys. Rev. Lett.* **67**, 1727 (1991).
23. B. Julsgaard, A. Kozhekin, and E. S. Polzik, *Nature* **413**, 400 (2001).
24. J. M. Geremia, J. K. Stockton, and H. Mabuchi, *Science* **304**, 270 (2004).
25. L. M. Duan, M. D. Lukin, J. I. Cirac, and P. Zoller, *Nature* **414**, 413 (2001).
26. A. S. Parkins, P. Marte, P. Zoller, and H. J. Kimble, *Phys. Rev. Lett.* **71**, 3095 (1993).
27. Q. A. Turchette, R. J. Thompson, H. J. Kimble, *Appl. Phys. B* **60**, S1 (1995).
28. L.-M. Duan and H. J. Kimble, *Phys. Rev. Lett.* **92**, 127902 (2004).
29. H. Mabuchi and H. J. Kimble, *Opt. Lett.* **19**, 749 (1994); D. W. Vernooy, A. Furusawa, N. Ph. Georgiades, V. S. Ilchenko, and H. J. Kimble, *Phys. Rev. A* **57**, 2293 (1998); F. Treussart, V. S. Ilchenko, J. F. Roch, J. Hare, V. Lefevre-Seguin, J. -M. Raimond, and S. Haroche, *Eur. Phys. J. D* **1**, 235 (1998).
30. T. J. Kippenberg, S. M. Spillane, D. K. Armani, and K. J. Vahala, *Appl. Phys. Lett.* **83**, 797 (2003).
31. D. K. Armani, T. J. Kippenberg, S. M. Spillane, and K. J. Vahala, *Nature* **421**, 925 (2003).
32. O. Painter, R. K. Lee, A. Scherer, A. Yariv, J. D. O'Brien, P. D. Dapkus, and I. Kim, *Science* **284**, 1819 (1999).
33. A. Kiraz, C. Reese, B. Gayral, L. D. Zhang, W. V. Schoenfeld, B. D. Gerardo, P. M. Petroff, E. L. Hu, and A. Imamoglu, *J. Opt. B-Quantum S. O.* **5**, 129 (2003).
34. M. Keller, B. Lange, K. Hayasaka, W. Lange, and H. Walther, *Appl. Phys. B* **76**, 125 (2003).
35. A. B. Mundt, A. Kreuter, C. Russo, C. Becher, D. Leibfried, J. Eschner, F. Schmidt-Kaler, and R. Blatt, *Appl. Phys. B* **76**, 117 (2003).
36. H. Mabuchi, Q. A. Turchette, M. S. Chapman, and H. J. Kimble, *Opt. Lett.* **21**, 1393 (1996).
37. H. Mabuchi, J. Ye, and H. J. Kimble, *Appl. Phys. B* **68**, 1095 (1999).
38. C. J. Hood, T. W. Lynn, A. C. Doherty, A. S. Parkins, and H. J. Kimble, *Science* **287**, 1447 (2000).
39. P. W. H. Pinkse, T. Fischer, P. Maunz, and G. Rempe, *Nature* **404**, 365 (2000).
40. A. C. Doherty, T. W. Lynn, C. J. Hood, and H. J. Kimble, *Phys. Rev. A* **63**, 013401 (2000).
41. A. C. Doherty and K. Jacobs, *Phys. Rev. A* **60**, 2700 (1999).
42. T. Fischer, P. Maunz, P. W. H. Pinkse, T. Puppe, and G. Rempe, *Phys. Rev. Lett.* **88**, 163002 (2002).
43. D. A. Steck, K. Jacobs, H. Mabuchi, T. Bhattacharya, and S. Habib, quant-ph/0310153.
44. P. Horak, G. Hechenblaikner, K. M. Gheri, H. Stecher, and H. Ritsch, *Phys. Rev. Lett.* **79**, 4974 (1997); V. Vuletic and S. Chu, *Phys. Rev. Lett.* **84**, 3787 (2000).
45. J. Ye, D. W. Vernooy, H. J. Kimble, *Phys. Rev. Lett.* **83**, 4987 (1999); S. J. van Enk, J. McKeever, H. J. Kimble, and J. Ye, *Phys. Rev. A* **64**, 013407 (2001); P. Maunz, T. Puppe, I. Schuster, N. Syassen, P. W. H. Pinkse, and G. Rempe, *Nature* **428**, 50, 2004.
46. T. Pellizzari, *Phys. Rev. Lett.* **79**, 5242 (1997); S. J. van Enk, H. J. Kimble, J. I. Cirac and P. Zoller, *Phys. Rev. A* **59**, 2659 (1999).
47. A. C. Doherty, A. S. Parkins, S. M. Tan, and D. F. Walls, *J. Opt. B-Quantum S. O.* **1**, 475 (1999).
48. L.-M. Duan, A. Kuzmich, and H. J. Kimble, *Phys. Rev. A* **67**, 032305 (2003).
49. "Cavity QED 'by the Numbers'," J. McKeever, J. R. Buck, A. D. Boozer, and H. J. Kimble, *Phys. Rev. Lett.* (submitted, 2004); available as quant-ph/0403121.

50. K. Vahala, *Nature* **424**, 839 (2004).
51. K. Srinivasan, P. E. Barclay, M. Borselli, and O. Painter, quant-ph/0309190.
52. A. Blais, R.-S. Huang, A. Wallraff, S. M. Girvin, and R. J. Schoelkopf, cond-mat/0402216 (2004).

Quantum Information Processing with Trapped Neutral Atoms

P. S. Jessen,[1,3] I. H. Deutsch,[2] and R. Stock[2]

Quantum information can be processed using large ensembles of ultracold and trapped neutral atoms, building naturally on the techniques developed for high-precision spectroscopy and metrology. This article reviews some of the most important protocols for universal quantum logic with trapped neutrals, as well as the history and state-of-the-art of experimental work to implement these in the laboratory. Some general observations are made concerning the different strategies for qubit encoding, transport and interaction, including trade-offs between decoherence rates and the likelihood of two-qubit gate errors. These trade-offs must be addressed through further refinements of logic protocols and trapping technologies before one can undertake the design of a general-purpose neutral-atom quantum processor.

KEY WORDS: Quantum information processing; neutrals atoms; controlled collisions; optical lattice.

PACS: 03.67.Lx; 32.80.Pj; 34.50.-s.

1. INTRODUCTION

An important lesson from 20th-century information science is that "information is physical". One cannot understand the power of algorithms, communication protocols or other information processing tasks separately from the physical description of the devices that perform them. In particular, quantum systems allow the implementation of new types of logic that cannot be efficiently simulated on classical systems governed by laws based

[1]Optical Sciences Center, University of Arizona, Tucson, AZ 85721, USA; E-mail: poul.jessen@optics.arizona.edu
[2]Department of Physics and Astronomy, University of New Mexico, Albuquerque, NM 87131, USA.
[3]To whom correspondence should be addressed.

on local realism. This has allowed a whole new field to emerge—quantum information science—whose ultimate vision is the construction of a universal quantum computer capable of executing any algorithm that can be described by a quantum evolution.

Exactly what features give quantum computers their power is still a subject of debate, but certain ingredients are generally agreed upon as essential:

- A many-body system whose Hilbert space has scalable tensor product structure.
- The ability to prepare a fiducial quantum state.
- A universal set of quantum operations capable of implementing an arbitrary quantum map.
- A method to read-out the quantum state.
- A dissipative mechanism to remove the entropy associated with unavoidable errors in a fault-tolerant manner.

Since they were proposed in their original form, we have learned that some of the so-called "DiVincenzo Criteria"(1) can be relaxed. For example, universal quantum maps need not be unitary and may instead have irreversible quantum measurements at their core, as shown by proposals for linear optics quantum computation,(2) quantum computation via teleportation,(3) and the so-called "one-way quantum computer" in which conditional measurements are performed on an entangled "cluster state".(4) Such developments highlight an important fact: the roadmap to a universal quantum computer is still evolving, and the "best" way to accomplish a computational task will depend on the strengths and weaknesses of the physical system at hand. Even so, the essential ingredient is clear: quantum control of a many-body system,(5) including both reversible unitary evolution and irreversible quantum measurement. Robust, high fidelity execution of these tasks is the goal of all physical implementations of quantum information processing (QIP).

Given these preliminaries, it is clear that atomic, molecular and/or optical (AMO) systems offer unique advantages for QIP. More than in any other subdiscipline, the quantum optics community has explored the foundations of quantum mechanics in the laboratory, including detailed studies of the processes of measurement and decoherence, entanglement and the violation of Bell's inequalities. In appropriately designed dilute systems, coherence times can be very long and decades of research in spectroscopy, precision metrology, laser cooling, and quantum optics has produced a large toolbox with which to manipulate them and drive their quantum dynamics. Indeed, atom- and ion-based atomic clocks are arguably the best controlled, most quantum coherent devices available, and present a strong motivation to consider the use of similar systems for QIP.

2. SURVEY

Proposals to use neutral atoms as the building blocks of a quantum computer followed closely after the first demonstration of quantum logic in ion traps.[6] Laser cooling of ions and neutrals was initially developed as an enabling technology for precision metrology. Both systems were known to have long coherence times but also complementary features that lead to radically different approaches to, e.g. atomic clock design. Because ions are charged they can be tightly confined in deep traps and observed for very long times, but the strong Coulomb repulsion limits the number of ions that can be precisely controlled in a single trap. In contrast, neutral atoms usually interact only at very short range and can be collected in large ensembles without perturbing each other, a clear advantage for both metrology and QIP, On the downside, traps for neutrals are shallow compared to ion traps, and the atom/trap field interaction invariably perturbs the atomic internal state. In QIP, one must balance an intrinsic conflict – qubits must interact with each other and with external control fields that drive the quantum algorithm, while at the same time the system must couple only weakly to the noisy environment which leads to decoherence. In an ion trap the Coulomb interaction leads to collective modes of center-of-mass motion, which can be used as a "bus" for coupling qubits together.[6] However, control of a strongly coupled many-body system becomes increasingly complex as the system size grows, and will likely require the use of intricate multitrap designs to overcome the difficulty of working with even a handful of ions in a single trap.[7] Also, the strong interactions can have a parasitic effect by coupling the ionic motion to noisy electric fields such as those associated with patch potentials on the trap electrodes.[8] Neutral atoms in the electronic ground state, in contrast, couple weakly to each other and to the environment, and so offer a different compromise between coupling vs. control complexity and decoherence.

The generally weak- and short-range coupling between neutrals makes the introduction of non-separable two-qubit interactions *the critical element* of neutral atom QIP. Brennen *et al.*[9] and Jaksch *et al.*[10] realized independently that this might be achieved by encoding qubits in the hyperfine ground manifold of individual atoms trapped in optical lattices,[11] and using the state-sensitive nature of the trap potential to bring the atomic center-of-mass wavepackets together for controlled interactions mediated by either optical dipole–dipole coupling[9] or ground state collisions.[10] Further ideas include a proposal for fast quantum gates based on interactions between Rydberg atoms,[12] and another based on magnetic spin–spin interaction.[13] These developments occurred against a backdrop of steady progress in the technologies for cooling, trapping and manipulating neutrals, in particular in optical lattices. Early work that helped inspire proposals for QIP include the demonstration of Raman sideband cooling to the lattice vibrational ground state,[14] the generation of vibrational Fock- and delocalized Bloch-states,[15] and tomographic reconstruction of the atomic internal[16] and center-of-mass state.[17] At the same time theoretical work indicated that loading an optical lattice from a Bose–Einstein condensate can induce a transition to a Mott-insulator state with nearly perfect, uniform occupation of the lattice sites.[18] A series of ground-breaking experiments by the group of Bloch and Hänsch have recently demonstrated, in short order, first the Mott-insulator transition,[19] followed by coherent splitting and transport of atomic wavepackets,[20] and finally controlled ground–ground state collisions and the generation of entanglement in an ensemble consisting of short strings of atoms.[21] Other elements of neutral atom QIP have been pursued in a number of laboratories, including patterned loading of optical lattices,[22] addressing of individual lattice sites,[23] and alternative trap technologies such as magnetic microtraps,[24] and arrays of optical tweezers traps.[25,26]

2.1. Neutral Atom Traps

Implementation of neutral atom QIP is closely tied to the development of suitable traps. Neutral atom traps in general rely on the interaction of electric or magnetic dipole moments with AC and/or DC electromagnetic fields. Magnetic traps have found wide use in the formation of quantum degenerate gases, but tend to be less flexible than optical traps in terms of the atomic states that can be trapped, and therefore have not been as widely considered for QIP. For this reason, we concentrate on optical traps created by the dynamical (AC) Stark effect in far detuned, intense laser fields. In principle, these traps suffer from decoherence caused by the spontaneous scattering of trap photons, but in practice the rate can be suppressed to a nearly arbitrary degree through the use of intense trap light tuned very far from atomic resonance. Proposals for QIP typically have considered alkalis (e.g. Rb or Cs), which are easy to laser-cool and have nuclear spin so qubits can be encoded in long-lived hyperfine ground states. For these atomic species trap detunings are always much larger than the excited state hyperfine splitting. In this limit, the optical potential can be written in the compact form,[27] $U(\mathbf{x}) = U_s(\mathbf{x}) - \mu \cdot \mathbf{B}_{\text{fict}}(\mathbf{x})$, where $U_s(\mathbf{x})$

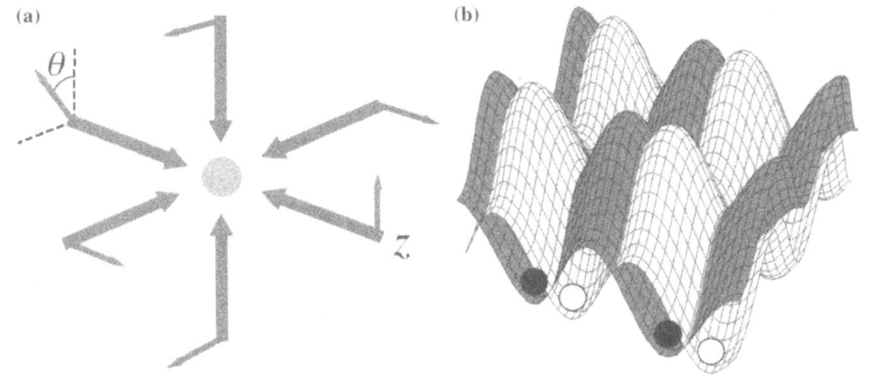

Fig. 1. Schematic of a 3-D optical lattice. (a) Two pairs of linearly polarized beams provide transverse confinement, and the beams along z in the lin-θ-lin configuration provide longitudinal confinement in σ_+ and σ_- standing waves. (b) Potential surfaces for the atom in different magnetic sublevels, described in the text, shown here as in gray and white, are moved along the z-axis through a rotation of the angle θ between polarization vectors for controlled collisions.

is a scalar potential (independent of the atomic spin) proportional to the total laser intensity, and $\mathbf{B}_{\text{fict}}$ is a *fictitious* magnetic field that depends on the *polarization* of the trap light, and $\boldsymbol{\mu} = g_F \mu_B \mathbf{F}$, where $\mathbf{F}$ is the total angular momentum (electron plus nuclear) and g_F is the Landé g-factor. For trap detunings much larger than the excited state fine structure $\mathbf{B}_{\text{fict}} \to 0$, and the potential is always purely scalar.

This description is the foundation for designing QIP protocols. To illustrate this point we consider how to bring atoms together for controlled interactions in a one-dimensional (1-D) optical lattice consisting of a pair of counterpropagating plane waves whose linear polarizations form an angle θ (Fig. 1). Choosing the z-axis along the lattice beams, the optical potential is given by $U_s(\mathbf{x}) = 2U_0(1 + \cos\theta \cos 2kz)$, $\mu_B \mathbf{B}_{\text{fict}} = U_0 \sin\theta \sin 2kz\, \mathbf{e}_z$, where U_0 is the light shift in a single, linearly polarized lattice beam and k the laser wave number. For $\sin\theta \neq 0$ there is a gradient of the fictitious $\mathbf{B}$-field near the minima of the scalar potential $U_s(\mathbf{x})$, which separates the different magnetic sublevels as in a Stern–Gerlach apparatus and causes the trap minima for hyperfine substates $|F, \pm m_F\rangle$ to move in opposite directions along z. A closer inspection of the full lattice potential shows that the trap minima move by $\pm\lambda/2$ for every 2π increase of the polarization angle θ. Thus, a pair of atoms in, e.g. $|F, m_F\rangle$ and $|F, -m_F\rangle$, trapped in neighboring wells at $\theta = \pi/2$, can be superimposed by rotating the lattice polarization to $\theta = \pi$, and separated again by further polarization rotation.

2.2. Quantum Logic

The basic design of a QIP protocol in the standard quantum circuit model involves a choice of qubit encoding, initialization method, single- and two-qubit gates, and read-out method. Of these mutually dependent design elements, the implementation of unitary two-qubit entangling gates poses the most fundamental challenge. One well-known example of a universal two-qubit gate is the controlled-phase (CPhase) gate, which maps the two-qubit logical basis state $|1\rangle|1\rangle \rightarrow -|1\rangle|1\rangle$, and leaves the others unchanged. In fact, any gate based on a diagonal two-qubit Hamiltonian can be converted to CPhase by single-qubit rotations, provided that the energy shifts are non-separable, $\Delta E = E_{11} + E_{00} - (E_{10} + E_{01}) \neq 0$, and the duration of the interaction is $\tau = \pm\pi\hbar/\Delta E$. If noise and/or decoherence introduces errors at a rate γ then we can estimate the minimum error probability of such a gate, $P_{\mathrm{error}} = 1 - \mathrm{e}^{-\gamma\tau} \approx \pi\hbar\gamma/\Delta E$. The quantity $\Delta E/\gamma$ is thus a key figure of merit of the gate operation, with a clear physical interpretation; it is the *spectral resolvability* of the coupled two-qubit states.

Because of their short range, neutral–atom interactions are best understood in terms of controlled collisions. To implement high-fidelity quantum logic these collisions must be state-dependent, but at the same time they must not cause scattering into states outside the computational basis. In atomic systems, these requirements are generally in conflict, but can be reconciled through appropriate choices of qubit encoding and trap geometry. Jaksch *et al.* proposed to use elastic *s*-wave collisions of atoms in the electronic ground state.[10] In this protocol, the main concern is to suppress inelastic collisions caused by the Heisenberg spin–exchange interaction that preserves only the *total* magnetic quantum number, but not that of the individual atoms. Jaksch *et al.* solved this problem by encoding qubits in the stretched states $|1\rangle = |F_+, m_F = F_+\rangle$, $|0\rangle = |F_-, m'_F = F_-\rangle$, where $F_\pm = I \pm 1/2$. Because $g_{F_\pm} = \pm 1/F$ these states move in opposite directions in a lattice of the type discussed in Sec. 2.1. Rotating the lattice polarization angle from $\theta = 0$ to π will then cause an atom in the state $|0\rangle$ and moving to the right to collide with an atom in the state $|1\rangle$ and moving to the left, i.e., the two qubits interact only if the state is $|0\rangle|1\rangle$ and not otherwise. In that case $\Delta E = E_{01} \neq 0$ and a CPhase can be achieved. Furthermore, because *s*-wave scattering conserves $m_F + m'_F$ (to good approximation) and neither m_F nor m'_F can increase, this collision *must* be elastic.

Several additional protocols for two-qubit interactions have been proposed. For example, Charron *et al.*[28] and Eckert *et al.*[29] considered encoding qubits in the ground and first excited center-of-mass vibrational states of trapped atoms, and to couple atomic qubits in neighboring traps by lowering the intervening potential barrier until tunneling causes atoms in the excited states to couple via s-wave collisions. Brennen *et al.* considered collisions of nearby but non-overlapping wavepackets associated with different internal states in different potentials.[9] This gives greater flexibility to design elastic but state-dependent interactions, but requires resonant and/or longer range forces than the $1/r^6$ van der Waals potential between ground state atoms. Brennen *et al.* proposed to use the $1/r^3$ electric dipole–dipole interactions created when an off-resonant laser field mixes the ground-state manifold with excited electronic states. These excited states will spontaneously emit photons and cause errors, but the rate saturates to that of the two-atom super-radiant state when the atoms are separated by less than a wavelength, while the dipole–dipole interaction continues to increase with decreasing atomic separation. Thus, for very tightly localized wavepackets in close proximity, the dipole–dipole interaction can be nearly coherent. Relatively long-range interactions provide yet another strategy to implement quantum logic with neutrals.[12] If atoms are excited into high-lying Rydberg states one can induce very large dipole moments by applying a static electric field. The interaction between two such dipoles is large enough to provide useful level shifts even if atoms are separated by several microns. In one possible protocol, qubits are encoded in the magnetic field-insensitive "clock doublet", $|1\rangle = |F_+, m_F = 0\rangle, |0\rangle = |F_-, m_F = 0\rangle$. To execute a two-qubit gate the atoms are excited by a laser tuned to the transition from the logical state $|1\rangle$ to a Rydberg level. If the atoms are not too far separated the Rydberg dipole–dipole interaction is strong enough to shift the two-atom, doubly excited state out of resonance and prevent it from becoming populated, a phenomenon referred to as "dipole-blockade". Since the blockade occurs only for the $|1\rangle|1\rangle$ logical state it can be used to achieve a CPhase.

2.3. Experimental Progress

Efforts to implement neutral atom QIP in the laboratory represent a natural but challenging extension of existing tools to prepare, control and measure the quantum state of trapped neutrals. A number of experiments have demonstrated several of the key components that go into QIP, and very recently some of these have been combined for the first time to demonstrate control and entanglement in a neutral-atom many body system. In this section, we briefly review progress in three main areas: initialization of the qubit register, implementation of single- and two-qubit gates, and methods to address individual qubits.

Optical lattices typically confine atoms tightly on the scale of an optical wavelength (the Lamb–Dicke regime), and lend themselves readily to the use of Raman sideband cooling. In a first demonstration, Hamann *et al.* initialized 98% of a 10^6-atom ensemble in a single spin- and vibrational-ground state of a sparsely filled 2-D lattice,(14) and subsequent work has achieved a somewhat lesser degree of state preparation in nearly filled 3-D lattices.(30) These laser cooling-based approaches are relatively simple to implement and will work in any tightly confining trap geometry, but when used in a lattice will produce a random pattern of vacant and occupied sites. Sparse, random filling may suffice for ensemble-based investigations of quantum logic,(31) but falls short of the requirements of full-scale lattice-based QIP.

Better filling and initialization can be achieved by loading a 3-D lattice from a high-density Bose–Einstein condensate and driving the atom/lattice through a superfluid to Mott insulator phase transition.(18) The group of Bloch and Hänsch at MPQ in Münich used this approach as a starting point for a series of proof of principle experiments to establish the viability of the Jaksch *et al.* collisional protocol.(10) As the first step, Greiner *et al.* successfully demonstrated the transition to an "insulator" phase consisting of individual ^{87}Rb atoms localized in the ground state of separate potential wells.(19) Mandel *et al.* then explored spin-dependent coherent transport in the context of interferometry.(20) This was done by preparing atoms in the logical-$|0\rangle$ state, transferring them to an equal superposition of the states $|0\rangle$ and $|1\rangle$ with a microwave $\pi/2$-pulse, and "splitting" them into two wavepackets by rotating the laser polarization vectors. The "which way information" was then erased with a final $\pi/2$-pulse and the atoms released from the lattice, allowing the separated wavepackets of each atom to overlap and interfere as in a two-slit experiment. Inhomogeneities across the ensemble were at least partially removed through a spin–echo procedure using additional π-pulses. In this fashion, the experiment achieved fringe visibilities of 60% for separations of three lattice sites, limited by quantum phase-errors induced by magnetic field noise, vibrational heating and residual inhomogeneities. Finally, Mandel *et al.* performed a many-body version of this experiment in a nearly filled lattice,(21) where the majority of atoms underwent collisional interactions with their neighbors according to the Jaksch *et al.* protocol. For appropriate collision-induced phase shifts this will lead to the formation of chains of entangled atoms, which cannot then be disentangled again by "local" operations such as the final $\pi/2$-pulse. In the experiment, a periodic disappearance and reappearance of interferometer fringe visibility was clearly observed as a function of interaction time and corresponding degree of entanglement. Technical limitations, in particular the inability to perform single qubit measurements, have so far made it difficult to obtain quantitative estimates for the size and degree of entanglement of these cluster states, or to extract the fidelity of the underlying CPhase interaction.

The experiments just described are essentially multiparticle interferometry, and illustrate how proof-of-principle and optimization of a gate protocol can be achieved with ensemble measurements. To proceed toward universal QIP it will be necessary to develop an ability to manipulate and read out the state of individual atomic qubits. In principle this can be accomplished by performing single-qubit rotations with focused Raman beams rather than microwave fields, and single-qubit readout with focused excitation beams and/or high-resolution fluorescence imaging. However, the necessary optical resolving power will be nearly impossible to achieve in current lattices whose sites are separated by roughly $0.5\,\mu\text{m}$. There are several possible ways around this problem: the lattice can be formed by a CO_2 laser so individual sites are $5\,\mu\text{m}$ apart and resolvable with a good optical microscope,[(23)] or a conventional lattice can be loaded with a pattern where atoms occupy only every nth well.[(22)] Alternatively, one might use other trapping geometries, such as arrays of very tightly focused optical tweezers-type traps. Schlosser *et al.* has shown that a few such traps can be formed in the focal plane of a single high-NA lens, and that the trap lens can be used at the same time to achieve spatially resolved detection of fluorescence.[(25)] This work used the ability to detect single atoms, in combination with a phenomenon known as "collisional blockade", to load individual traps with exactly one atom each. Much larger arrays of such traps have been demonstrated using microfabricated arrays of high-NA microlenses,[(26)] but this approach has yet to demonstrate the loading and detection of one atom per trap.

3. LESSONS LEARNED AND FUTURE RESEARCH

The seminal experiments by the Münich group have demonstrated the feasibility of coherent spin transport and entanglement via controlled collisions, but also served to highlight some of the fundamental limitations of the particular protocol employed. To implement high-fidelity collisional gates one must achieve a spin-dependent phase shift, while at the same time restrict the interaction to a single collisional channel so as to prevent scattering outside the computational basis. Jaksch *et al.* accomplished this with their stretched-state encoding, but at the cost of being maximally sensitive to magnetic field- and trap noise which was already a limiting factor in the Münich experiments. Moreover, in a filled lattice the protocol leads to large entangled chains rather than the isolated two-qubit interactions required in the standard quantum circuit model.

It is of course conceivable that one might switch between-noise protected encodings and encodings suitable for collisions during the course of a computation, but such an approach would be cumbersome. Our group is now exploring an alternative, by developing new methods to accurately control collisions between cold atoms in tight traps. As in the original proposal by Brennen *et al.*, we consider logical basis states $|0\rangle = |F_+, m_F\rangle$ and $|1\rangle = |F_-, -m_F\rangle$ for which Zeeman and AC Stark shifts are close to identical. With such encodings the logical states move on identical optical potentials and are never split into separated wavepackets. This provides excellent immunity against noise, but at a cost: in a two-qubit interaction all four logical states interact. The challenge is then to engineer a collision to produce a non-separable phase shift without inelastic scattering. The possibilities of coherent control by directly manipulating the center-of-mass wave packets for atoms in tight traps offer new avenues to reach this goal. A particularly promising approach is to consider *resonant* interactions between atoms in spatially separated traps that can then be used to pick out and strengthen a single elastic channel and suppress off-resonance inelastic processes.

Stock *et al.* have studied the resonant interaction that occurs when a molecular-bound state is AC Stark shifted into resonance with a center-of-mass vibrational state of the two-atom system.(32) These "trap-induced shape resonances" show up as avoided crossings in the energy spectrum as a function of the trap separation, as shown in Fig. 2. The energy gaps indicate the strength of the resonance and become substantial when the scattering length associated with the collision is on the order of the trapped wave packet's width. At this point, the two-atom interaction energy is a non-negligible fraction of the vibrational energy. The Münich experiments used ^{87}Rb atoms for which the relevant scattering length is $\sim 100\ a_0$, and a shallow lattice potential where the trapped wave packet width was $\sim 1200\ a_0$, resulting in a negligible energy gap of order $10^{-22}\,\hbar\omega$. If we choose to work instead of ^{133}Cs, the relevant scattering length lies in the range from 280 a_0 to 2400 a_0, which is comparable to the $\sim 200\ a_0$ wave packet width in a moderately deep lattice. In this case, the trap-induced shape resonance will be significant, and should provide a new and flexible mechanism for designing quantum logic protocols. Additional flexibility and control can in principle be introduced by tuning the scattering length via optically or magnetically induced Feshbach resonances, as demonstrated in several BEC experiments.(33)

The Jaksch *et al.* proposal and Münich experiments together provide proof-of-principle that the most important components of QIP can be achieved with trapped neutral atoms, but are still far from a full quantum computer architecture. Spin-dependent trapping forces are at the

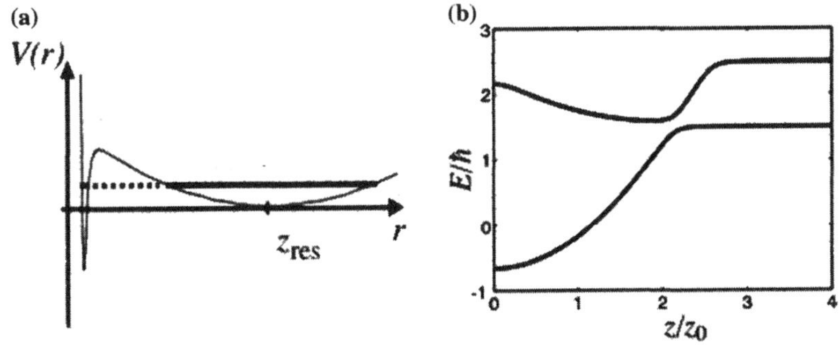

Fig. 2. (a) Sum of the harmonic trapping potential and chemical-binding potential (gray line), as a function of the relative coordinate r along a line through the two trap minima. The trap eigenstate can become resonant with a molecular-bound state at a critical separation Δz_{res}. (b) The energy spectrum as a function of separation between traps Δz (in units of the trap ground state width z_0) shows the energy shift of the molecular-bound state due to the harmonic trapping potential and the avoided crossings associated with the trap-induced resonance.

heart of the protocol, and the trap detuning therefore can be at most comparable to the excited state fine structure. The resulting photon scattering ultimately leads to motional heating, decoherence, and even the occasional loss of an atom. It is, therefore, necessary to explore mechanisms for re-cooling and replacing atoms, and to provide a supply of fresh ancilla atoms as required for error correction. Most importantly, trapping architectures must be developed that allow efficient, programmable transport and qubit interaction, along with individual qubit manipulation and readout. Long-period or pattern loaded(22) lattices or arrays of tweezers traps are one step in this direction, as is recent work on microwave spectroscopy in micro-magnetic traps.(34) Protocols based on Rydberg atoms provide additional freedom to design a workable QIP architecture.(12) Because of the longer range of the interaction there is in principle no need for spin-dependent transport, and trap fields can therefore be detuned much further from resonance. This should effectively remove one important source of heating and decoherence. However, the approach raises new challenges related to the coherent control of Rydberg atoms, e.g. accurate and highly coherent π-pulses between ground and Rydberg levels. Rydberg atoms are also highly susceptible to background DC and AC electric fields, as well as to spontaneous decay and perturbation by thermal blackbody radiation.

As the Review and Discussion in this article illustrates, both the details and overall architecture of a hypothetical neutral atom quantum processor continues to evolve. Every known approach involves tradeoffs between conflicting requirements, and much additional research is required before we can hope to identify a winning strategy. In addition, new paradigms are being developed, inspired by the physical constraints of the particular implementations under study. An excellent example is the "one-way quantum computer" of Raussendorf and Briegel, in which the type of cluster states generated in the Münich experiments become a resource for computation rather than a liability.[4] Whether this protocol can be made fault-tolerant is a subject of continued research. Indeed, fault tolerance is the ultimate goal of any QIP implementation, and it will eventually be necessary to consider in detail how it might be achieved in the context of concrete logic protocols and architectures. Optical lattices and similar traps that allow blocks of physical qubits to be encoded and manipulated in parallel provide an attractive architecture for error correction. More speculatively, error correction based on topological codes might be implemented in a lattice geometry[35] and lead to a very robust fault-tolerant architecture. Which, if any of these ideas ultimately turn out to be practical remains to be seen. Clearly, information is still physical.

ACKNOWLEDGMENTS

Research at UA was partly supported by the Defense Advanced Research Projects Agency (DARPA) under Army Research Office (ARO) Contract No. DAAD19-01-1-0589, and by the National Science Foundation under Contract No. ITR-0113538. Research at UNM was partly supported by the National Security Agency (NSA) and the Advanced Research and Development Activity (ARDA) under Army Research Office (ARO) Contract No. DAAD19-01-1-0648 and by the Office of Naval Research under Contract No. N00014-00-1-0575.

REFERENCES

1. D. P. DiVincenzo, *Fortschr. Phys.* **48**, 771 (2000).
2. E. Knill, R. Laflamme, and G. J. Milburn, *Nature* **409**, 46 (2001).
3. M. A. Nielsen, *Phys. Lett. A* **308**, 96 (2003).
4. R. Raussendorf and H. J. Briegel, *Phys. Rev. Lett.* **86**, 5188 (2001).
5. R. Blume-Kohout, C. M. Caves, and I. H. Deutsch, *Found. Phys.* **32**, 1641 (2002).
6. Q. A. Turchette *et al.*, *Phys. Rev. Lett.* **81**, 3631 (1998).
7. D. Kielpinski, C. Monroe, and D. J. Wineland, *Nature* **417**, 709 (2002).
8. Q. A. Turchette *et al.*, *Phys. Rev. A* **61**, 063418 (2000).
9. G. K. Brennen *et al.*, *Phys. Rev. Lett.* **82**, 1060 (1999); G. K. Brennen, I. H. Deutsch, and C. J. Williams, *Phys. Rev. A* **65**, 022313 (2002).

10. D. Jaksch *et al.*, *Phys. Rev. Lett.* **82**, 1975 (1999).
11. P. S. Jessen and I. H. Deutsch, *Adv. At. Mol. Opt. Phys.* **37**, 95 (1996).
12. D. Jaksch *et al.*, *Phys. Rev. Lett.* **85**, 2208 (2000)
13. L. You and M. S. Chapman, *Phys. Rev. A* **62**, 052302 (2000)
14. S. E. Hamann *et al.*, *Phys. Rev. Lett.* **80**, 4149 (1998).
15. M. Morinaga *et al.*, *Phys. Rev. Lett.* **83**, 4037 (1999); M. BenDahan *et al.*, *Phys. Rev. Lett.* **76**, 4508 (1996).
16. G. Klose, G. Smith, and P. S. Jessen, *Phys. Rev. Lett.* **86**, 4721 (2001).
17. S. H. Myrskog, e-print quant-ph/0312210.
18. D. Jaksch *et al.*, *Phys. Rev. Lett.* **81**, 3108 (2000).
19. M. Greiner *et al.*, *Nature (Lond.)* **415**, 39 (2002).
20. O. Mandel *et al.*, *Phys. Rev. Lett.* **91**, 010407 (2003).
21. O. Mandel *et al.*, *Nature (Lond.)* **425**, 937 (2003).
22. S. Peil *et al.*, *Phys. Rev. A* **67**, 051603 (2003).
23. R. Scheunemann *et al.*, *Phys. Rev. A* **62**, 051801 (2000).
24. R. Folman *et al.*, *Adv. At. Mol. Opt. Phys.* **48**, 263 (2002).
25. N. Schlosser *et al.*, *Nature (Lond.)* **411**, 1024 (2001).
26. R. Dumke *et al.*, *Phys. Rev. Lett.* **89**, 097903 (2002).
27. I. H. Deutsch, and P. S. Jessen, *Phys. Rev. A* **57**, 1972 (1998).
28. E. Charron *et al.*, *Phys. Rev. Lett.* **88**, 077901 (2002).
29. K. Eckert *et al.*, *Phys. Rev. A* **66**, 042317 (2002).
30. M. T. Depue *et al.*, *Phys. Rev. Lett.* **82**, 2262 (1999); A. J. Kerman *et al.*, *Phys. Rev. Lett.* **84**, 439 (2000).
31. I. H. Deutsch, G. K. Brennen, and P. S. Jessen, *Fortschr. Phys.* **48**, 925 (2000).
32. R. Stock, E. L. Bolda, and I. H. Deutsch, *Phys. Rev. Lett.* **91**, 183201 (2003).
33. S. Inouye *et al.*, *Nature (Lond.)* **392**, 151 (1998); T. Weber *et al.*, *Science* **299**, 232 (2003); A. Widera *et al.*, e-print cond-mat/0310719.
34. P. Treutlein *et al.*, quant-ph/0311197.
35. E. Dennis, A. Kitaev, A. Landahl, and J. Preskill, *J. Mat. Phys.* **43**, 4452 (2002).

The Road to a Silicon Quantum Computer

J. R. Tucker[1,3] and T.-C. Shen[2]

We discuss prospects for building a silicon-based quantum computer with phosphorous donor qubits. A specific architecture is proposed for initial demonstrations; and the advantages and difficulties of this approach are described along with a plan for systematic development and calibration of the individual components.

KEY WORDS: Qubits; quantum computer; phosphorous donors; STM lithography.

PACS: 73.21.-b; 73.23.Hk; 81.16.-c; 85.35.-p; 85.40.Ry.

1. INTRODUCTION

Phosphorous donors in silicon present a unique opportunity for solid-state quantum computation.[(1)] Electrons spins on isolated Si:P donors have very long decoherence times of ~60 ms in isotopically purified ^{28}Si at 7 K.[(2)] By contrast, electron spin dephasing times in GaAs are orders-of-magnitude shorter due spin–orbit interaction; and the background nuclear spins of the III–V host lattice cannot be eliminated by isotope selection. Finally, the Si:P donor is a self-confined, perfectly uniform single-electron quantum dot with a *non-degenerate* ground state. A strong Coulomb potential breaks the 6-valley degeneracy of the silicon conduction band near the donor site, yielding a substantial energy gap of ~15 meV to the lowest excited state[(3)] as required for quantum computation.

[1]Department of Electrical and Computer Engineering, University of Illinois, Urbana-Champaign, 1406 West Green Street, Urbana, Illinois 61801. E-mail: jrtucker@uiuc.edu

[2]Department of Physics, Utah State University, Logan, Utah 84322. E-mail: tcshen@cc.usu.edu

[3]To whom correspondence should be addressed.

Kane's original proposal[1] envisions encoding quantum information onto the nuclear spin 1/2 states of ^{31}P qubits in a spinless $I=0$ ^{28}Si lattice. Nuclear spin relaxation times for P donors are extremely long when the electron spin is polarized, many hours at LHe temperature and far longer below. The Kane architecture employs an array of top-gates to manipulate the ground state wavefunctions of the spin-polarized electrons at each donor site in a high magnetic field $B \sim 2$ T, at very low temperature $T \sim 100$ mK. "A-gates" above each donor tune single-qubit NMR rotations *via* the contact hyperfine interaction; and "J-gates" between them induce an indirect two-qubit nuclear exchange interaction *via* overlap of the spin-polarized electron wavefunctions. Although the nuclear spin offers unlimited decoherence times for quantum information processing, the technical problems of dealing with nuclear spins through the electrons are exceedingly difficult. A modified version of the Kane architecture was soon proposed using the spin of the P donor electron as the qubit.[4] In this scheme, A-gates would modulate the electron's g-factor by polarizing its ground state into Ge-rich regions of a SiGe heterostructure for selective ESR rotations, while two-qubit electron exchange is induced through wavefunction overlap.

At its inception, the most difficult challenge to realizing Kane's proposal was thought to be the positioning of P donors into silicon with atomic accuracy. By coincidence, we published a process for doing this only a few weeks later.[5] Previous work had shown that the low-energy electron beam of a scanning tunneling microscope (STM) could be used to remove hydrogen from H-terminated silicon surfaces with atomic precision.[6] By dosing the resulting pattern with phosphine gas, individual PH_3 precursor molecules can be selectively adsorbed onto the STM-exposed dangling bonds, followed by low-temperature silicon overgrowth to incorporate the P atoms as activated donors. Over the last three years, both our group and the Australian Centre for Quantum Computer Technology have been actively engaged in implementing these techniques as the 'bottom-up' route to a silicon quantum computer. Thus far, linear arrays of individual PH_3 precursor molecules have been accurately positioned onto hydrogen-terminated Si(1 0 0)-2×1 surfaces,[7] and their P atoms incorporated into the surface crystal structure *via* low-temperature anneal.[8] Ultra-low-temperature silicon overgrowth has also been developed by fabricating unpatterned P δ-layers from self-ordered PH_3 molecular precursors.[9,10] The saturated surface yields an ultra-dense 2D electron gas with a carrier density of $\sim 1.5 \times 10^{14}$ cm^{-2}, corresponding to full activation of a $\sim$1/4ML PH_3 adlayer. Magnetoresistance measurements show metallic conductivity of $\sim$1–2 kΩ/sq., and 2D weak-localization effects with an electron phase coherence (Thouless) length $L_\phi \sim 150$ nm.[11,12] Close agreement of results between two groups indicates a robust, high quality process.

All of the techniques needed to place P qubits into silicon are now in place. To realize a silicon quantum computer (Si QC), P donors must be integrated with: (1) single-electron transistors (SETs) to perform spin-to-charge state readout and (2) top-gate arrays for accurate control of bound electron wavefunctions and nearest-neighbor exchange. Both of these tasks are very difficult. Here, we will outline an approach based on independent development and scaling of these two major components prior to combining them into a working prototype, as in conventional integrated circuits.

2. THE 3-SPIN "UNIVERSAL EXCHANGE"QUBIT

The choice of architecture will be crucial to a first demonstration of Si QC. DiVincenzo *et al.*[13] have shown that the nearest-neighbor Heisenberg exchange can be universal for composite 3-spin qubits; and this appears to offer the simplest possible implementation. Figure 1 illustrates a 3-donor qubit with logic states encoded onto the $S=1/2$, $S_z=1/2$ subspace of the bound electrons. Logic zero is represented by spins 1 and 2 in the singlet state S, and spin 3 up. Logic one is a linear combination of triplet states T_+ and T_0 for spins 1 and 2, with spin 3 down and up, respectively, that preserves the overall spin quantum numbers. Initializing to logic zero is achieved by cooling the system in a large magnetic field to polarize spin 3, while inducing an even greater exchange coupling, $J_{\text{initial}} > 2\mu_B B \gg k_B T$, between spins 1 and 2 to produce the equilibrium spin singlet. Typical parameters for 10^{-6} initialization error at $T =$ 100 mK are $B \sim 1$ T and $J_{\text{initial}} \sim 200\,\mu$eV. In this electron-spin qubit, the

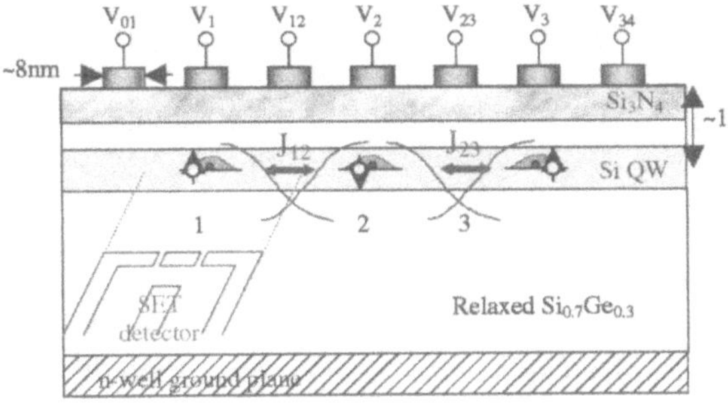

Fig. 1. Sketch for a composite 3-spin "universal exchange" qubit of Si:P donors in a Kane-type architecture with the integrated SET readout. A strained Si quantum well on relaxed SiGe prevents ionization of the donors during gate operations.

hyperfine interaction with ^{31}P nuclei is suppressed by the large magnetic field, adding only a rapidly oscillating spin-flip component of order $\sim 10^{-3}$ to electron wavefunctions outside the encoded subspace.

One-qubit operations are implemented with the nearest-neighbor exchange in four or fewer steps, eliminating the need for gated ESR rotations. The two-qubit CNOT operation can also be performed by the nearest-neighbor exchange within a 1D array, in 19 steps.(13) This large overhead for "universal exchange" may not be optimal for a large-scale quantum computer, but the internal structure of the 3-spin qubit is uniquely suited to an initial demonstration.

As indicated in Fig. 1, this adaptation to the Kane architecture employs A-gates over each donor site in addition to the J-gates between nearest neighbors. Although this is an exchange-only qubit, calculations of the potential landscape show that both sets of gates are needed to span the required range of J-coupling. During entanglement operations, the programmed exchange energy sets a time scale for the gate pulses:

$$t_{2\pi} = \frac{h}{J} \approx \frac{4 \times 10^{-9}\mathrm{s}}{J(\mu\,\mathrm{eV})}. \tag{1}$$

Assuming that gate response can be tailored on a ~ 10 ps scale, a fidelity standard of 10^{-4} would require exchange energies of ~ 10 neV in order to lengthen gate pulses into the ~ 0.1 μs range. On the other hand, controlling the magnitude of exchange at the ~ 10 neV level is likely to be problematic, keeping in mind that $k_B T \sim 8.6$ μeV at $T = 100$ mK (not to mention voltage noise on the gates!). Initial plans call for a larger 'on-state' exchange energy in the range $J_{on} \sim 0.1$–1 μeV, implying a pulse length fidelity of $\sim 10^{-3}$ sufficient for testing basic entanglement operations. The 'off-state' coupling will need to be $J_{off} \sim 10^{-12}$ eV or less.

Including the initialization requirement $J_{initial} \sim 200$ μeV, the electron exchange energy must be controlled over at least eight orders-of-magnitude. A complete set of potential calculations has recently been carried out for the 3-donor, 7-gate array shown in Fig. 1, employing realistic SiGe heterolayer structures and gate parameters.(14) Combined with our previously published simulations of P donor exchange in an applied parabolic potential,(15) these results indicate that the structure in Fig. 1 should have all of the capabilities needed to initialize logic 0, transform the spin entanglement to logic 1, and read out the result. Similar calculations also show that individual electrons can be transferred to empty neighboring sites, carrying quantum information around a larger array.

The most important property of the 3-spin qubit is the ability to perform a detailed calibration before attempting high-frequency gate operations. An integrated SET is included for singlet/triplet readout on electrons 1 and 2, *via* gate-induced charge transfer in the spin singlet state. The exchange energy J_{12} can thus be measured *inside the qubit* for the "on" configuration of gate voltages that couples electrons 1 and 2 and isolates electron 3. Readout is accomplished by rapidly adjusting the gate voltages to apply a potential difference between coupled donors 1 and 2 on a time scale fast compared to the electron spin lifetime. Our simulations show a sharp threshold for spin-singlet charge transfer into the doubly- occupied D^- state at a potential difference roughly equal to the on-site repulsion $U \sim 42$ meV for P donors in the Hubbard model.[15] The spin-triplet remains unpolarized until ionization takes place at ~20% higher potential difference, providing a well-defined window for spin-state readout. By reinstating the 'on' gate voltages and slowly ramping the magnetic field between measurements, the equilibrium singlet/triplet crossover can be identified at $B_{cr} = J_{12}/2\mu_B$ yielding a direct experimental measurement of exchange energy. The J_{23} parameter space could also be characterized by repositioning the SET to monitor charge on donor 2. By comparing data of this type with simulations of the gate operations, it should be possible to produce a fully calibrated 3-spin qubit ready for high-frequency entanglement.

3. SET SPIN-STATE READOUT

The planar SET shown in Fig. 1 can be patterned into the P δ-layer by STM along with the individual P donor qubits, in the same lithographic step.[5,16] The entire SET structure—source, drain, Coulomb island and capacitive gate—is comprised of ultra-dense 2D metallic regions grown into the silicon crystal from PH_3 molecules at saturation coverage. The gaps between these metallic areas serve as planar tunnel junctions, with a maximum built-in potential comparable to the ~44 meV binding energy of an isolated P donor. The ~8 nm width of the tunnel junctions is dictated by the relatively large Bohr radius, a_B ~2.5 nm, of the P donor ground state—large enough to permit modulation of tunnel barriers with a top-gate. Recently, we fabricated our first electrical device between pre-implanted contacts in the form of an Aharonov–Bohm ring to demonstrate electron wave interference across a ~50 nm diameter.[17] The Australian Centre has now applied *ex situ* contacts to a 90 nm-wide STM-patterned line, and demonstrated the sharp 2D to 1D crossover in magnetoresistance.[12] We believe that an integrated SET will be essential to development of Si QC, and efforts to realize it are currently underway.

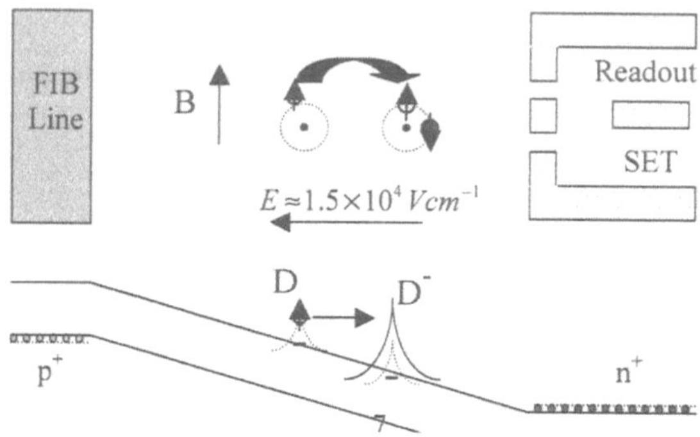

Fig. 2. Sketch for a planar implementation of singlet/triplet-spin-state readout.

Figure 2 illustrates a planar circuit for developing singlet/triplet state detection without top-gates. Here, the P-donor SET is combined with a focused-ion beam (FIB) counter electrode. Two individual P "qubits" are separated by $10a_B \sim 25$ nm and positioned at a distance of ~100 nm from the SET in the same lithographic step. Because the pre-implanted Ga^+ FIB line is p-type, a large built-in electric field will be generated along the axis of the P-donor "molecule" in the direction of the ultra-dense n-type SET. The magnitude of this field is $\sim 1.3 \times 10^4$ V/cm for a total separation of ~800 nm, close to our theoretical estimate for the critical field, $E_{crit} \sim 1.5 \times 10^4$ V/cm, needed to induce singlet-state charge transfer into the D^- configuration at $10a_B$ donor separation.(15) Relatively small voltages can then be applied to the FIB line to induce reversible polarization in the equilibrium singlet state at magnetic fields below $J/2\ \mu_B$, and no polarization in the triplet state above, yielding an experimental value for exchange energy.

The absence of amorphous tunnel barriers in the P-donor SET can be expected to eliminate the large "telegraph noise" seen in conventional SETs as a series of random discrete changes in the voltage threshold of the Coulomb blockade.(18) This enhanced stability, and greatly reduced $1/f$ noise, should make it possible to develop spin-state detection in a simplified planar circuit, and then use that capability to calibrate the exchange interactions inside a Si QC prototype.

4. DEVELOPING TOP-GATES

Gate arrays of ~15 nm pitch are needed to control the potential at each donor site and the exchange coupling between them. Fabrication

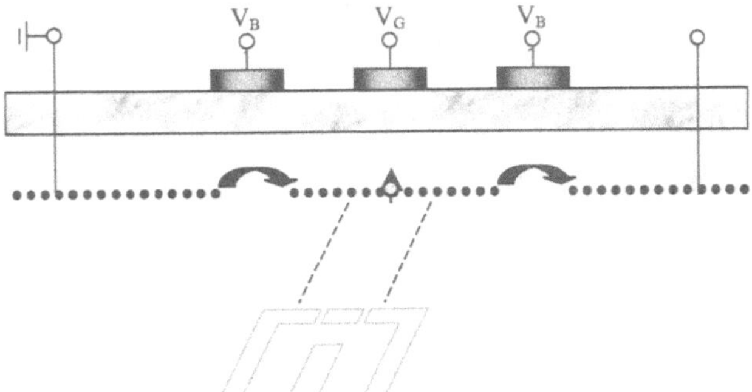

Fig. 3. Sketch for one section of a "single-electron pump" with ultra-dense P-donor islands for scaling gate pitch and registration down to single P qubits.

techniques for doubling a larger pitch defined by e-beam lithography into the ~15 nm range are currently under development; but the gates must also be registered to the underlying P donor qubits with ~2 nm accuracy. For the anticipated exchange energies $J_{\text{on}} \sim 0.1$–1 μeV, operating frequencies will be in the ~100 MHz range:

$$\frac{J}{h} \approx 240 \text{ MHz} \times J(\mu\text{eV}). \tag{2}$$

In that case, an initial fidelity standard of $\sim 10^{-3}$ will require a well-characterized gate response at ~100 GHz. This combined set of requirements for gate dimensions, registration, and frequency is unlikely to be realized in entanglement experiments on qubit arrays, since this is an all-or-nothing proposition.

A systematic strategy for gate development can be implemented, however, by employing metallic P donor islands as the elements of a charge-coupled device (CCD) array. In this approach, a series of Coulomb islands would be inserted into an STM-patterned line, each coupled to its nearest neighbors by a planar tunnel junction. At first, the islands would contain large numbers of donors in the form of an ultra-dense metallic 'quantum dot'. Figure 3 illustrates the smallest such unit, a 3-gate array to control both tunnel barriers and the dot potential for one island. As control is established over single-electron transport through a series of these structures, both islands and gates would be scaled continuously toward single P qubits. The frequency response of the gate arrays can also be tested at each stage. The final goal here is a 7-gate "single-electron pump"[19] that can control current flow through three P donor islands—the same configuration as the 3-spin qubit.

5. CONCLUSION

Most of the processes needed to fabricate a 3-spin "universal exchange" qubit with Si:P donors have been separately demonstrated. Integrating them will be difficult, but strategies are available for systematic development and testing of the individual components. Hurdles include registration of gates to P-donor qubits, the quality of ultra-thin gate dielectrics, and the material perfection of SiGe heterolayers; but there appear to be no show-stoppers in fabrication. High-frequency entanglement operations will entail additional issues of measurement, refocusing, and control.

In spite of these difficulties, it is important to appreciate the intrinsic ~60 ms electron spin dephasing time in the Si:P system and its potential for large scale integration. The non-degenerate P-donor ground state is large enough to permit electronic control, but small enough to yield level splittings that prevent state mixing. Deforming these very robust wavefunctions and coupling them to the nearest neighbors can realize all of the basic operations of quantum information processing. The same set of gates can move this quantum information along linear arrays, providing the basis for a scalable architecture.

Finally, a silicon-based quantum computer can have the tremendous advantage of on-chip MOS control circuits. A tunneling MOS transistor with metal silicide source/drain has been proposed and demonstrated at ~25 nm gate length by one of us,[20,21] and refined to ~20 nm × 25 nm overall dimensions by a University of California Berkeley, group.[22] These tunneling transistors can operate at $T = 0$ as well as at room temperature; and they can be integrated with gates and P-donor SETs in a low-temperature process. The potential for large-scale Si QC is unparalleled in the solid state, if it can be realized.

ACKNOWLEDGMENTS

The authors wish to thank the following colleagues who have made important contributions to our joint efforts: Y.-C. Chang, M. Feng, R.-R. Du, M. A. Zudov, J. S. Kline, S. J. Robinson, R. Chan, J.-Y. Ji, A. Fang, and J. C. Kim. This work is based on research supported by the DARPA QuIST Program under Contract No. DAAD19-01-1-0324 and the U.S. National Science Foundation under ITR Grant EIA-01-21568.

REFERENCES

1. B. E. Kane, *Nature* **393**, 133 (1998).
2. A. M. Tyryshkin, S. A. Lyon, A. V. Astashkin, and A. M. Raitsimring, *Phys. Rev. B* **68**, 193207 (2003).
3. D. K. Wilson and G. Feher, *Phys. Rev.* **124**, 1068 (1961).
4. R. Vrijen, E. Yablonovitch, K. Wang, H. W. Jiang, A. Balandin, V. Roychowdhury, T. Mor, and D. DiVincenzo, *Phys. Rev. A* **62**, 12306 (2000).
5. J. R. Tucker and T.-C. Shen, *Solid State Electronics* **42**, 1061 (1998).
6. T.-C. Shen, C. Wang, G. C. Abeln, J. R. Tucker, J. W. Lyding, Ph. Avouris, and R. E. Walkup, *Science* **268**, 1590 (1995).
7. J. L. O'Brien, S. R. Schofield, M. Y. Simmons, R. G. Clark, A. S. Dzurak, N. J. Curson, B. E. Kane, N. S. McAlpine, M. E. Hawley, and G. W. Brown, *Phys. Rev. B* **64**, 161401(R) (2001).
8. S. R. Schofield, N. J. Curson, M. Y. Simmons, F. J. Rueß, T. Hallam, L. Overbeck, and R. G. Clark, *Phys. Rev. Lett.* **91**, (2003).
9. T.-C. Shen, J.-Y. Ji, M. A. Zudov, R.-R. Du, J. S. Kline, and J. R. Tucker, *Appl. Phys. Lett.* **80**, 1580 (2002).
10. L. Overbeck, N. J. Curson, M. Y. Simmons, R. Brenner, A. R. Hamilton, S. R. Schofield, and R. G. Clark, *Appl. Phys. Lett.* **81**, 3197 (2002).
11. M. A. Zudov, C. L. Yang, R. R. Du, T.-C. Shen, J.-Y. Ji, J. S. Kline, and J. R. Tucker, http://arXiv.org/abs/cond-mat/0305482.
12. F. J. Rueß, L. Oberbeck, M. Y. Simmons, K. E. J. Goh, A. R. Hamilton, T. Hallam, N. J. Curson, and R. G. Clark (unpublished).
13. D. P. DiVincenzo, D. Bacon, J. Kempe, G. Burkard, and K. B. Whaley, *Nature* **408**, 339 (2000).
14. Y.-C. Chang and J. R. Tucker (unpublished).
15. A. Fang, Y. C. Chang, and J. R. Tucker, *Phys. Rev. B* **66**, 155331 (2002).
16. J. R. Tucker and T.-C. Shen, *Int. J. Circuit Theory and Appl.* **28**, 553 (2000).
17. J. R. Tucker and T.-C. Shen, *Ext. Abstracts of the 2nd Int. Workshop on Quantum Dots for Quantum Computing*, University of Notre Dame, August 2003, pp. 44–45.
18. N. M. Zimmerman, W. H. Huber, A. Fujiwara, and Y. Takahashi, *Appl. Phys. Lett.* **79**, 2188 (2001).
19. H. Pothier, P. Lafarge, P. F. Orfila, C. Urbina, D. Esteve, and M. H. Devoret, *Physica B* **169**, 1598 (1991).
20. J. R. Tucker, C. Wang, and P. S. Carney, *Appl. Phys. Lett.* **65**, 618 (1994).
21. C. Wang, J. P. Snyder, and J. R. Tucker, *Appl. Phys. Lett.* **74**, 1174 (1999).
22. J. Kedzierski, P. Xuan, V. Subramanian, E. Anderson, J. Bokor, T.-J. King, and C. Hu, *IEDM* (2000).

Controlling Spin Qubits in Quantum Dots

Hans-Andreas Engel,[1] L. P. Kouwenhoven,[2] Daniel Loss[3] and C. M. Marcus[4]

We review progress on the spintronics proposal for quantum computing where the quantum bits (qubits) are implemented with electron spins. We calculate the exchange interaction of coupled quantum dots and present experiments, where the exchange coupling is measured via transport. Then, experiments on single spins on dots are described, where long spin relaxation times, on the order of a millisecond, are observed. We consider spin-orbit interaction as sources of spin decoherence and find theoretically that also long decoherence times are expected. Further, we describe the concept of spin filtering using quantum dots and show data of successful experiments. We also show an implementation of a read out scheme for spin qubits and define how qubits can be measured with high precision. Then, we propose new experiments, where the spin decoherence time and the Rabi oscillations of single electrons can be measured via charge transport through quantum dots. Finally, all these achievements have promising applications both in conventional and quantum information processing.

KEY WORDS: Spin qubits; coupled quantum dots; spin filter; spin read out.

PACS: 03.67.Lx; 03.67.Mn; 73.23.Hk; 85.35.Be.

1. INTRODUCTION

The spin degree of freedom promises many applications in electronics.[1–3] Prominent experiments have shown injection of spin-polarized currents

[1]Department of Physics and Astronomy, University of Basel, Klingelbergstrasse 82, CH-4056 Basel, Switzerland.

[2]Department of NanoScience and ERATO Mesoscopic Correlation Project, Delft University of Technology, P.O. Box 5046, 2600 GA Delft, The Netherlands.

[3]Department of Physics and Astronomy, University of Basel, Klingelbergstrasse 82, CH-4056 Basel, Switzerland.

[4]Department of Physics, Harvard University, Cambridge, Massachusetts 02138, USA.

into semiconductor material,[4,5] long spin dephasing times in semiconductors (approaching microseconds),[6] ultrafast coherent spin manipulation,[7] as well as phase-coherent spin transport over distances of up to 100 μm.[6] Irrespective of spin, the charge of the electrons can be used to control single electrons by confining them in quantum dot structures, which leads to striking effects in the Coulomb blockade regime.[8] The Loss and DiVincenzo proposal[9] combines these two fields of research and uses the spin of electrons confined on quantum dots as spin qubits for quantum computation. This proposal comprises two-qubit quantum gates relying on the exchange interaction of coupled quantum dots and comprises spin-to-charge conversion for efficient read-out schemes, satisfying all theoretical requirements for quantum computing. This quantum computer proposal, based on exchange interaction, can be mapped from electron spins on dots to nuclear spins of P atoms in Si, as shown by Kane[10] (see article in this issue).

The spin qubit proposal[9] addresses the central issues for building a quantum computer. However, for a concrete implementation of spin qubits, a more detailed theoretical and experimental understanding of spins on quantum dots is required. This demand has led to many new theoretical and experimental investigations on quantum dots, which also address interesting aspects of physics on their own. In this article we will review some of these recent results.

1.1. Quantum Dots

In this article we consider semiconductor quantum dots. These are structures where charge carriers are confined in all three spatial dimensions. The dot size, typically between 10 nm and 1 μm,[8] is on the order of the Fermi wavelength in the host material. The confinement of the quantum dots is usually achieved by electrical gating of a two-dimensional electron gas (2DEG), possibly combined with etching techniques, see Figs. 1, 2(a), and 5(b). Small dots have charging energies in the meV range, resulting in quantization of charge on the dot (Coulomb blockade). This allows precise control of the number of electrons and of the spin ground state on the dot. Such a control of the number of electrons in the conduction band of a quantum dot (starting from zero) has been achieved with GaAs heterostructures, e.g., for vertical dots[11] and lateral dots.[12,13] Quantum dots have various tunable parameters. These include geometry, energy spectrum, coupling between dots, etc. which open up many possibilities by providing a versatile system for manipulation of electronic states, in particular the spin state. Further, the electronic dot-orbitals are highly sensitive to external magnetic and electric fields,[8,11] since the magnetic length

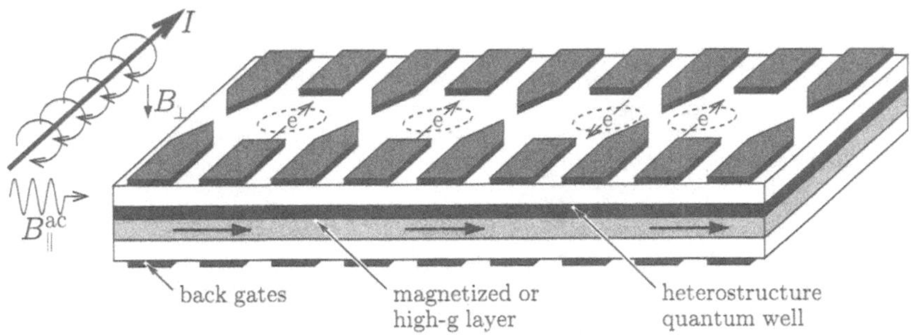

Fig. 1. An array of quantum dots (circles) is defined by gate electrodes (dark gray) which confine the electrons. For spin manipulations, electrons can be moved by changing the gate voltage, pushing the electron wave function into the magnetized or high-*g* layer, allowing for spatially varying Zeeman splittings. Alternatively, local magnetic fields can be achieved by a current-carrying wire (indicated on the left of the dot array). Then, the electron in each dot is subject to a distinct Zeeman splitting. This can be used for one-qubit gates, since only relative spin rotations are sufficient. Further, the spins can be addressed individually with ESR pulses of an oscillating in-plane magnetic field which is in resonance with a particular Zeeman splitting. These mechanisms allow single-spin rotations in different spatial directions. For gate operations on two qubit spins, their exchange coupling can be controlled by lowering the tunnel barrier between the dots (see Sec. 2). Here, the two rightmost dots are drawn schematically as tunnel-coupled. Note that only electrical switching is required to control spin dynamics and quantum computation with such a device.

corresponding to fields of $B \approx 1\,\mathrm{T}$ is comparable to typical dot sizes. In coupled quantum dots, Coulomb blockade effects,[14] tunneling between neighboring dots,[8,14,15] and magnetization[16] have been observed as well as the formation of a delocalized single-particle state[17,18] and coherent charge oscillations.[19]

1.2. Quantum Computing with Spin Qubits

The interest in quantum computing[20,21] derives from the hope to outperform classical computers using new quantum algorithms. These algorithms make use of the quantum computer's abilities to exist in a quantum superpositions of its "binary" basis states $|0\cdots 00\rangle$, $|0\cdots 01\rangle$, $|0\cdots 10\rangle$,..., and to perform unitary time evolutions $U|\Psi_{\rm in}\rangle = |\Psi_{\rm out}\rangle$ for computation. The basis states can be realized by concatenating several quantum bits (qubits) which are states in the Hilbert space spanned by $|0\rangle$ and $|1\rangle$. A natural candidate for the qubit is the electron spin because every spin 1/2 encodes exactly one qubit. Such spin qubits on quantum dots are good candidates for realizing a quantum computer.[9] We consider the five criteria of DiVincenzo's checklist[22] which must *all* be satisfied for any physical implementation of a quantum computer. We briefly discuss

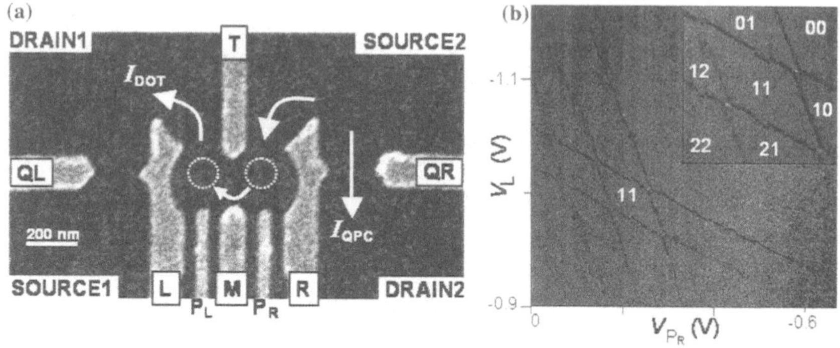

Fig. 2. (a) Double dot structure with a single electron in each dot, shown as scanning electron micrograph of the metallic surface gates.[12] The circles indicate the two quantum dots and the arrows show the possible current paths. A bias voltage, V_{DOT} , can be applied between source 2 and drain 1, leading to current through the dots. A bias voltage, V_{SDi} between source and drain $i = 1, 2$ yields a current, I_{QPC}, through the corresponding QPC. (b) Charge stability diagram (honeycomb)[15] of the double quantum dot, measured with QPC-R.[12] A modulation (0.3 mV at 17.77 Hz) is applied to gate L, and dI_{QPC}/dV_L is measured with a lock-in amplifier and plotted versus V_L and V_{PR}. The bias voltages are $V_{SD2} = 100\,\mu V$ and $V_{DOT} = V_{SD1} = 0$. The inset shows a magnification of the honeycomb pattern for the first few electrons in the double dot. The labels "$n_L n_R$" indicate the number of electrons in the left and right dot, and the double dot is completely empty in the region "00."

that these criteria are satisfied for spins qubits.[9,23] These criteria provides us with a good starting point for going into the details of concrete parts of the actual implementation of spin qubits. In the following sections we then show where specific theories and current experiments give new insight into the realization of spin qubits.

(i) *A scalable system with well characterized qubits* is required. To speed up calculations using a quantum computer, one needs a large number of qubits, i.e., on the order of 10^5. This requirement is achievable for spin qubits, since producing arrays of quantum dots is feasible with state-of-the-art techniques for defining nanostructures in semiconductors. Further, the electron's spin 1/2 provides a natural qubit, setting $|0\rangle \equiv |\uparrow\rangle$ and $|1\rangle \equiv |\downarrow\rangle$.

(ii) *The state of the qubits must be initialized* to a known value at the beginning of a computation. To initialize spin qubits, one can apply a large magnetic field $g\mu_B B \gg kT$ that allows them to relax to the thermal ground state. Alternatively, one can inject polarized electrons into the dot by using spin-polarizing materials[4,5] or by using a spin filter[24] which we describe in Sec. 4.

(iii) *Long decoherence times, much longer than the gate operation time,* is the most difficult criterion to satisfy for many quantum computer proposals. Here, the current knowledge about the spin qubits is very promising. Gate operation times well below one ns are in principle feasible.[23] Using theoretical estimates and experimental data on spin flip times, the expected decoherence times can reach ms (see Sec. 3). Thus, the decoherence times could be eight orders of magnitude larger than the gate operation times.

(iv) With *a universal set of quantum gates*, any quantum algorithm can be implemented by controlling a particular unitary evolution of the qubits. It is sufficient to have single-qubit gates and a universal two-qubit gate (e.g., XOR or square root of SWAP). Single qubit gates can be produced by controlling the local magnetic field, the local g factor (or g tensor), or local Overhauser field, which, e.g., can be achieved with a semiconductor heterostructure and electrical gating,[23,25,26] (see Fig. 1). To build two-qubit gates, one can use the exchange interaction which arises when two neighboring dots are tunnel coupled, which can again be controlled via gate voltages.[9,27] We describe the exchange interaction of coupled dots in Sec. 2.

(v) *Qubit read out* determines the result at the end of the computation by measuring specific qubits. There are several proposals for measuring the spin in quantum dots, most of them rely on transferring the information from the spin to the charge state,[9] e.g., by making use of the Pauli principle,[24,28,29] via the spin-orbit interaction,[30] or by making use of the Zeeman splitting.[29] We discuss concrete read-out schemes for spin qubits in Sec. 5 and address experiments[31] where single-shot read out has been achieved.

2. TWO COUPLED QUANTUM DOTS AS QUANTUM GATES

We now consider a pair of spin qubits which are coupled by the exchange interaction, which results from the combination of the Coulomb interaction and the Pauli exclusion principle. Two electrons in coupled quantum dots and in the absence of a magnetic field have a spin-singlet ground state, while the first excited state in the presence of sufficiently strong Coulomb repulsion is a spin triplet. Higher excited states are separated from these two lowest states by an energy gap, given either by the Coulomb repulsion or the single-particle confinement. The low-energy dynamics of such a system is described by the effective Heisenberg spin Hamiltonian,

$$H_S(t) = J(t)\,\mathbf{S}_1 \cdot \mathbf{S}_2, \tag{1}$$

where $J(t)$ describes the exchange coupling between the two spins $\mathbf{S}_1$ and $\mathbf{S}_2$ and is given by the energy difference between the triplet and the singlet, $J = E_{T_0} - E_S$. After a pulse of $J(t)$ with $\int_0^{\tau_s} dt J(t)/\hbar = \pi \pmod{2\pi}$, the time evolution $U(t) = T \exp(i \int_0^t H_s(\tau) d\tau/\hbar)$ corresponds to the SWAP operator U_{sw}, whose application leads to an interchange of the two spin states. While U_{sw} is not sufficient for quantum computation, any of its square roots, say $U_{sw}^{1/2}|\phi\chi\rangle = (|\phi\chi\rangle + i|\chi\phi\rangle)/(1+i)$, turns out to be a *universal* quantum gate. It can be used, together with single-qubit rotations, to assemble any quantum algorithm.[9]

We consider a system of two coupled quantum dots in a 2DEG, containing one (excess) electron each (see Fig. 2(a)). The dots are arranged in a plane such that the electrons can tunnel between the dots, leading to an exchange interaction J between the two spins, which we now calculate. We model this system of coupled dots with the Hamiltonian $H = \sum_{i=1,2} h_i + C + H_Z = H_{orb} + H_Z$. The single-electron dynamics in the 2DEG (xy-plane) is defined with the Hamiltonian h_i, containing the quartic confinement potential

$$V(x, y) = \frac{m\omega_0^2}{2}\left[\frac{1}{4a^2}\left(x^2 - a^2\right)^2 + y^2\right] \tag{2}$$

with inter-dot distance $2a$, effective Bohr radius $a_B = \sqrt{\hbar/m\omega_0}$, and effective mass m. Separated dots ($a \gg a_B$) are thus modeled as two harmonic wells with frequency ω_0, consistent with experiments where the low-energy spectrum of single dots indicates a parabolic confinement.[11] A magnetic field $\mathbf{B} = (0, 0, B)$ is applied along the z-axis, which couples to the electron spins through the Zeeman interaction H_Z and to the charges through the vector potential $\mathbf{A}(\mathbf{r}) = (B/2)(-y, x, 0)$. In almost depleted regions, like few-electron quantum dots, the screening length λ can be expected to be much larger than the screening length in bulk 2DEG regions (where it is 40 nm for GaAs). Thus, for small quantum dots, say $\lambda \gg 2a \approx 40$ nm, we consider the bare Coulomb interaction $C = e^2/\kappa|\mathbf{r}_1 - \mathbf{r}_2|$, where κ is the static dielectric constant.

Now we consider only the two lowest orbital eigenstates of H_{orb}, leaving us with one symmetric (spin singlet) and one antisymmetric (spin triplet) orbital state. The spin state for the singlet is $|S\rangle = (|\uparrow\downarrow\rangle - |\downarrow\uparrow\rangle)/\sqrt{2}$, while the triplet spin states are $|T_0\rangle = (|\uparrow\downarrow\rangle + |\downarrow\uparrow\rangle)/\sqrt{2}$, $|T_+\rangle = |\uparrow\uparrow\rangle$, and $|T_-\rangle = |\downarrow\downarrow\rangle$. For $kT \ll \hbar\omega_0$, higher-lying states are frozen out and H_{orb} can be replaced by the effective Heisenberg spin Hamiltonian (Eq. (1)). To calculate the triplet and singlet energies, we use the analogy between atoms and quantum dots and make use of variational methods similar to the ones in molecular physics. Using the Heitler–London ansatz with the ground-state single-dot orbitals, we find,[27]

$$J = \frac{\hbar\omega_0}{\sinh\left(2d^2\,\frac{2b-1}{b}\right)}\left\{\frac{3}{4b}\left(1+bd^2\right)\right.$$
$$\left.+c\sqrt{b}\left[e^{-bd^2}\,I_0\left(bd^2\right)-e^{d^2(b-1/b)}\,I_0\left(d^2\,(b-1/b)\right)\right]\right\} \quad (3)$$

with zeroth order Bessel function I_0, dimensionless distance $d = a/a_B$ between the dots, magnetic compression factor $b = \sqrt{1+\omega_L^2/\omega_0^2}$, and Larmor frequency $\omega_L = eB/2mc$. In Eq. (3), the first term arises from the confinement potential, while the terms proportional to the parameter $c = \sqrt{\pi/2}(e^2/\kappa a_B)/\hbar\omega_0$ result from the Coulomb interaction C; the exchange term is recognized by its negative sign. We are mainly interested in the weak coupling limit $|J/\hbar\omega_0| \ll 1$, where the ground-state Heitler–London ansatz is self-consistent. We plot $J(B)$ (Eq. (3)) in Fig. 3(a) and observe the singlet–triplet crossing, where the sign of J changes from positive to negative (for the parameters chosen in Fig. 3(a) at $B \approx 1.3\,\mathrm{T}$). Finally, J is suppressed exponentially, $\propto \exp(-2d^2b)$, either by compression of the electron orbitals through large magnetic fields ($b \gg 1$), or by large distances between the dots ($d \gg 1$), where in both cases the orbital overlap of the states in the two dots is reduced. The Heitler–London result (Eq. (3)) was refined by taking higher levels and double occupancy of the dots into account (implemented in a Hund–Mullikan approach), which leads to qualitatively similar results,[27] in particular concerning the singlet–triplet crossing. These results have been confirmed by numerical calculations which take more single-particle levels into account.[32]

A characterization of a double dot can be performed with transport measurements. We describe transport through a double quantum dot, using a master equation approach.[34] We calculate differential conductance $G = \mathrm{d}I/\mathrm{d}V_{SD}$ as a function of the bias voltage $V_{SD} = \Delta\mu/e$ in the sequential tunneling and cotunneling regime. We obtain the main peak of the Coulomb blockade diamond and its satellite peaks. Since the positions of these peaks are related to the interdot tunnel splitting and to the singlet–triplet splitting J, one can determine these values in a standard transport experiment. Further, our model can be checked independently, since we also predict which satellite peaks have positive or negative values of G and since we describe structures inside the Coulomb blocked diamonds which are due to a combined effect of cotunneling and sequential tunneling.[34] When we measure transport properties of a structure resembling a single dot, we observe features as would be expected for a double dot.[33] This indicates that a double dot in formed within our structure.

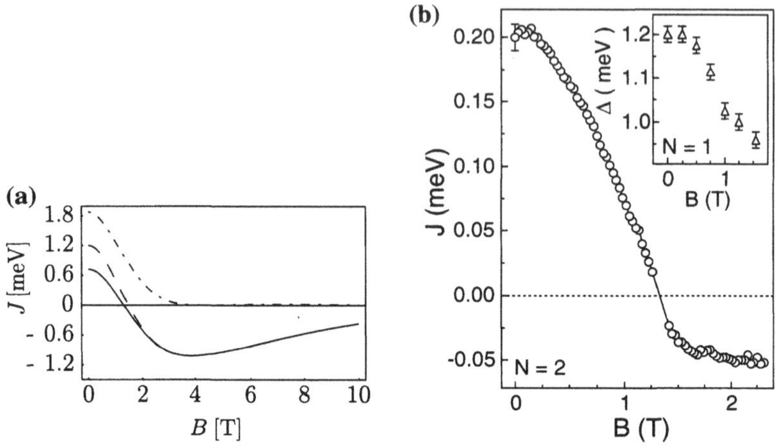

Fig. 3. The exchange coupling J (solid line) for GaAs quantum dots a function of the magnetic field B. (a) Theoretical prediction (Eq. (3)) for a double dot with confinement energy $\hbar\omega = 3\,\text{meV}$, inter-dot distance $d = a/a_B = 0.7$, and $c = 2.42$.[27] For comparison, the short-range Hubbard result $J = 4t^2/U$ (dashed–dotted line) and the extended Hubbard result $J = 4t^2/U + V$ (dashed line) are plotted. (b) Experimentally observed exchange coupling J via transport measurements.[33] Although a single dot structure was used, the measurements show double dot features, indicating that a double dot is formed within the structure. The dependence on magnetic field B is in agreement with the theoretical predictions, in particular, J can be tuned through zero near $B = 1.3\,\text{T}$.

We can then extract the B-dependent exchange coupling from our data which again is in agreement with theoretical predictions for double dots (see Fig. 3(b)). That singlet-triplet crossings occur in single dots is established experimentally.[35]

In further experiments, we measured a double quantum dot with tunable tunnel couplings. Spectroscopy of the double dot states was performed using a quantum point contact (QPC) as a local charge sensor. From the charge distribution on the double dot, we can deduce charge delocalization as a function of temperature and strength of tunnel coupling. Conversely, we can measure the tunnel coupling t as function of the voltage applied on a gate in the coupling region. We find that the tunneling coupling is tunable from $t = 0$ to 22 μeV when the gate voltage is increased.[18]

For few-electron quantum dots, the charging energies of a double quantum dot can be tuned such that there is only a single electron in each dot. The number of electrons on the dots can be controlled by simultaneously measuring the charge distribution with a QPC charge sensor[12] (see Fig. 2), or by measuring transport through the double dot.[36]

3. SPIN RELAXATION

The lifetime of an electron spin is described by the following two time scales. The (longitudinal) spin relaxation time T_1 describes the time scale of a spin-flip process when the electron is aligned along the external magnetic field. The spin decoherence time T_2 is the lifetime of a coherent superposition $\alpha|\uparrow\rangle + \beta|\downarrow\rangle$. Since quantum gate operations require coherence of the underlying qubits, they must be carried out on times shorter than T_2. We note that $T_2 \leq 2T_1$ and typically even $T_2 \ll T_1$,[(37)] thus from the sole knowledge of T_1, no lower bound for T_2 can be deduced. Therefore, it is of interest to investigate the interactions leading to decoherence (as we do now) and to find ways of measuring the decoherence time T_2 in an experiment (see Sec. 6).

For spins on quantum dots, one possible source of spin relaxation and decoherence is spin-orbit interaction. Calculations show that phonon-assisted spin-flip times[(38,39)] in quantum dots are unusually long. This is so because the spin-orbit coupling in two-dimensions (2D) is linear in momentum, both for Dresselhaus and Rashba contributions. Due to this linearity, the effective magnetic field due to spin-orbit fluctuates transversely to the external magnetic field (in leading order). This implies that $T_2 = 2T_1$ for spin-orbit interaction[(40)] and thus long decoherence times are expected. Another source of decoherence is the hyperfine coupling between electron spin and nuclear spins in a quantum dot,[(27,41,42)] since all naturally occurring Ga and As isotopes have a nuclear spin $I = 3/2$. It is known that such decoherence can be controlled by a large magnetic field or by polarizing the nuclear spins, i.e., by creating an Overhauser field.[(27)]

The spin relaxation time T_1 of single electron spins on quantum dots was measured in recent experiments. One way to assess T_1 is to measure transport through the dot while applying double-step pulses to the gate voltage of the dot. First, the dot is emptied and filled again with one electron with a random spin. Then, the electron is held in the dot during a time t_h. Finally, the gate voltage is tuned such that the electron can tunnel out of the dot and contribute to a current, but only if it is in the excited spin state. Thus, the (time-averaged) current will be proportional to the probability of having an excited spin on the dot after time t_h; this probability decays on the time scale of T_1. In these experiments, the limited current sensitivity puts an upper bound on t_h. Since T_1 turned out to be longer than this bound, one was not able to measure T_1. Still, it is possible to obtain a lower bound of for T_1 and $\approx 100\,\mu s$ was obtained for triplet to singlet transitions[(43)] and for $N = 1$ Zeeman levels.[(44)] Using a charge read-out device (see Sec. 5), single tunneling events can be observed. This allowed us to measure T_1 directly and $T_1^{\rm exp} = 1\,\mathrm{ms}$ was obtained at $B = 8\,\mathrm{T}$.[31] We now compare this value with theoretical predictions.[40] We assume a GaAs dot with Dresselhaus spin–orbit interaction $H_{\rm SO} = \beta(-p_x\sigma_x + p_y\sigma_y)$, with quantum well thickness d = 5 nm, and with lateral size quantization energy $\hbar\omega_0 = 1.1\,\mathrm{meV}$, corresponding to a Bohr radius $a_{\rm B} = 32\,\mathrm{nm}$. The material parameters are the dielectric constant $\kappa = 13.1$, coupling constant of deformation potential $\Xi_0 = 6.7\,\mathrm{eV}$, piezoelectric constant $h_{14} = -0.16\,\mathrm{C/m^2}$, sound velocity s_j for branch j, namely $s_1 = 4.73 \times 10^5\,\mathrm{cm/s}$ and $s_2 = s_3 = 3.35 \times 10^5\,\mathrm{cm/s}$, sample density $\rho_c = 5.3 \times 10^3\,\mathrm{kg/m^3}$, and effective mass $m^* = 0.067 m_e$. The remaining unknown parameter is the spin–orbit length $\lambda_{\rm SO} = \hbar/m^*\beta$. It can be extracted from (independent) weak antilocalization measurements,[45] where $\lambda_{\rm SO} \approx 9\,\mu\mathrm{m}$ was found. Taking the Zeeman splitting used in the measurement of $T_1^{\rm exp}$, we obtain[40] $T_{1,\rm SO}^{\rm th} \approx 750\,\mu\mathrm{s}$, with an error of 50% due to the uncertainty of the value of the Zeeman splitting. There is some additional uncertainty on the value of $\lambda_{\rm SO}$ which depends on electron density and growth of the sample. For example, we find $\lambda_{\rm SO} \approx 17\,\mu\mathrm{m}$ in other samples,[46] which would indicate a longer T_1 time since $T_1 \propto \lambda_{\rm SO}^2$.[40] Within these uncertainties we find an agreement between experiments and theory, $T_1^{\rm exp} \approx T_{1,\rm SO}^{\rm th}$. Moreover, the predicted B-dependence[40] of $1/T_1$ agrees well with the experiment,[31] where a plateau is seen around $B \sim 10\,\mathrm{T}$. From this we can conclude that the spin-phonon mechanism is the dominant source for spin relaxation (and not hyperfine interaction). Since $T_2 = 2T_1$ for spin–orbit interaction[40] and since there is no difference between decoherence and relaxation for hyperfine interaction,[41,42] we can expect spin decoherence times T_2 to be on the order of milliseconds.

4. SPIN FILTER

An important device for spintronics is a spin filter which selectively transmits electrons with respect to their spin orientation. For quantum computation with spin qubits, such a spin filter can be used for initialization and read out (see Secs. 1.2 and 5). We proposed to use a quantum dot attached to in- and outgoing current leads as a spin filter.[24] The direction of polarization of this spin filter can be tuned electrically by changing the gate voltage on the quantum dot. We now describe the operational principle of such a spin filter and present experimental implementations.[47–49]

Our spin filter proposal[24] requires a lifted spin-degeneracy on the dot with a Zeeman splitting $\Delta_z = |\mu_{\rm B} g B|$. For two electrons on the dot, we assume a singlet ground state with energy $E_{\rm S}$, while the lowest-lying triplet state has a higher energy $E_{\rm T_+}$. Let us consider the sequential tunneling transition where the number of electrons on the dots changes from 1 to 2. The bias between the leads at chemical potentials $\mu_{1,2}$ is $\Delta\mu = \mu_1 - \mu_2 > 0$. For small bias and low temperatures such that $\Delta\mu$, $kT < \min\{\Delta_z, E_{T_+} - E_S\}$, only ground state transitions are energetically allowed, i.e., $|\uparrow\rangle \leftrightarrow |S\rangle$. Thus, only spin down electrons can tunnel through the dot (see Fig. 4(a)). We calculate the current through the dot using the standard tunneling Hamiltonian approach in the Coulomb blockade regime[8] and the master equation for the reduced density matrix of the dot.[24] The current in first order in tunneling is the sequential tunneling current I_s,[8] which is spin-$\downarrow$ polarized. The second-order contribution is the cotunneling current I_c[50] which involves a virtual intermediate state, where energy conservation can be violated for a short time. Thus, our energetic argument does not hold here and the cotunneling current I_c contains a spin-$\uparrow$ component, reducing the

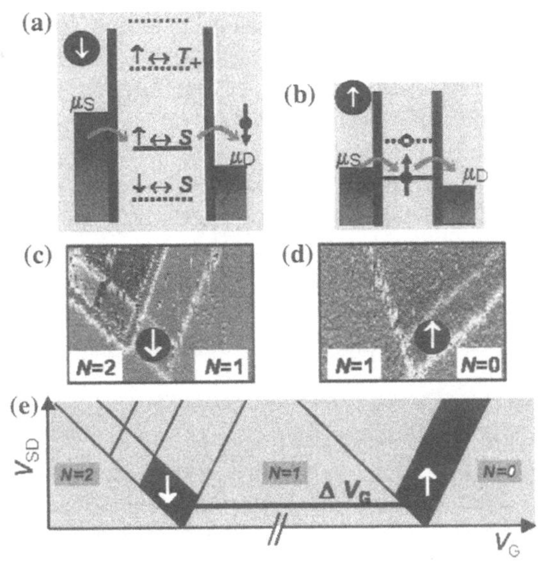

Fig. 4. Spin filter in the sequential tunneling regime.[24] (a) and (b) Operation principle of the spin filter. (a) Regime where the only allowed $1 \leftrightarrow 2$ electron transitions are $\uparrow \leftrightarrow S$ due to energy conservation, thus only spin-$\downarrow$ electron pass through the dot (see text). (b) The $0 \leftrightarrow 1$ electron transition provides a spin filter for spin-$\uparrow$ electrons. (c) and (d) The experimentally measured dI/dV_{SD} is plotted as function of bias voltage V_{SD} and gate voltage V_G at $B_{\|} = 12\,T$.[47] In the region labeled "$\downarrow$" only spin-down electrons pass through the dot while in the region "$\uparrow$" only spin-up electrons. (e) Analyzing all transitions between the dot states $|0\rangle$, $|\uparrow\rangle$, $|\downarrow\rangle$, $|S\rangle$, $|T_0\rangle$, and $|T_\pm\rangle$, the predicted dI/dV_{SQ} is shown schematically and agrees with the experimental data. This indicates that the current is spin polarized in the regimes labeled by arrows.[47]

efficiency of the spin-filtering effect. For $\Delta_z < E_{T_+} - E_S$, the ratio of spin-polarized to unpolarized current is[24]

$$I_s(\downarrow)/I_c(\uparrow) \sim \frac{\Delta_z^2}{(\gamma_1 + \gamma_2)\max\{k_B T, \Delta\mu\}}, \tag{4}$$

where γ_l is the tunneling rate between lead l and the dot. In the sequential tunneling regime we have $\gamma_l < k_B T$, $\Delta\mu$, thus, the ratio Eq. (4) is large and the spin-filter is efficient. We implemented this spin filter with a single quantum dot in the few electron regime.[47] The measured currents agree well with the theoretical predictions (see Fig. 4).

Spin filtering properties of both open[48] and Coulomb blockaded[49] quantum dots were measured directly in a polarizer-analyzer geometry, where the spin polarization of current emitted from the dot (polarizer) was detected using a QPC at $g = 0.5e^2/h$ (analyzer).[51] These polarizer and analyzer elements were coupled by transverse focusing with the use of a small magnetic field applied perpendicular to the sample plane shown in the inset of Fig. 5(b). The collector voltage at the QPC shows a focusing peak when the distance between emitter and collector is an integer multiple of the cyclotron diameter. Measuring at the focusing peak, we find that in the presence of an in-plane field of a few Tesla or more, the current through the quantum dot (which is strongly coupled to leads) is indeed spin polarized. For the case of open dots,[48] the direction of polarization can be readily tuned from along to against the applied in-plane field, see Fig. 5. However, for the closed dots, reversed spin filtering was not observed though ground-state peak motion was seen.[49] More work is needed to clarify this departure from expectation.

5. READ-OUT OF A SINGLE SPIN

At the end of every (quantum) computation, one reads out the result of the computation. For this it is sufficient to determine the state of some qubits which are either in state $|\uparrow\rangle$ or in state $|\downarrow\rangle$ (we do not need to measure a coherent superposition). However, it is very hard to detect an electron spin by directly coupling to its tiny magnetic moment (on the order of μ_B). This difficulty is overcome by converting the spin information into charge information, which is then measured (we describe implementations below). Ideally, the qubit state can be determined in a single measurement, referred to as single shot read out. In general, however, there are some errors associated with the measurement, thus the preparation and measurement of the qubit need to be performed not only once but n times.

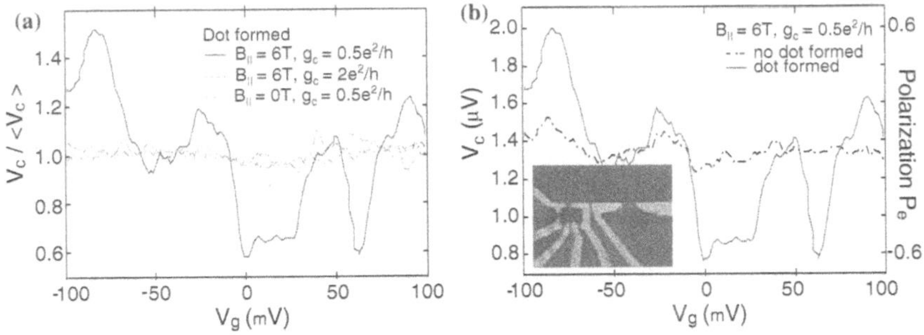

Fig. 5. The spin polarization of current through a quantum dot is detected with an analyzer setup. The polarization is measured via the collector voltage (at the focusing peak, see text). The polarization of the current though the quantum dot in a magnetic field fluctuates as function of gate voltage. The fluctuations in the collector voltage only occur when the emitter forms a quantum dot, the collector is spin-sensitive, and an in-plane magnetic field is applied.[48] (a) Comparison of normalized focusing peak height as a function of V_g at $B_{\|}=6$T for a spin-selective collector, $g_c=0.5e^2/h$, at $B_{\|}=6$T for an unpolarized collector, $g_c=2e^2/h$, and at $B_{\|}=0$ with $g_c=0.5e^2/h$. Dividing by average peak height, $\langle V_c \rangle$, normalizes for changes in focusing efficiency. (b) Focusing peak height at $B_{\|}=6$T with spin-selective collector, $g_c=0.5e^2/h$, comparing an emitter which is a point contact at $2e^2/h$ and an emitter which is a quantum dot with both leads at $2e^2/h$. The inset shows a micrograph of the measured device, where the dot on the left and the QPC on the right side.[48]

We now determine n by assuming that the measurement has two possible outcomes, $A_\uparrow$ or $A_\downarrow$. Then, for an initial qubit state $|\uparrow\rangle$, with probability $p_\uparrow$ the outcome is $A_\uparrow$, which we would interpret as "qubits was in state $|\uparrow\rangle$." However, with probability $1-p_\uparrow$, the outcome is $A_\downarrow$ and one might incorrectly conclude that "qubit was in state $|\downarrow\rangle$". Conversely, the initial state $|\downarrow\rangle$ leads with probability $p_\downarrow$ to $A_\downarrow$ and with $1-p_\downarrow$ to $A_\uparrow$. How many times n do the preparation of a qubit in the *same* initial state and subsequent measurement need to be performed until the state of the qubit is known with some given infidelity α (n-shot read out)? We model the read out process with a positive operator valued measure (POVM) and find from a statistical analysis that we need[29]

$$n \geq z_{1-\alpha}^2\left(\frac{1}{e}-1\right), \tag{5}$$

$$e = \left(\sqrt{p_\uparrow p_\downarrow} - \sqrt{(1-p_\uparrow)(1-p_\downarrow)}\right)^2 \tag{6}$$

with the quantile (critical value) $z_{1-\alpha}$ of the standard normal distribution function, $\Phi(z_{1-\alpha})=1-\alpha=(1/2)\left[1+\text{erf}(z_{1-\alpha}/\sqrt{2})\right]$. We interpret e as *measurement efficiency*,[29] since it is a single parameter $e\in[0, 1]$ which tells us if n-shot read out is possible. For $p_\uparrow = p_\downarrow = 1$, the efficiency is maximal, $e = 100\%$, and single-shot read out is possible ($n = 1$). When the measurement outcome is independent of the qubit state, i.e., $p_\downarrow = 1 - p_\uparrow$ (e.g., $p_\uparrow = p_\downarrow = (1/2)$), the state of the qubit cannot be determined and the efficiency is $e = 0\%$. For the intermediate regime, $0\% < e < 100\%$, the state of the qubit is known after several measurements, with n satisfying Eq. (5). In the more general case, the state of a register with k different qubits should be determined with infidelity β. The probability that the state of all qubits is determined correctly is $1 - \beta = (1 - \alpha)^k$. One could expect that the required n grows dramatically with k. Fortunately this is not the case, from Eq. (5) we find that $n \geq 2(1/e - 1) \log k/\beta$ is sufficient.

For the actual implementation of the spin qubit read out, the most prominent idea is to transfer the qubit information from spin to charge,[9,24,28,29,31,52,53] which can then be accessed experimentally with sensitive voltage or current measurements. A straightforward concept yielding a potentially 100% reliable measurement requires a "spin-filter"[24] which allows only, say, spin-up but no spin-down electrons to pass through, as it is described in Sec. 4. For performing a measurement of a spin in a quantum dot, the spin filter is connected between this dot and a second ("reference") dot. The charge distribution on this system can be detected with sensitive electrometers[54] by coupling the dots to a quantum point contact[12,55] or to a single-electron transistor (SET).[56] Then, if the spin had been up, it would pass through the spin filter into the second dot and a change in the charge distribution would be measured, while there is no change for spin down[9]. Instead of a spin filter, one can use different Zeeman splittings on qubit and reference dot or make use the Pauli principle to read out the spin qubit via charge detection.[29]

Finally, we consider the qubit dot coupled to a lead instead of a reference dot. For Zeeman splittings larger than temperature, one can tune the dot levels such that only the excited spin state, $|\downarrow\rangle$, can tunnel into the leads[28] with rate γ_{out} (spin $\uparrow$ electrons can tunnel only *onto* the dot). Such a tunneling event changes the number of electrons on the dot and produces a pulse in the QPC current, whose duration must exceed t_{m} to be detected, until a spin $\uparrow$ electron tunnels onto the dot with rate γ_{in}. After waiting a time t to detect such a signal, we have $p_\uparrow = 1$ and $e = p_\downarrow = (1 - e^{-t\gamma_{\mathrm{out}}})e^{-t_{\mathrm{m}}\gamma_{\mathrm{in}}}$. We implemented this scheme experimentally.[31] Accounting also for finite T_1 and temperature, we obtain $p_\uparrow = 92\%$ and $p_\downarrow = 70\%$. This means that the measurement efficiency is $e = 41\%$, which is already very close to single-shot read out. For example, after 16 measurements, one knows the state of a 10 qubit register with an error smaller than 10^{-4}. Further, this single spin detection scheme made it possible to determine the T_1 times of electron spins on quantum dots,[31] see Sec. 3.

6. DETECTION OF SINGLE-SPIN DECOHERENCE

As it was seen in Sec. 3, it is an important research goal to measure the decoherence time T_2 of single spins on quantum dots. For this, we now describe how to extract the decoherence time T_2 from the sequential tunneling current through a quantum dot, in the presence of an applied electron spin resonance (ESR) field producing spin-flips on the dot.[28] We assume that the Zeeman splitting on the dot is $g\mu_B B > \Delta\mu, k_B T$, while the Zeeman splitting in the leads is different, such that the effect of the ESR field on the leads can be neglected. This can be achieved, e.g., by using materials with different g-factors for the dot and the leads. We derive the master equation and find the stationary reduced density matrix of the quantum dot in the basis $|\uparrow\rangle$, $|\downarrow\rangle$, $|S\rangle$ (with corresponding energies $0 = E_\uparrow < E_\downarrow < E_S$). We can assume that the triplet is higher in energy and does not contribute to the sequential tunneling current. In the regime $E_S > \mu_1 > E_S - g\mu_B B > \mu_2$, the current is blocked in the absence of the ESR field due to energy conservation. We calculate the the stationary current and find[28]

$$I(\omega) \propto \frac{V_{\downarrow\uparrow}}{(\omega - g\mu_B B)^2 + V_{\downarrow\uparrow}^2}, \tag{7}$$

where the width of the resonance at $\omega = g\mu_B B$ is given by the total spin decoherence rate $V_{\downarrow\uparrow} = (W_{S\uparrow} + W_{S\downarrow})/2 + 1/T_2$. Here, $W_{S\sigma}$ denotes the rate for the transition from the state $|\sigma\rangle = |\uparrow\rangle, |\downarrow\rangle$ to the singlet $|S\rangle$ due to electrons tunneling from the leads onto the dot. Therefore, the inverse of the observed line width $1/V_{\downarrow\uparrow}$ represents a lower bound for the intrinsic single-spin decoherence time T_2. For finite temperatures and in the linear response regime $\Delta\mu < kT$, the current has roughly the standard sequential tunneling peak shape $\cosh^{-2}[(E_S - E_\downarrow - \mu)/2k_B T]$ as a function of the gate voltage $V_{\text{gate}} \propto \mu = (\mu_1 + \mu_2)/2$, while the width of the resonance in Eq. (7) as a function of ω remains unaffected.

The spin of a quantum dot in the presence of an ESR field shows coherent Rabi oscillations. It is possible to observe these Rabi oscillations of a single spin via time-averaged currents when ESR pulses are applied. Then, the time-averaged current $\bar{I}(t_p)$ as a function of the pulse length t_p exhibits the Rabi oscillations of the spin-state of the dot,[28] see Fig. 6. Observing such Rabi oscillations of a single spin would be a significant achievement, since this implied an working implementation of a one qubit gate.

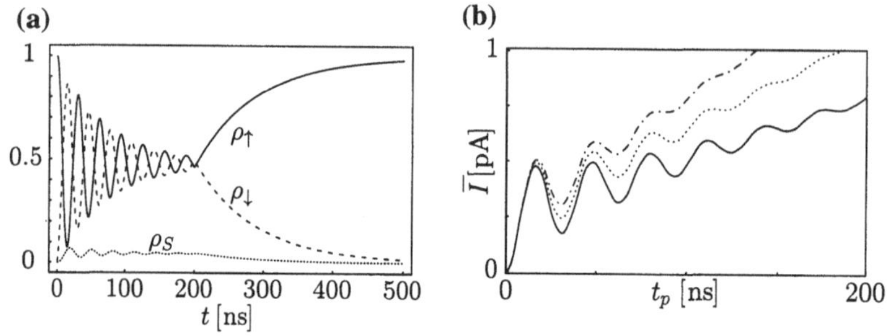

Fig. 6. Single spin Rabi oscillations, generated by ESR pulses of length t_p, are observable in the time-averaged current $I(t_p)$ through a quantum dot.[28] We take the amplitude of ESR field as $B_x^0 = 20\,\mathrm{G}$ (and $g = 2$), and $\Delta\mu > kT$, $\gamma_1 = 2 \times 10^7\,\mathrm{s}^{-1}$, $\gamma_2 = 5\gamma_1$, $T_1 = 1\,\mu\mathrm{s}$, and $T_2 = 150\,\mathrm{ns}$. (a) Evolution of the density matrix ρ, where a pulse of length $t_p = 200\,\mathrm{ns}$ is switched on at $t = 0$, obtained via integration of master equation. (b) Time-averaged current $\bar{I}(t_p)$ (solid line) for a pulse repetition time $t_r = 500\,\mathrm{ns}$. We also show the current where γ_1 and γ_2 are increased by a factor of 1.5 (dotted) and 2 (dash–dotted). Calculating the current contributions analytically, we obtain $\bar{I}(t_p) \propto 1 - \rho_\uparrow(t_p)$, up to a background contribution $\bar{I}_{bg}$ for times $t < t_p$, which is roughly linear in t_p. Thus, the current $\bar{I}$ probes the spin state of the dot at time t_p and therefore allows one to measure the Rabi oscillations of a single spin.[28]

7. CONCLUSIONS

We described the basic requirements for building a quantum computer with spin qubits. We addressed several concrete implementation issues for spin qubits, namely coupling between quantum dots, spin relaxation and decoherence measurements, spin filter devices, and single-spin read out setups. For all these issues, we reviewed theoretical and experimental results. These results give further insight in the details of quantum computing with spin qubits.

ACKNOWLEDGMENTS

We acknowledge G. Burkard, L. DiCarlo, J.M. Elzerman, J.A. Folk, V. Golovach, R. Hanson, A. Khaetskii, R. Potok, P. Recher, L.M.K. Vandersypen, L.H. Willems van Beveren, and D. Zumbühl. This work was supported by ARO, DARPA, FOM, NCCR Nanoscience, and Swiss NSF.

REFERENCES

1. G. A. Prinz, *Phys. Today* **45**(4), 58 (1995); *Science* **282**, 1660 (1998).
2. S. A. Wolf, D. D. Awschalom, R. A. Buhrman, J. M. Daughton, S. von Molnár, M. L. Roukes, A. Y. Chtchelkanova, and D. M. Treger, *Science* **294**, 1488 (2001).

3. *Semiconductor Spintronics and Quantum Computation* D. D. Awschalom, D. Loss, and N. Samarth (eds.) (Springer, Berlin, 2002).
4. R. Fiederling, M. Keim, G. Reuscher, W. Ossau, G. Schmidt, A. Waag, and L. W. Molenkamp, *Nature* **402**, 787 (1999).
5. Y. Ohno, D. K. Young, B. Beschoten, F. Matsukura, H. Ohno, and D. D. Awschalom, *Nature* **402**, 790 (1999).
6. J. M. Kikkawa, I. P. Smorchkova, N. Samarth, and D. D. Awschalom, *Science* **277**, 1284 (1997); J. M. Kikkawa and D. D. Awschalom, *Phys. Rev. Lett.* **80**, 4313 (1998); D. D. Awschalom and J. M. Kikkawa, *Phys. Today* **52**(6), 33 (1999).
7. J. A. Gupta, R. Knobel, N. Samarth, and D. D. Awschalom, *Science* **292**, 2458 (2001).
8. L. P. Kouwenhoven, G. Schön, and L. L. Sohn, in *Mesoscopic Electron Transport*, Vol. **345** of *NATO Advanced Study Institute, Series E*, edited by L. L. Sohn, L. P. Kouwenhoven, and G. Schön (Kluwer Academic Publishers, Dordrecht, 1997).
9. D. Loss and D. P. DiVincenzo, *Phys. Rev. A* **57**, 120 (1998); Ph. 9701055.
10. B. E. Kane, *Nature* **393**, 133 (1998).
11. S. Tarucha, D. G. Austing, T. Honda, R. J. van der Hage, and L. P. Kouwenhoven, *Phys. Rev. Lett.* **77**, 3613 (1996).
12. J. M. Elzerman, R. Hanson, J. S. Greidanus, L. H. Willems van Beveren, S. De Franceschi, L. M. K. Vandersypen, S. Tarucha, and L. P. Kouwenhoven, *Phys. Rev. B* **67**, 161308R (2003).
13. M. Ciorga, A. S. Sachrajda, P. Hawrylak, C. Gould, P. Zawadzki, Y. Feng, and Z. Wasilewski, *Physica E* **11**, 35 (2001).
14. F. R. Waugh, M. J. Berry, D. J. Mar, R. M. Westervelt, K. L. Campman, and A. C. Gossard, *Phys. Rev. Lett.* **75**, 705 (1995); C. Livermore, C. H. Crouch, R. M. Westervelt, K. L. Campman, and A. C. Gossard, *Science* **274**, 1332 (1996).
15. W. G. van der Wiel, S. De Franceschi, J. M. Elzerman, T. Fujisawa, S. Tarucha, and L. P. Kouwenhoven, *Rev. Mod. Phys.* **75**, 1 (2003).
16. T. H. Oosterkamp, S. F. Godijn, M. J. Uilenreef, Y. V. Nazarov, N. C. van der Vaart, and L. P. Kouwenhoven, *Phys. Rev. Lett.* **80**, 4951 (1998).
17. R. H. Blick, D. Pfannkuche, R. J. Haug, K. V. Klitzing, and K. Eberl, *Phys. Rev. Lett.* **80**, 4032 (1998); *ibid.* **81**, 689 (1998); T. H. Oosterkamp, T. Fujisawa, W. G. van der Wiel, K. Ishibashi, R. V. Hijman, S. Tarucha, and L. P. Kouwenhoven, *Nature* **395**, 873 (1998); I. J. Maasilta and V. J. Goldman, *Phys. Rev. Lett.* **84**, 1776 (2000).
18. L. DiCarlo, H. J. Lynch, A. C. Johnson, C. M. Marcus, M. P. Hanson, and A. C. Gossard, cond-mat/0311308 (2003).
19. T. Hayashi, T. Fujisawa, H.-D. Cheong, Y.-H. Jeong, and Y. Hirayama, *Phys. Rev. Lett.* **91**, 226804 (2003).
20. A. Steane, Rep. *Prog. Phys.* **61**, 117 (1998).
21. M. A. Nielsen, and I. L. Chuang, *Quantum Computation and Quantum Information* (Cambridge U. Press, New York, 2000).
22. D. P. DiVincenzo, *Fortschr. Phys.* **48**, 771 (2000).
23. G. Burkard, H.-A. Engel, and D. Loss, *Fortschr. Phys.* **48**, 965 (2000); G. Burkard and D. Loss, Ch. 8 in Ref. 3.
24. P. Recher, E. V. Sukhorukov, and D. Loss, *Phys. Rev. Lett.* **85**, 1962 (2000).
25. G. Salis, Y. Kato, K. Ensslin, D. C. Driscoll, A. C. Gossard, and D. D. Awschalom, *Physica E* **16**, 99 (2003).
26. Y. Kato, R. C. Myers, D. C. Driscoll, A. C. Gossard, J. Levy, and D. D. Awschalom, *Science* **299**, 1201 (2003).
27. G. Burkard, D. Loss, and D. P. DiVincenzo, *Phys. Rev. B* **59**, 2070 (1999).
28. H.-A. Engel, D. Loss, *Phys. Rev. Lett.* **86**, 4648 (2001); *Phys. Rev. B* **65**, 195321 (2002).

29. H.-A. Engel, V. Golovach, D. Loss, L. M. K. Vandersypen, J. M. Elzerman, R. Hanson, and L. P. Kouwenhoven, cond-mat/0309023.
30. L. S. Levitov and E. I. Rashba, *Phys. Rev. B* **67**, 115324 (2003).
31. J. M. Elzerman, *et al.* (unpublished).
32. X. Hu and S. Das Sarma, *Phys. Rev. A* **61**, 062301 (2000).
33. D. M. Zumbühl, *et al.* (unpublished).
34. V. N. Golovach and D. Loss, cond-mat/0308241.
35. B. Su, V. J. Goldman, J. E. Cunningham, *Phys. Rev. B* **46**, 7644 (1992); R. C. Ashoori, H. L. Stormer, J. S. Weiner, L. N. Pfeiffer, K. W. Baldwin, and K. W. West, *Phys. Rev. Lett.* **71**, 613 (1993); T. Schmidt, M. Tewordt, R. H. Blick, R. J. Haug, D. Pfannkuche, and K. V. Klitzing, *Phys. Rev. B* **51**, 5570 (1995).
36. I. H. Chan, P. Fallahi, A. Vidan, R. M. Westervelt, M. Hanson, A. C. Gossard, cond-mat/0309205.
37. A. Abragam and B. Bleaney, *Electron Paramagnetic Resonance of Transition Ions* (Clarendon Press, Oxford, 1970).
38. A. V. Khaetskii and Y. V. Nazarov, *Phys. Rev. B* **64**, 125316 (2001).
39. S. I. Erlingsson and Y. V. Nazarov, *Phys. Rev. B* **66**, 155327 (2002).
40. V. Golovach, A. Khaetskii, and D. Loss, cond-mat/0310655.
41. A. Khaetskii, D. Loss, and L. Glazman, *Phys. Rev. Lett.* **88**, 186802 (2002).
42. A. Khaetskii, D. Loss, and L. Glazman, *Phys. Rev. B* **67**, 195329 (2003).
43. T. Fujisawa, D. G. Austing, Y. Tokura, Y. Hirayama, S. Tarucha, *Nature* **419**, 278 (2002).
44. R. Hanson, B. Witkamp, L. M. K. Vandersypen, L. H. Willems van Beveren, J. M. Elzerman, L. P. Kouwenhoven, *Phys. Rev. Lett.* **91**, 196802 (2003).
45. D. M. Zumbühl, J. B. Miller, C. M. Marcus, K. Campman, and A. C. Gossard, *Phys. Rev. Lett.* **89**, 276803 (2002). Note that their λ'_{SO} is related to ours by $\lambda_{SO} = 2\lambda'_{SO}$.
46. A. G. Huibers, J. A. Folk, S. R. Patel, C. M. Marcus, C. I. Duruöz, and J. S. Harris, Jr., *Phys. Rev. Lett.* **83**, 5090 (1999).
47. R. Hanson, L. M. K. Vandersypen, L. H. Willems van Beveren, J. M. Elzerman, I. T. Vink, and L. P. Kouwenhoven, cond-mat/0311414.
48. J. A. Folk, R. M. Potok, C. M. Marcus, and V. Umansky, *Science* **299**, 679 (2003).
49. R. M. Potok, J. A. Folk, C. M. Marcus, V. Umansky, M. Hanson, and A. C. Gossard, *Phys. Rev. Lett.* **91**, 016802 (2003).
50. D. V. Averin and Yu. V. Nazarov, in *Single Charge Tunneling*, eds. H. Grabert and M. H. Devoret, *NATO ASI Series B*, Vol. 294 (Plenum Press, New York, 1992).
51. R. M. Potok, J. A. Folk, C. M. Marcus, and V. Umansky, *Phys. Rev. Lett.* **89**, 266602 (2002).
52. B. E. Kane, N. S. McAlpine, A. S. Dzurak, R. G. Clark, G. J. Milburn, H. B. Sun, and H. Wiseman, *Phys. Rev. B* **61**, 2961 (2000).
53. M. Friesen, C. Tahan, R. Joynt, and M. A. Eriksson,*Phys. Rev. Lett.* **92**, 037901 (2004).
54. M. Devoret, D. Estève, and Ch. Urbina, *Nature* (London) **360**, 547 (1992).
55. M. Field, C. G. Smith, M. Pepper, D. A. Ritchie, J. E. F. Frost, G. A. C. Jones, and D. G. Hasko, *Phys. Rev. Lett.* **70**, 1311 (1993).
56. W. Lu, Z. Q. Ji, L. Pfeiffer, K. W. West, and A. J. Rimberg, *Nature* **423** (6938), 422 (2003).

Spin-based Quantum Dot Quantum Computing in Silicon

Mark A. Eriksson,[1] Mark Friesen,[1] Susan N. Coppersmith,[1] Robert Joynt,[1] Levente J. Klein,[1] Keith Slinker,[1] Charles Tahan,[1] P. M. Mooney,[2] J. O. Chu,[2] and S. J. Koester[2]

The spins of localized electrons in silicon are strong candidates for quantum information processing because of their extremely long coherence times and the integrability of Si within the present microelectronics infrastructure. This paper reviews a strategy for fabricating single electron spin qubits in gated quantum dots in Si/SiGe heterostructures. We discuss the pros and cons of using silicon, present recent advances, and outline challenges.

KEY WORDS: Quantum computation; quantum dot; silicon; silicon–germanium; spin; quantum well.

PACS: 03.67.Pp; 03.67.Lx; 85.35.Be; 73.21.La.

1. INTRODUCTION

The seminal paper by Loss and DiVincenzo(1) outlined essential components of quantum dot quantum computing (QDQC): (1) spin qubits in single electron dots, (2) qubit initialization by thermalization in a magnetic field, (3) qubit rotations performed using electron spin resonance (ESR), (4) two-qubit gates enabled by electrostatic control of exchange coupling in neighboring dots, and (5) readout by spin-charge transduction. Subsequent theoretical work has shown that two-qubit gates can be sufficiently fast (sub-nanosecond),(2,3) and that these same interactions can be harnessed for single-qubit rotations,(4,5) albeit with some encoding overhead. The most challenging aspect of scalable QDQC is fast readout: spin-dependent

[1]Physics Department, University of Wisconsin-Madison, Madison, WI 53706 USA.
[2]IBM Research Division T.J. Watson Research Center Yorktown Heights, NY 10598 USA.

tunneling schemes have been proposed,[6,7] as well as microwave-enabled, fast initialization and readout in a closed dot.[8]

Quantum dots in semiconductors have a long history, much of which is reviewed in the excellent book.[9] An important step forward for quantum computing was the realization of dots in GaAs containing controlled numbers of electrons as few as 0 and 1.[10–12] Spin spectroscopy has been performed in dots, indicating that they are indeed viable candidates for qubits.[13–15] More recently, the ability to readout a single spin inside a quantum dot was demonstrated by Elzerman *et al.*[16] These and other important advances are reported in the paper by Engel, Kouwenhoven, Loss, and Marcus of this volume.

Many techniques developed in atomic physics can be directly adapted for quantum dots, at least in principle. Examples include readout and initialization,[8] as well as a recent proposal by Lukin and coworkers to enable long-range interactions between quantum dots.[17] However, the flexibility of quantum dots comes at the price of embedding the qubits in a solid matrix, with consequent issues related to decoherence. For this reason, materials properties are crucial for quantum dot-based devices. A major motivation for the development of the silicon quantum dot architecture is that the materials properties of silicon result in unusually long electron spin coherence times.

2. STRAINED SILICON QUANTUM DOT QUBITS

Here we outline the main challenges to QDQC in silicon, and we describe solutions for many of these problems. We discuss six critical areas: growth of strained silicon, silicon two-dimensional electron gas (2DEG) based quantum dots, valley degeneracies and their consequences, tolerance to impurities, decoherence, and bandwidth concepts at both high and low frequency limits.

2.1. Strained Silicon Growth

Unlike the AlGaAs system, SiGe structures inherently involve strain, as the lattice parameter of Ge is 4% larger than that of Si. Thus, $Si_{0.75}Ge_{0.25}$ strain-relaxed buffer layers provide a template for silicon growth resulting in silicon with biaxial tensile strain of about 1%. As a result, the cubic symmetry of Si is broken and the six conduction band valleys are no longer degenerate. In the case of biaxial tensile strain, the two perpendicular Δ_2 valleys having electrons with a light in-plane effective mass are lower in energy than the four in-plane Δ_4 valleys with a heavy in-plane effective mass, and the energy level of the lowest two Δ_2 valleys is lower than that of the conduction band in bulk SiGe. Thus, the quantum well formed in such a strained Si layer is occupied by light effective mass electrons. Because of the large energy splitting of the conduction band valleys, intervalley scattering is also reduced, resulting in higher electron mobility.

One challenge in attaining high mobility Si/SiGe heterostructures is to minimize the threading dislocation density arising from the lattice mismatch between Si and SiGe. Since bulk SiGe substrates are not available, structures with strained Si layers having a high mobility two-dimensional electron gas are achieved by first growing a strain-relaxed SiGe buffer layer on a Si(001) substrate, which provides a "virtual substrate" for the growth of a pseudomorphic Si layer under biaxial tensile strain. When a $Si_{0.7}Ge_{0.3}$ layer is grown directly on Si(001), strain-induced roughening occurs, leading to the random nucleation of misfit dislocations and a threading dislocation density on the order of $10^{10}\,cm^{-2}$.[(18)] In contrast, at lower mismatch strain, e.g., SiGe $x = 0.15$, the surface remains flat and dislocation nucleation takes place by a multiplication mechanism that results in much lower threading dislocation densities. The strain-relaxed buffer layer typically used for modulation-doped field-effect transistors (MODFETs) is a thick structure in which the Ge concentration is increased linearly or in small steps up to 25 or 30% plus a thick uniform composition $Si_{0.7}Ge_{0.3}$ layer. Grading allows dislocation nucleation to occur at low mismatch strain and threading dislocation densities are reduced to the 10^5–$10^8\,cm^{-2}$ range, depending on the grading rate and growth conditions.[(18)] A strained Si quantum well is grown on this virtual substrate and is then modulation doped by capping with a thin intrinsic alloy layer, followed by a P-doped alloy layer, and finally a thin Si layer as shown in Fig. 1(a). For some experiments, the strained Si quantum well is grown with isotopically pure ^{28}Si.

To obtain high mobility, scattering must be minimized. Scattering is induced by local changes in electric field and strain, as well as interfacial roughness on short length scales. Increasing the setback of the donors from the well decreases Coulomb scattering, increasing mobility until other scattering mechanisms are dominant. It has been shown that threading dislocation densities that exceed $3 \times 10^8\,cm^{-2}$ reduce the electron mobility in modulation-doped strained Si.[(19)] Additionally, the strained Si layer must be below the critical thickness for misfit dislocation formation at the Si/SiGe interface to avoid scattering.[(20)] Roughening of the surface of the SiGe virtual substrate, the so-called cross-hatch roughness, is inherent in the strain relaxation process.[(18)] This roughness appears as interface roughness in the pseudomorphic layer structure that forms the 2DEG,

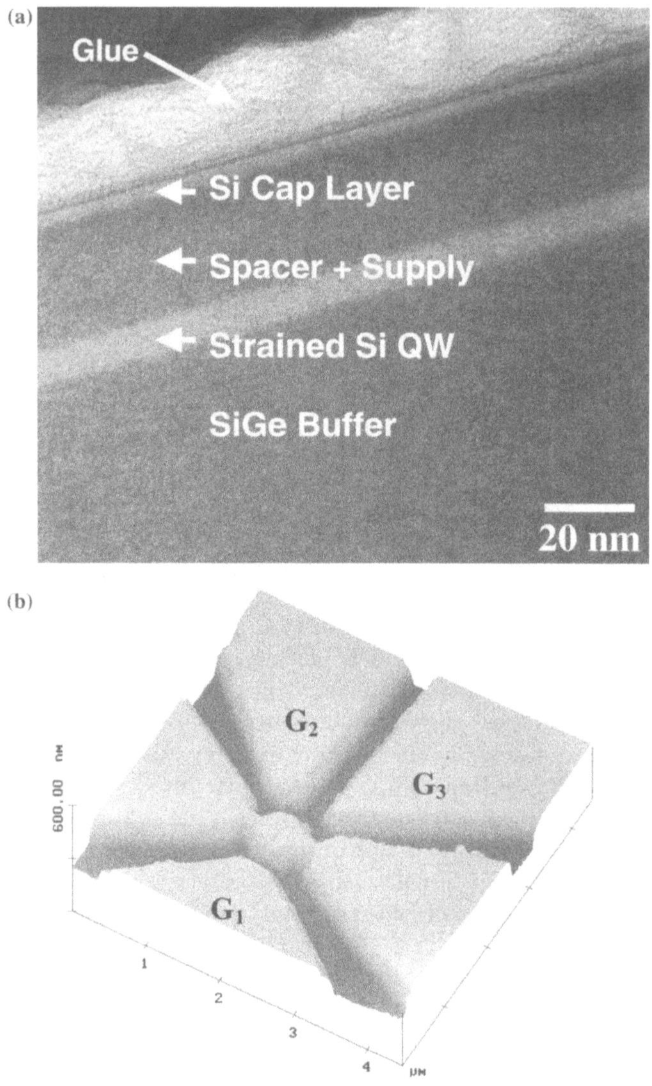

Fig. 1. (a) Cross sectional TEM image of the modulation-doped layer structure. The strained Si quantum well is grown on a uniform composition $Si_{0.7}Ge_{0.3}$ layer, which is grown on top of a step-graded buffer layer (not shown). The spacer and supply layers are also $Si_{0.7}Ge_{0.3}$. (b) AFM image of an etched dot fabricated from the heterostructure shown in (a). Three 2DEG side gates are visible.

specifically the strained Si quantum well. Fortunately, the length scale of this roughness is long enough that it does not reduce the electron mobility.[21] Low temperature electron mobility in the range of $1–6 \times 10^5\,\text{cm}^2/\text{V s}$ has been achieved in modulation-doped strained Si/SiGe structures grown epitaxially by various growth methods.[22–25]

In addition to low temperature 2DEG and quantum dot formation, these developments are also critical for non-cryogenic applications. The room temperature electron mobility in modulation-doped strained Si structures is typically $2000–2800\,\text{cm}^2/\text{V s}$, about 3–5 times that in n-type Si metal-oxide semiconductor field-effect transistors (MOSFETs). Thus, faster transistors are anticipated using strained Si structures, provided the device dimensions remain favorable. High-speed modulation-doped field-effect transistors (MODFETs) have been fabricated with Si/SiGe layer structures.[26] Recently MODFETs having 70–79 GHz f_T and record 194 GHz f_{Max} at room temperature were reported.[27]

2.2. Quantum Dots

A critical challenge for single-electron strained Si dots is the fabrication of high quality Schottky contacts on Si/SiGe heterostructures. Although it is relatively easy to fabricate large barrier Schottky contacts on silicon, it is challenging to create ultra-low leakage contacts on Si/SiGe heterostructures, due to the proximity of high P doping (typically $> 10^{19}\,\text{cm}^{-3}$) near the interface of the gate electrode.[28] Possible alternative approaches are the use of dielectric films to create metal–insulator–semiconductor (MIS) structures and the relocation of the P-doped supply layer underneath the Si quantum well. Bottom doping has been demonstrated by MBE growth techniques, but for CVD this is extremely difficult to achieve due to memory effects associated with phosphorous doping from PH_3.[28]

A second strategy is to avoid metal top-gates entirely, and instead to use 2DEG side gates, separated from the active region of the device by etch trenches. We have observed Coulomb blockade in such quantum dots with multiple gates to independently control the tunneling to the leads as well as the overall electron occupation of the quantum dot.[29] Quantum dots are fabricated by electron beam lithography and subsequent CF_4 reactive-ion etching. An AFM image of such a device is shown in Fig. 1(b). The electron density in the 2DEG from which the dot was formed is $4 \times 10^{11}\,\text{cm}^{-2}$ and the mobility is $40{,}000\,\text{cm}^2/\text{V s}$ at 2 K. Ohmic contacts to the 2DEG are formed by a Au/Sb alloy. Figure 2 shows a Coulomb blockade trace at $T = 1.8\,\text{K}$. Control of the dot electron population and the lead resistances is achieved with three separately

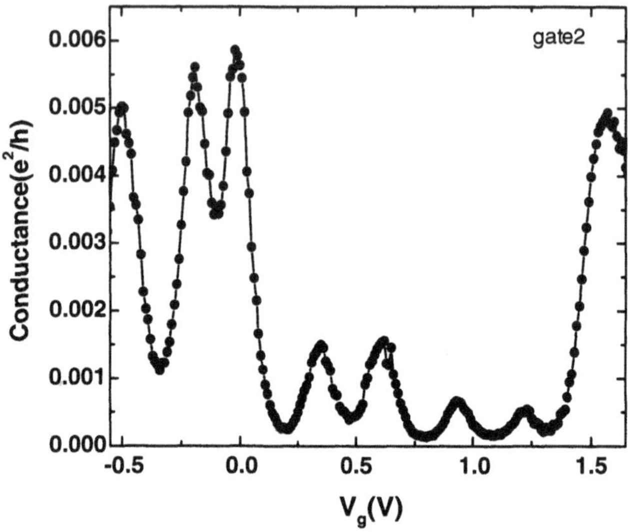

Fig. 2. Coulomb blockade trace for the dot shown in Fig. 1(b). The temperature was 1.8 K, and the gate voltage V_g was applied to gate 2.

tunable gates. Each gate is fabricated from the same 2DEG from which the quantum dot is created. Such in-plane coupling of one 2DEG to another has been used to monitor the electron population in GaAs quantum dots.[(12)] Here we have inverted this idea and used the 2DEG–2DEG coupling to control the dot.

Etched gates are very effective for individual dots, and can likely be used by themselves to create two coupled quantum dots. However, due to the relatively large size of 2DEG side gates it is likely that truly metallic top-gates will be required to couple many dots together. A second important challenge is the achievement of low charge noise. Switching events in the dot shown in Fig. 1(b) occur on the time scale ~ 1 h (Note added in proof: recent advances have extended this time to more than 10 hours between switching events). Improving this charge noise is an important goal. It is known that the charge noise in some types of silicon quantum dots, for example, oxide confined dots, can be extremely low, allowing repeatable measurement over very long time periods.[(30)]

2.3. Valley States

As described above, strain in Si/SiGe heterostructures reduces the sixfold silicon valley degeneracy to twofold. This remaining twofold valley degeneracy is a potential complication in two-qubit gates mediated

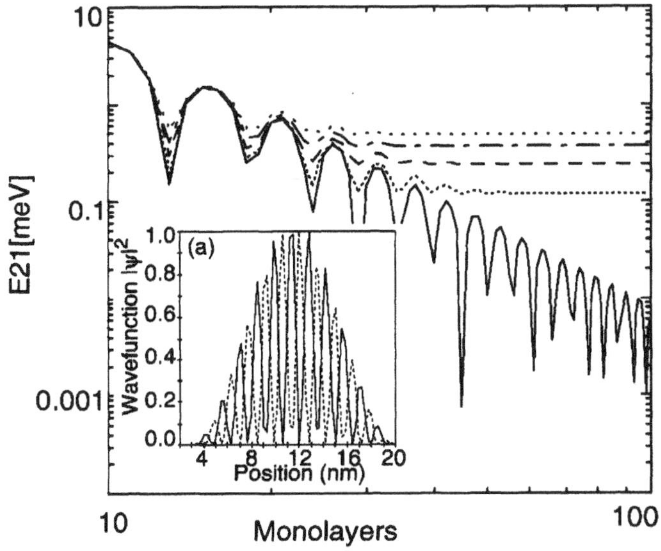

Fig. 3. Computed valley splitting E_{21} versus number of atomic monolayers in the quantum well. Solid line corresponds to zero applied field; oscillations reflect transitions of valley ground state from even to odd symmetry. Dotted and dashed lines correspond to finite applied E fields between 1–4 mV/nm. Inset: even and odd symmetry traces of tight-binding coefficients for a pair of valley split ground states.

by inter-dot exchange coupling.(1) Fortunately, the twofold valley degeneracy in strained silicon quantum wells is split by the quantum well confinement potential. As shown in Fig. 3, recent work has demonstrated that the valley splitting can be engineered both by varying the well width and by applying electric fields, and that the splitting, in some cases, can be quite large.(31) The valley splitting is most easily understood in the infinite square well limit. Because the valley minimum is not at the center of the Brillouin zone ($k = 0$), the electron wavefunctions experience atomic-scale modulations. In a semiconductor with two degenerate valleys, the oscillations of the two lowest energy states have very similar envelopes, but are out of phase by 90°. For an infinite square well, the energy eigenstates are linear combinations of four different k-values, yielding an energy splitting that decays as $(\text{width})^{-3}$; the splitting for a 4 nm well is 1.5 meV. In nonzero electric field the potential is asymmetric, and the energy difference between the two lowest energy states increases with increasing external electric field.(32) Typical modulation doped heterostructures experience internal electric fields of order 10 mV/nm. In calculations involving isolated electrons in a quantum well, we obtain splittings larger than 1 meV—quite a large energy.(31)

A potential concern is whether the exchange coupling between two qubits will oscillate uncontrollably as a function of position, in analogy with donor-bound electrons in silicon.[33] In fact, the situation in strained silicon quantum dots is quite different. Because of strain, the charge density in the plane of the quantum well does not have atomic-scale oscillations. Perpendicular to the quantum well the oscillations self-align because of the presence of the strong confining heterostructure potential. Because the length scale of the quantum well potential is so much shorter than that of variations in the quantum well plane, the Born–Oppenheimer approximation is appropriate and immediately shows that the oscillations in charge density perpendicular to the quantum well plane follow any slow variations in quantum well width and position.[34]

It is important to mention that it is also possible to view valley degeneracy as a resource. It may be possible to use valley states rather than spin states to store quantum information. Such states would be charge qubits with little difference in charge distribution for the two states, possibly leading to weak decoherence. In a different approach, using spin, conceivably one could access all low-lying eigenstates to form a four-dimensional qubit Hilbert space.

2.4. Tolerance to Impurities

For scalable QDQC, it is important that quantum dot exchange couplings be tolerant to the presence of low concentrations of impurities. It is well known that charged impurities such as ionized donors cause scattering effects which limit the 2DEG mobility.[20,35] These charges also have some effect on electrostatic control of qubit gate operations. However, a more important issue from the standpoint of decoherence and scalability is the influence of neutral dopant impurities like P in Si and Si in GaAs. Such impurities can potentially act as renegade qubits, siphoning off quantum information in an uncontrolled way. When the exchange coupling between qubit and impurity becomes large enough, fault-tolerant quantum error correction schemes[36,37] are no longer effective. Such donor-bound spins are abundantly present in the modulation-doped supply layer of semiconductor 2DEGs, and they also occur at low densities throughout the sample.

To investigate this issue, we have computed the qubit-impurity exchange coupling J for two cases:[38] an impurity in the supply layer, and an impurity in or near the quantum well. In the first case, due to the potentially large numbers of neutral donors in the supply layer, the important quantity is the distance between the supply layer and the quantum well. We obtain the following minimum set-back distances between the quantum well and the supply layer: 18 nm for Si in GaAs and 8 nm for P in Si. These numbers are only weakly dependent on the details of the structure or on the choice of fault-tolerance schemes and error correction coding, because of the exponential dependence of the exchange coupling on qubit-impurity separation. The results do depend on choice of barrier materials and, in particular, on the height of the quantum well barriers. Fortunately, these distances are consistent with typical experimental set-backs of ~20 nm for GaAs and ~10 nm for Si.

We have also studied the effect of impurities in or very near the quantum well. We find that impurity spins near the quantum well pose a threat to qubits at a distance of ~100 nm for Si in GaAs and ~60 nm for P in Si. The results are somewhat sensitive to specific details of the qubit confinement potential. A crucial observation, from the perspective of scalability, is that the computed impurity danger zones are approximately equal to the radius of a single electron dot. That is, a single impurity can only kill one, or at most two qubits in a 1D array. Therefore, a modest amount of parallel connectivity would enable scalable computations, provided the impurity density is somewhat smaller than the qubit density. We can estimate this critical impurity density by assuming that only impurities near the quantum well can trap electrons. For Si in GaAs the critical density is about $1.0 \times 10^{15}\,\mathrm{cm}^{-3}$ (assuming a 25-nm quantum well), while for P in Si the density is $1.6 \times 10^{16}\,\mathrm{cm}^{-3}$ (assuming a 6-nm quantum well). Both of these impurity density limits are achievable in good materials.

2.5. Decoherence

Silicon-based quantum dots have the compelling attribute that the spin coherence time T_2 can be very long. The reasons for long coherence times are the availability of the spin-zero ^{28}Si isotope, use of which greatly reduces relaxation via nuclear spins (hyperfine coupling), and silicon's small spin–orbit coupling (SOC), which suppresses phonon and SOC-based decoherence mechanisms.

Each electron spin **S** interacts with all nuclear spins $\mathbf{I}_i$ with which it overlaps spatially. As pointed out in Refs. 39 and 40, in the QC operational regime the external magnetic field **B** will exceed 100 gauss, so that electron spin-flips accompanied by one nuclear spin flip are not allowed energetically. This suppresses the single-spin-flip mechanism considered in Refs. 41 and 42, and the dominant ESR relaxation mechanism becomes spectral diffusion.(40) Measurements(43) and theory(40) have made it clear that in the spectral diffusion regime the relaxation rate increases with the density of the nuclear moments.

Natural Si has isotopic fractions 95.33% spin 0 (mostly ^{28}Si and a small fraction ^{30}Si) and 4.67% spin 1/2(^{29}Si). Spins in naturally occurring silicon have T_2 in the range 0.1–1 ms at low temperatures, and the dominant source of decoherence is coupling to the ^{29}Si nuclei.(44) (One must note here that the decay of spin echoes is not purely exponential, and more than one time scale may enter.) Measured T_2 values for low doped isotopically purified ^{28}Si are substantially longer.(45) In 1958 Gordon and Bowers (44) observed a T_2 of around 0.5 ms at 1.4 K for phosphorus-doped isotopically pure ^{28}Si, versus 0.24 ms for similarly doped natural Si:P. Tyryshkin *et al.*(45) recently compared T_2 times with different doping levels in isotopically pure silicon. Below 12 K the relaxation time T_2 in isotopically pure ^{28}Si was as large as 3 ms. Furthermore, by comparing different doping levels and attributing the remaining linewidths to a magnetic dipole–dipole interaction of neighboring phosphorus donors (via instantaneous diffusion),(46,47) they extrapolated from their data $T_2 = 62$ ms (at 7 K) for isolated donors in ^{28}Si. The presence of non-zero nuclear spin isotopes clearly results in shorter electron spin coherence times.

Electron spins in qubits can dephase even while in contact with a bath at zero temperature, because the quantum computer itself is not in equilibrium, and excited states are populated. Because of spin–orbit coupling, there is an effective spin–phonon coupling, and spins can flip by spontaneous emission of phonons. This process contributes to the relaxation rate T_1^{-1}, which is usually a lower bound to the decoherence rate T_2^{-1}. Generally, transition rates are proportional to $(g-2)^2$. This leads to very long T_1 and T_2 values in Si donor states,(48) and these favorable numbers are expected to extend also to dot-confined electrons.(49) Additional spin–orbit mixing due to the Rashba field,(50) typically prevalent in asymmetrically doped semiconductor heterostructures, is also expected to be quite weak in silicon.(51) Thus, spectral diffusion should be the predominant decoherence limiting mechanism in silicon QDQC.

For electrons in a Si/SiGe 2D electron gas, Tyryshkin *et al.*(52) have measured $T_2 = 2.98\,\mu$s in a sample with a phosphorus delta-doping layer above the well that was illuminated and thermally annealed. The relatively short decoherence time is due to the increase in scattering mechanisms in a mobile, 2D electron system, as explained recently in the context of Rashba spin–orbit coupling and the D'yakanov/Perel' (DP) spin-relaxation mechanism.(53) Confinement of the electrons laterally in a quantum dot suppresses the dominant 2DEG relaxation mechanism, greatly increasing the coherence time.(54–60) Since there should be very few phosphorus donors within the well to contribute to magnetic dipole–dipole driven instantaneous diffusion, quantum dots fabricated in isotopically pure small ^{28}Si quantum wells in principle should have better coherence times.

2.6. Bandwidth Issues

It is important to note that bandwidth in quantum computing is limited on both the high and the low end. At low frequencies, decoherence forms a fundamental, yet device dependent limit. As we have seen, the natural decoherence timescale for a SiGe QDQC should be $\sim 10\,\text{ms}$. Threshold theorems for fault-tolerant quantum computing vary according to qubit architectures and coding sophistication.[37] In particular, schemes have been devised for local gates,[61] and can even be extended to 1D arrays with nearest-neighbor coupling.[62] Nonetheless, the exact probability threshold for fault tolerance in a QDQC is not available yet. Somewhat arbitrarily, we estimate it here as 10^{-6}–10^{-4}. Thus, quantum gate operations must be at least as fast as 0.01 to $1\,\mu\text{s}$ in SiGe. Furthermore, fault-tolerance requires that readout and initialization steps must be performed at these same speeds. This does not imply that spins cannot be read out at much slower speeds, only that high speeds are required for scalability.

High-bandwidth constraints include non-adiabatic gating effects,[63,64] and sensitivity limits for readout. Based on shot noise analysis, the upper bound[65,66] on detection sensitivity for charge induced on the island of an optimized rf-SET (the lowest noise detector currently available) is about 4×10^{-6} $e/\sqrt{\text{Hz}}$. Simulations suggest that fast readout and initialization in SiGe can meet the stringent high and low-bandwidth criteria,[8] but experimental confirmation of this result is required.

The technical criteria for scalable QDQC are also challenging. On the low-bandwidth side, computations should be completed at speeds consistent with laboratory or human timescales (probably less than days!). If the necessary structures can be built, this limit is not a problem for solid state QC implementations. However, high-bandwidth technical limits are set by control and measurement electronics. We mention here a single example, discussed in Refs. 67 and 68. In these papers, we investigated the control sensitivity of the exchange coupling J to voltage pulses ΔV for particular SiGe devices. We found that, because of the exponential dependence of J on ΔV in typical architectures, small fluctuations in ΔV produced relatively large errors in J. As consistent with fault tolerant computing, the total exchange pulse (consisting of J integrated over pulse time Δt) should have an error less than 10^{-3}–10^{-2}.[69] However, the accuracy of control electronics is strongly sensitive to bandwidth, in terms of both the height and width of the pulse. As shown in Fig. 4, using specially designed

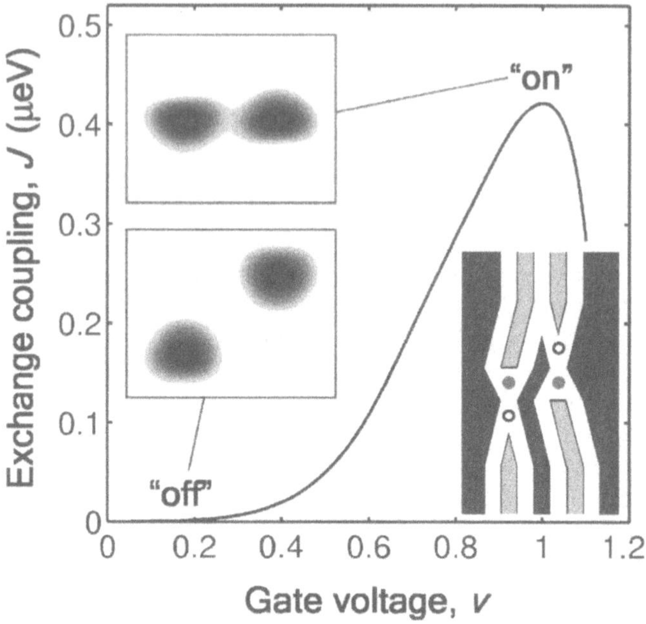

Fig. 4. Computed exchange coupling as a function of reduced gate voltage. The pseudo-digital technique allows a flat-top working point to replace the usual exponential dependence of $J(v)$. Lower right: pseudo-digital top-gate scheme for coupled double dots, each with a single electron. Upper left: computed electron densities for "off" and "on" configurations.

"pseudo-digital" dot architectures greatly decreases the electronics requirements in these simulations.

3. OUTLOOK FOR QUANTUM DOTS

As described above, recent advances point to a promising future for QDQC. Nonetheless, important challenges remain. A major goal for silicon 2DEG-based quantum dots is the fabrication of highly tunable, coupled dots, and the demonstration of spin measurement. A challenge for quantum dots in all materials is connectivity: is it possible to develop systems that are more highly connected than linear arrays? It is well known that such connectivity is an important aid in algorithm and error correction development, and there is an inverse relation between connectivity and resource requirements and operating timescales. Recent advances show that coupling between dots need not be highly localized.[70] Perhaps the greatest challenge is the development of long-distance couplings between qubits. Although not required, such couplings would be a great benefit to QDQC. Cross-fertilization from other QC disciplines should play an important role in meeting this challenge.

ACKNOWLEDGMENTS

We thank Steve Lyon, Alex Rimberg, and the University of Wisconsin solid state quantum computing group for many fruitful discussions. We acknowledge support from ARDA, ARO, and NSF.

REFERENCES

1. D. Loss and D. P. DiVincenzo, *Phys. Rev. A* **57**, 120 (1998).
2. G. Burkard, D. Loss, and D. P. DiVincenzo, *Phys. Rev. B* **59**, 2070 (1999).
3. X. D. Hu and S. Das Sarma, *Phys. Rev. A* **61**, 62301 (2000).
4. D. Bacon *et al.*, *Phys. Rev. Lett.* **85**, 1758 (2000).
5. D. P. DiVincenzo *et al.*, *Nature* **408**, 339 (2000).
6. P. Recher, E. V. Sukhorukov, and D. Loss, *Phys. Rev. Lett.* **85**, 1962 (2000).
7. H. A. Engel and D. Loss, *Phys. Rev. Lett.* **86**, 4648 (2001).
8. M. Friesen *et al.*, *Phys. Rev. Lett.* **92**, 037901 (2004).
9. L. L. Sohn, L. P. Kouwenhoven, and G. Schön, eds., *Mesoscopic Electron Transport*, NATO ASI Ser. E, Vol. 345. Kluwer, 1997.
10. R. C. Ashoori *et al.*, *Phys. Rev. Lett.* **68**, 3088 (1992).
11. M. Ciorga *et al.*, *Phys. Rev. B* **61**, R16315 (2000).
12. J. M. Elzerman *et al.*, *Phys. Rev. B* **67**, 161308 (2003).
13. T. Fujisawa *et al.*, *Nature* **419**, 278 (2002).
14. R. M. Potok *et al.*, *Phys. Rev. Lett.* **91**, (2003).
15. R. Hanson *et al.*, *Phys. Rev. Lett.* **91**, 196802 (2003).
16. Elzerman, *et al.* (unpublished).
17. L. I. Childress, *et al.*, *Phys. Rev. A* **69**, 042302 (2004).
18. P. M. Mooney, *Mater. Sci. Eng. R-Rep.* **17**, 105 (1996).
19. K. Ismail, *J. Vac. Sci. Technol. B* **14**, 2776 (1996).
20. K. Ismail *et al.*, *Phys. Rev. Lett.* **73**, 3447 (1994).
21. R. M. Feenstra and M. A. Lutz, *J. Appl. Phys.* **78**, 6091 (1995).
22. K. Ismail *et al.*, *Appl. Phys. Lett.* **66**, 842 (1995).
23. F. Schaffler *et al.*, *Semicond. Sci. Tech.* **7**, 260 (1992).
24. Y. J. Mii *et al.*, *Appl. Phys. Lett.* **59**, 1611 (1991).
25. T. Okamoto *et al.*, *Phys. Rev. B* **69**, 041202 (2004).
26. K. Ismail *et al.*, *IEEE Electr. Device Lett.* **EDL-14**, 348 (1993).
27. S. J. Koester *et al.*, *Electronics Lett.* **39**, 1684 (2003).
28. P. M. Mooney and J. O. Chu, *Annu. Rev. Mater. Sci* **30**, 335 (2000).
29. L. Klein *et al.*, *Appl. Phys. Lett.* **84**, 4074 (2004).
30. N. M. Zimmerman *et al.*, *Appl. Phys. Lett.* **79**, 3188 (2001).
31. T. B. Boykin *et al.*, *Appl. Phys. Lett.* **84**, 115 (2004).
32. L. J. Sham and M. Nakayama, *Phys. Rev. B* **20**, 734 (1979).
33. B. Koiller, X. D. Hu, and S. Das Sarma, *Phys. Rev. B* **66**, (2002).

34. S. N. Coppersmith, (unpublished).
35. F. Stern and S. E. Laux, *Appl. Phys. Lett.* **61**, 1110 (1992).
36. P. W. Shor, in *Proceedings of the 35th Annual Symposium on Fundamentals of Computer Science* (IEEE Press, Los Alamitos, 1996), pp. 56–65.
37. A. M. Steane, *Phys. Rev. A* **68**, 042322 (2003).
38. S. Liao, S. N. Coppersmith, M. A. Eriksson, and M. Friesen, (unpublished).
39. R. de Sousa and S. Das Sarma, *Phys. Rev. B* **67**, 033301 (2003).
40. R. de Sousa and S. Das Sarma, *Phys. Rev. B* **68**, 115322 (2003).
41. A. V. Khaetskii, D. Loss, and L. Glazman, *Phys. Rev. Lett.* **88**, 186802 (2002).
42. A. Khaetskii, D. Loss, and L. Glazman, *Phys. Rev. B* **67**, 195329 (2003).
43. M. Chibi and J. Hirai, *J. Phys. Soc. Jpn.* **33**, 730 (1972).
44. J. P. Gordon and K. D. Bowers, *Phys. Rev. Lett.* **1**, 368 (1958).
45. A. M. Tyryshkin *et al.*, *Phys. Rev. B* **68**, 193207 (2003).
46. J. P. Klauder and P. W. Anderson, *Phys. Rev.* **125**, 912 (1962).
47. W. B. Mims, *Phys. Rev.* **168**, 370 (1968).
48. G. Feher and E. A. Gere, *Phys. Rev.* **114**, 1245 (1959).
49. C. Tahan, M. Friesen, and R. Joynt, *Phys. Rev. B* **66**, 035314 (2002).
50. Y. A. Bychkov and E. I. Rashba, *J. Phys. C-Solid State* **17**, 6039 (1984).
51. Z. Wilamowski *et al.*, *Phys. Rev. B* **66**, 195315 (2002).
52. A. M. Tyryshkin, S. A. Lyon, W. Jantsch, and F. Schaeffler, preprint cond-mat/0304284.
53. C. Tahan and R. Joynt, "Spin relaxation in SiGe two-dimensional electron gases," preprint cond-mat/0401615.
54. A. V. Khaetskii and Y. V. Nazarov, *Phys. Rev. B* **61**, 12639 (2000).
55. A. V. Khaetskii and Y. V. Nazarov, *Phys. Rev. B* **64**, 125316 (2001).
56. B. I. Halperin *et al.*, *Phys. Rev. Lett.* **86**, 2106–2109 (2001).
57. Aleiner I. L. and V. I. Fal'ko, *Phys. Rev. Lett.* **87**, (2001).
58. D. M. Zumbuhl *et al.*, *Phys. Rev. Lett.* **89**, (2002).
59. E. Tsitsishvili and G.S. Lozano and A.O. Gogolin, "Rashba coupling in quantum dots: exact solution," preprint cond-mat/0310024.
60. C. Tahan, M.Friesen, and R. Joynt (unpublished).
61. D. Gottesman, *J. Mod. Optic* **47**, 333 (2000).
62. D. Aharonov and M. Ben-Or, in *Proc. 29th Ann. ACM Symp. on Theory of Computing*, p. 176, ACM, New York, 1998, preprints quant-ph/9611025, quant-ph/9906129.
63. J. Schliemann, D. Loss, and A.H. MacDonald, *Phys. Rev. B* **6308**, 085311 (2001).
64. X. D. Hu and S. Das Sarma, *Phys. Rev. A* **66**, 012312 (2002).
65. A. N. Korotkov and M. A. Paalanen, *Appl. Phys. Lett.* **74**, 4052 (1999).
66. M. H. Devoret and R. J. Schoelkopf, *Nature* **406**, 1039 (2000).
67. M. Friesen, R. Joynt, and M. A. Eriksson, *Appl. Phys. Lett.* **81**, 4619 (2002).
68. M. Friesen *et al.*, *Phys. Rev. B* **67**, 121301 (2003).
69. This criterion, different than used in our previous papers, represents the amplitude of error, consistent with the probability of error mentioned above (10^{-6}–10^{-4}) (D. Gottesman and E. Yablonovitch, private communications).
70. N. J. Craig *et al.*, *Science* **304**, 565 (2004).

Optically Driven Quantum Computing Devices Based on Semiconductor Quantum Dots

Xiaoqin Li,[1] Duncan Steel,[1,4] Daniel Gammon[2] and L. J. Sham[3]

This paper concerns optically driven quantum logic devices based on semiconductor quantum dots. It provides a brief review of recent theoretical and experimental progress towards building such devices and a description of a possible direction of further research. We consider both the exciton and the electron spin as a potential qubit. Quantum dot fabrication and single dot spectroscopy studies are briefly discussed followed by a description of experimental demonstrations of basic quantum logic operations. A scheme for a scalable quantum computer based on optical control of electron spins localized in quantum dots is described in detail. Important lessons as well as challenges for future research are summarized.

KEY WORDS: Quantum computing; quantum entanglement; semiconductor quantum dots; ultrafast optical spectroscopy.

PACS: 78.67.Hc; 42.50.Md; 03.67.Lx; 42.50.Hz.

1. INTRODUCTION

Semiconductor quantum dot (QD) studies have evolved from ensembles to single and coupled dots, opening up the possibility of building quantum devices based on the existing infrastructure for quantum optoelectronic semiconductor fabrication. It is believed that these dots may be used as the basic building blocks for future quantum information processing devices. The focus of this paper is on building optically driven quantum logic devices based on semiconductor QDs. We aim to provide a simple

[1]FOCUS, Harrison M. Randall Laboratory of Physics, The University of Michigan, Ann Arbor, MI 48109-1120.
[2]Naval Research Laboratory, Washington, DC 20375-5347.
[3]Department of Physics, University of California, San Diego, La Jolla, CA 92093-0319.
[4]To whom correspondence should be addressed. E-mail: dst@eecs.umich.edu

review of the rapid theoretical and experimental progress made in the past few years and discuss the future directions.

Following the first theoretical proposal[1] to use optical excitation in confined semiconductor systems for quantum computation, early optical spectroscopy studies of single quantum dots demonstrated the existence of isolated optical excitations[2,3] and provided the promising prospect that such localized excitations may be controlled individually and used as carriers of quantum information.[4] More detailed exciton-based proposals of implementation followed.[5–7] Following studies of the coherent nonlinear optical response[8] and coherent optical control[9] at the single dot level, an experimental breakthrough was made in 2001 when several groups demonstrated Rabi oscillations of excitons confined in single quantum dots using a variety of optical techniques.[10–13] Rabi oscillation of a qubit is an essential forerunner to arbitrary single qubit operations.

The essence of quantum computation and information processing lies in the generation of entanglement of qubits. Theoretical studies show that, while not scalable, two excitons in a single dot can be used to demonstrate simple quantum algorithms such as the Deutsch–Jozsa algorithm with pulse-shaping techniques for quantum operations.[7] Quantum coherence between arbitrary states within the computation basis was shown to persist for the lifetime of dipole transitions.[14,15] Entangled states of two polarized exciton states in a single dot were created and detected optically.[14,16] A key experimental demonstration of the capability of coherent control of excitons for conditional quantum operations is the demonstration of a simple two-bit quantum gate based on two excitons confined in a single quantum dot.[17]

A proposal for a scalable system is based on using single electron spins, with much longer coherence times than the above systems based on excitons, residing in closely spaced quantum dots with gate control of spins in separate dots.[18] There is considerable experimental effort currently directed at developing appropriate gate control of the spins. Optically driven electron spins confined in QDs is a promising alternative to electronic control for future quantum information processing since the speed for optical operations can easily exceed a THz.[19,20] The recent demonstration of fabrication of structures, where one extra electron is confined in a single dot,[21] has been a promising step towards achieving these proposals.

For charged quantum dots of the above type, optical excitation creates a charged exciton (trion), which can be used to rotate the electron spin.[22] This charged exciton state thus serves as an auxiliary state that provides optical access to the electron spin. Currently, the properties of negatively charged dots and, in particular, of trions are being studied intensely.[23]

In this review, we follow a particular framework for universal quantum computation[24] of defining qubits, the single qubit operations and the two-qubit conditional operations on demand, initialization and readout. We discuss the experimental demonstrations achieved so far based on exciton qubits, and describe proposals for the next generation of systems based on manipulation of spin qubits.

2. QUBITS IN SEMICONDUCTOR QUANTUM DOTS

Optical excitation of an electron into the empty conduction band of a semiconductor leaves a hole in the valence band. The electron and hole attract each other via the Coulomb interaction and form a bound-state exciton. The quantum dot serves as a trap with an effective potential for the electron and one for the hole as shown in Fig. 1(a). Because the band gap forms an excitation barrier, the effect of the Coulomb interaction of the large number of electrons in the ground state with the excited electron simply renormalizes its attraction with the hole by a dielectric constant and its mass in combination with the influence of the crystal potential.Unless there are too many excited electrons, the electron-hole pair

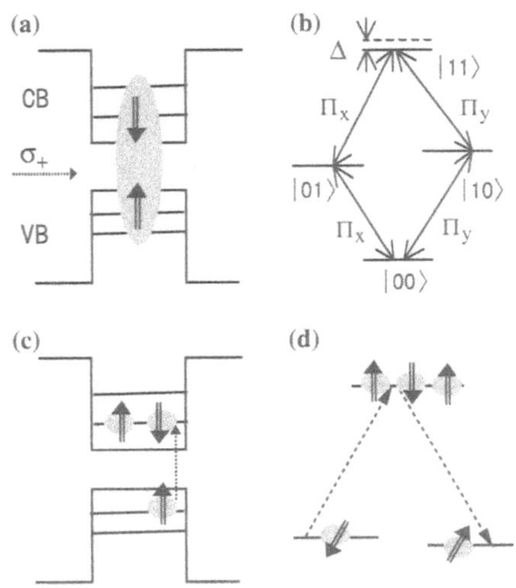

Fig. 1. Optical control of exciton-based qubits and spin-based qubits. (a) The single particle energy levels in a neutral dot indicating the optical excitation of an exciton qubit; (b) the excitation diagram of a two-exciton system confined in a single dot; (c) the single particle energy levels in a dot indicating the electron spin and the optically controlled exciton; (d) spin manipulation through the Raman process of the two lowest spin states and the trion state.

behaves remarkably like a positronium atom. The presence and absence of an exciton in a dot can serve as a qubit.

In an alternate scenario, one can dope each QD with a single electron. The spin of each localized electron then serves as a qubit. The state of the single electron spin is controlled via a stimulated Raman transition via the trion (electron-exciton) state[22] as illustrated in Fig. 1(c) and (d).

Laser pulses on the order of tens of femtoseconds are readily available in laboratories nowadays. The interaction energy of the excitons and the Zeeman energy of the spin levels are both of the order of meV. To minimize unintended dynamics of the unselected states in either the exciton or the spin qubit system, the pulse width cannot be arbitrarily short. Nevertheless, optically driven logic devices can be operated at clock speeds approaching the THz regime. Pulse shaping schemes[7,22] can be implemented to prevent the addition of errors in logical qubits. The clock speed of the operations is thus orders of magnitude faster than electrical or microwave control for the same spin qubit system.

Minimizing the errors in quantum operations caused by dephasing (and hence reducing the demand for error corrections) requires completing quantum operations before the system loses quantum coherence. The exciton coherence is limited by radiative recombination, whose time ranges from 100 ps in the large interface fluctuation dots[8] to 1 ns in the self-assembled dots.[25–27] For operations of ps duration, the number of possible operations is 10^2–10^3, which is currently believed to be insufficient for error correction. Shorter pulses are not viable, since the resultant increase in bandwidth results in excitation of other unintended states which compromises device performance.

The coherence time scale for electron spin based qubits is expected to be much longer, in the range of 1–100 μs.[28] Electron spins make better qubits obviously from this perspective. Optical manipulation of a single electron spin is still a challenging task at this point. Initial demonstrations towards building quantum dot logic gates were based on excitons. However, such optical manipulation of exciton-based qubits is in fact a significant component of what is needed to optically control a single electron spin, and hence is the penultimate demonstration leading to an optically driven spin-based quantum logic device.

3. FABRICATION OF QUANTUM DOTS AND SPECTROSCOPY STUDIES OF SINGLE DOTS

Using advanced semiconductor fabrication technologies such as molecular beam epitaxy (MBE), coherent (defect free) islands spontaneously form on an initially flat 2-dimensional layer in the Stranski–Krastanow growth mode of a highly strained system. Such a self-assembled mechanism is a rapid method to produce quantum dot arrays. Steady progress has been made in terms of fabricating quantum dots with more regular shape, size and positions.(29) This type of quantum dot is believed to be promising for practical applications involving a large number of quantum dots and has been widely studied. Many challenges remain but already the properties of single pairs of vertically coupled dots have been explored optically.(30) Other groups are attempting to develop laterally coupled dots, which may be necessary to scale to larger numbers of qubits. Considerable effort is needed, but researchers should eventually be able to provide designer QD molecules. For entanglement of two qubits from two dots on demand, it is better to have no interaction between two dots during quiescence but the two dots must be sufficiently close for the optical induction of qubit interaction such as via the optical RKKY interaction.(19)

Another model system studied extensively is the natural quantum dot formed due to interface fluctuations in narrow quantum wells following growth interruption. The monolayer high islands provide the necessary quantum potential to localize spins and excitons. The most direct evidence of such localized excitons is provided by images taken by near-field optical microscopes.(31–33) Besides near field microscopes, other approaches to achieve high spatial resolution include the use of shadow masks or mesas to limit the area of optical excitation. In this way single dot spectroscopy shows extremely sharp resonances.(2,3) Such spectra can only be observed at the single dot level—ensemble measurements show broad energy spectra due to inhomogeneous broadening as a result of fluctuations in size and shape of the large number of dots.

Nonlinear spectroscopy measurements using continuous wave (CW) lasers have confirmed that pure dephasing processes are reduced as a result of reduced interaction between localized excitons.(8) Furthermore, such CW spectroscopy studies have enabled detailed studies of biexciton formation in single dots, providing important information such as binding energy and nonradiative coherence times.(15) Transient nonlinear studies on self-assembled QDs have turned out to be quite challenging and have required wave-guide structures to enhance the interaction length in initial studies. Such measurements have revealed long dephasing times of a few hundred picoseconds.(25,26) In addition, optically induced entanglement of two orthogonally polarized exciton transitions in ensembles of InAs dots have been measured.(34)

4. OPTICAL COHERENT MANIPULATION OF EXCITON QUBITS

Universal computation may be achieved by quantum algorithms consisting of sequences made up from a set of arbitrary one-qubit operations and one entanglement creating two-qubit logic gate, such as a controlled-NOT (CNOT) gate or a square-root swap gate or a phase gate. A CNOT gate has now been demonstrated using two exciton qubits in a single quantum dot(17) based on the demonstration of Rabi oscillations and production of entangled states.(16)

We note that a two-bit system in a single quantum dot forms the basis for the device where the necessary coupling is provided by the enhanced Coulomb interaction as a result of quantum confinement. Fig. 1(b) shows the excitation level diagram of such a two-exciton system. This simplest two-bit system involves the crystal ground state ($|00\rangle$), two distinguishable excitonic states with orthogonal polarizations ($|01\rangle$ and $|10\rangle$) and the biexciton state ($|11\rangle$). It is important to note that the qubits are defined in the basis of the Bloch vectors of the exciton pseudo-spins. The value 1 (0) corresponds to the presence (absence) of an exciton with the Bloch vector pointing up (down).

An important feature evident in the excitation picture is the large binding energy of the biexciton state due to the three-dimensional quantum confinement in QDs. The binding energy varies with the size of QDs, typically in the range of 3–4 meV. The large binding energy implies the following: the excitation of one exciton leads to a different excitation energy of the other exciton, shifting it down by an amount equal to the binding energy. This feature gives rise to the characteristic conditional dynamics needed for building a CNOT gate.

Biexcitons confined in a single quantum dot have been identified optically, and Rabi oscillations of the exciton to biexciton transition have also been demonstrated.(17) An important achievement of such a demonstration is that the π pulse can serve as the operational pulse of a CNOT gate. The performance of this exciton-based CNOT gate can be examined by comparing the physical truth table shown in Fig. 2 with the ideal one. Like a classical gate, the truth table provides the population of each state at the output corresponding to a particular input. As an example, if the input state of the system is 10, after the gate operation, the populations in states (00, 01, 10, 11) are (0.14, 0.06, 0.17, 0.63) respectively, as opposed to the ideal (0, 0, 0, 1).

Quantum coherence and entanglement are critical to the superior performance of quantum logic devices compared to classical devices. The complete wavefunction immediately following a pulse that simultaneously excites both excitonic states can be written as $\psi = C_0|00\rangle + C_+|01\rangle +$

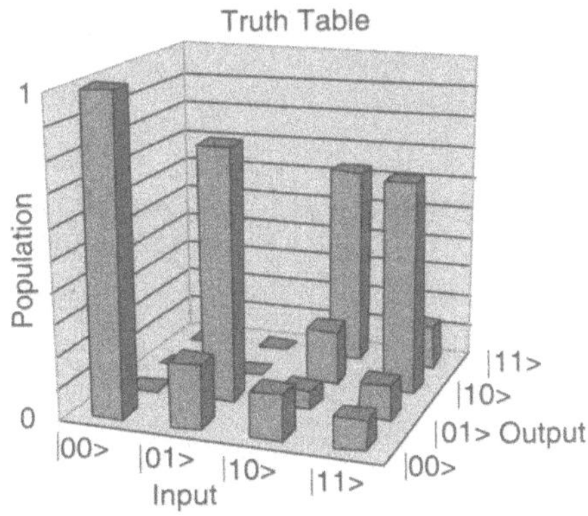

Fig. 2. Truth table of the numerically simulated CNOT gate using experimentally obtained parameters. In an ideal gate, the four highest bars are 1 and the others are 0.

$C_{-}|10\rangle + C_{+-}|11\rangle$. The existence and duration of quantum coherence between arbitrary states in the computational basis was first measured in the weak excitation regime in a series of nonlinear spectroscopy studies using CW lasers.[8,14,15] The creation of a specific entangled state requires laser pulses with pulse area (time-integrated electric field interaction) of $\sim\pi$. In an experiment aiming to create the Bell state $|01\rangle + |10\rangle$ using π pulses coupled to both orthogonal excitons, coefficients were estimated to be $C_0 = 0.48$, $C_+ = C_- = 0.62$, $C_{+-} = 0$, leading to entanglement entropy as high as ~0.7. The above discussion has assumed a pure quantum state following the optical excitation. Ref. 16 provides a more complete discussion of the mixed state entanglement. The temporal evolution of the non-radiative Raman coherence between states $|01\rangle$ and $|10\rangle$ was directly resolved in quantum beats measured in differential transmission (DT) geometry as shown in Fig. 3. The Raman coherence time was determined from the envelop decay of the beats, and was found limited by the lifetimes of the exciton transitions even in the strong field regime, thus showing that the potential coupling to other states under the high intensity optical field did not result in any detectable unintended dynamics or decoherence.

An extended definition of gate fidelity,[35] $\overline{Tr\left[\rho_P(t)\,\rho_I(t)\right]}$, measures the overlap between the gate-produced mixed state denoted by the physical density matrix r_{P} and the ideal one r_{I}, averaged over all possible initial states. Accounting for dephasing, the fidelity of the CNOT gate was calculated to be 0.7 assuming that the initial states can be prepared perfectly.[17] An ideal gate would have fidelity of 1 while any real gate would have fidelity between 0 and 1. Long operational pulses and short dephasing times due to fast recombination, a consequence of the large dipole moment in

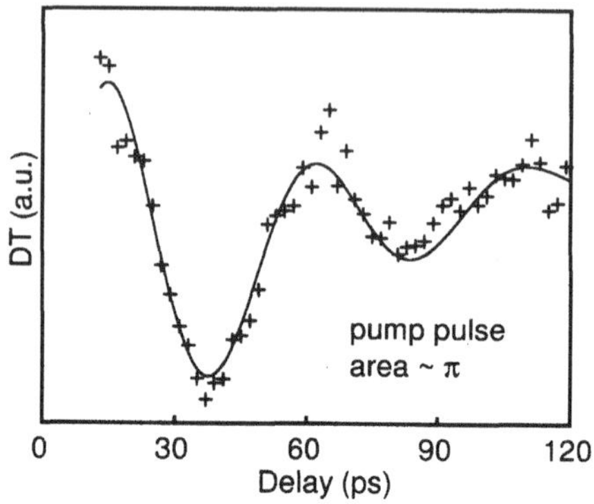

Fig. 3. An entangled state involving two polarized excitons confined in single dots was created and detected optically as evidenced by quantum beats between states $|01\rangle$ and $|10\rangle$ shown. The quantum coherence time between these two states is directly extracted from the decay of the envelope.

natural QDs, are two key issues that lead to fidelity below one for the gate. The combination of a short operational pulse (1 ps) and long population relaxation times (200 ps) with no pure dephasing would lead to a much improved gate with fidelity, (F), as high as 0.97 with the optimal separation between pulses.

In order to increase the quantum operations beyond one dot, interdot exciton interaction is required. One proposal is to use an electric field to increase the dipole–dipole interaction between two excitons in separate dots.[6] Dynamic control of the electric field means some sacrifice of the clock speed of the optical control. Another proposal is to put the dots in an optical cavity so that the cavity mode can serve as a data bus for the exciton qubits.[36,37]

5. OPTICAL CONTROL OF SPIN QUBITS IN QUANTUM DOTS

In this section we describe a scheme for a quantum computer that can be scaled up to many qubits based on optical control of the electron spin. The system consists of a 2-dimensional array of self-assembled quantum dots in a plane normal to the growth axis of an epitaxial cake of III–V semiconductors. Each dot is charged with one electron whose two spin states in the lowest orbital serve as the fundamental qubit.

A set of qubit operations required for universal quantum computation (consisting of arbitrary single qubit rotations and a suitable conditional two-qubit operation) are proposed using coherent optical control of off-resonance Raman processes which involve the spin states in a static magnetic field transverse to the growth direction. Since the shortest dot-center to dot-center distance will be tens of nanometers, current near-field optics capability is unable to address two neighboring qubits. To circumvent this problem, we have developed the zip-code concept of light covering an area containing a dozen or so of dots. Within each zip-code, frequency selection is used to carry out the quantum operations on one or two desired spins. The unintended dynamics of the other spins is eliminated by laser pulse shaping.

We classify the errors in quantum operations into two classes: the unintended dynamics due to the discrete states in the dots and the decoherence effects due to the coupling of the spin degrees of freedom with the environment defined as any system with a continuum of degrees of freedom such as the electromagnetic field and the solid vibrations. The direct spin dephasing time is long because of the weak spin flip terms. The optical decoherence incurred by the optical processes, potentially the strongest component of decoherence, is minimized by keeping the Raman processes off-resonance. The advantage of the optical control is then the fast (sub-nanosecond) clock speed in relation to very long spin dephasing time (possibly close to milliseconds).

5.1. Qubit Initialization

The system needs to be prepared initially to a specific state, say, all spins down. The qubit initialization is to be carried out by first splitting the two spin states in each dot with a constant magnetic field and then optical pumping of the spin up state to the trion state allowing time for relaxation into the spin down states.(38,39)

5.2. Measurement

Measurement can be carried out through optical recycling transitions between a specific electron spin state and the corresponding trion state under resonant optical excitation. The transition will occur only if the electron is in the targeted spin state. A photon will be emitted after each cycle until there is a spin flip process of the trion or electron, providing a burst of photons.(38)

5.3. Arbitrary One-qubit Operations

The use of coherent circularly polarized light propagating along the growth axis of the semiconductor layers (denoted as the z-axis) for the

Raman process shown in Fig. 1(d) can produce a single qubit rotation only about the z-axis. We have designed a method for a rotation about any axis through any angle by utilizing a static magnetic field pointing away from the z-axis and with sufficient Zeeman splitting so that two pulses of different frequencies can separately address the optical excitation (lowest energy trion) of each spin state without interfering with the other state.[22] The three parameters of an arbitrary rotation, such as the Euler parameters or the orientation of rotation axis relative to the magnetic field and the angle of rotation are determined by the relative phase and Rabi frequencies of the two applied pulses connecting the two spin states to the trion state and their common detuning from the trion state.

While we make use of the off-resonance Raman process in the adiabatic regime, our pulse design is conceptually from that for the population transfer using stimulated Raman adiabatic passage (STIRAP).[40] While both methods involve coherent control of the three-level system, STIRAP uses the dark optical dressed state to affect the population transfer whereas we use the other two dressed states to make a spin rotation. The "counter-intuitive" sequence of two pulses used in STIRAP requires the foreknowledge of the initial and the final state, which cannot be used as a qubit operation in quantum computing. Our design for a qubit operation makes use of the remaining two states and works for an arbitrary initial spin state.

5.4. Two-qubit Logic Gate

The principle is based on Ref. 19, optically creating a Heisenberg interaction between two spins in two nearest neighbor dots as illustrated in Fig. 4(a). The amount of rotation between the two antiparallel spin states shown in Fig. 4(b) can be controlled by the time duration of the optical pulse. For example, an effective $\pi/2$ pulse yields a qubit swap. However, to reduce the addressing problem because of the current focusing limit of the optical field, the intermediate state of the excited trion should be chosen to be the common bound states of the two dots. The dots are designed with the ground states of the electrons in different dots isolated from one another but with the first or higher excited states having overlapping wave functions between the nearest neighbors as shown in Fig. 4(b). Thus, for the single-qubit operation, the lowest trion used for the intermediate Raman state is isolated in a dot, whereas for the two-qubit operation, the excited trion state is designed to cover two and only two neighboring dots.

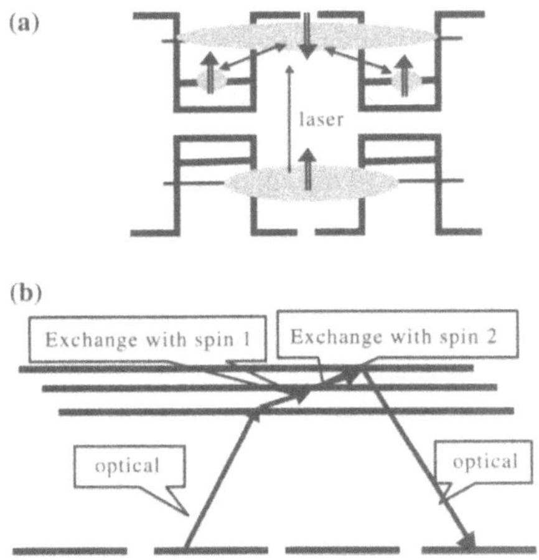

Fig. 4. Optical control of two spins in two separate dots. (a) The single particle energy level scheme; (b) the same process in terms of the four states of the two spins and the excited states with the addition of an exciton.

The excited trion energies of the two dots are brought to coincidence by the dynamic Stark shift (second-order ac electric field induced energy change) while the relevant energy levels from the other dots are kept away from the optical frequencies used. Under ideal operation conditions, the detuning from the intended intermediate state is always less than the energy differences between the intermediate state and other states. To refine the operation for less than ideal circumstances, we propose pulse shaping, to be discussed in Sec. 5.5.

5.5. The Possibility in Principle of Building a Large Quantum Computer

The problem of addressing specific single or two qubits belongs to the realm of the avoidance of unintended dynamics. Its solution is key to both specific quantum operations and to the scalability of the computer. In NMR,(41) a sequence of pulses in time is used to accomplish an operation. In quantum optics,(42) a pulse is shaped in the frequency domain to acquire a certain shape in the time domain without drastically increasing the operation time. The pulse shaping approach for quantum operations(7,43,44) in frequency space uses a broader bandwidth than the interdot and intradot energy differences to gain a short operation time. The idea is to use the flexibility of the pulses to return the unintended dynamics to a net zero at the end of the operations.

The use of the frequency space avoids the time consumed by pulse sequencing in NMR depending on the principle of spin echo, which the dot system cannot afford because of its much shorter decoherence time. The energy difference for unintended dynamics to arise is 0.1–1% of the laser bandwidth, one to two orders of magnitudes smaller than the NMR design. Our approach does not so much eliminate the unintended dynamics by interference to render the undesired level dark as to bring the recalcitrant electron back to its original state at the end of the pulse. It is possible to limit the overhead of pulse shaping to have a power-law dependence on the number of qubits.[7] We have simulated a few pulse-shaping operations to study their performance quality[7,43] but much work for the spin operations remains to be carried out.

5.6. Decoherence

There are two sources of decoherence. One is spin relaxation whose time is long, as discussed above, although the transverse relaxation time T_2 in a single dot has yet to be measured. The decoherence due to the optical processes during the operations involving the additional excitons could be fast and has to be ameliorated by design, such as detuning or pulse shaping.

6. SUMMARY

The experience of building quantum logic devices based on optically driven quantum dots has provided the basis for moving from the exciton qubit to the spin qubit. Many benchmark features predicted for zero-dimensional systems have been readily observable in the model system of natural dots formed by interface fluctuations once the technical challenges of nano-optical probing to enable the study of individual quantum dots were overcome. The body of work has shown that the quantum dots do indeed have the optical features of sharp energy level structure associated with atomic systems and that decoherence effects that are associated with the continuum states of higher-dimensional structures are not an issue in these systems, in general. Multiple reports of exciton Rabi oscillations demonstrated on different types of quantum dots verify the robustness of such qubit rotational operations. Evaluation of the performance of the these systems show surprisingly high fidelity for initial studies, and further progress in materials development and the application of optical pulse shaping and coherent control techniques will surely result in considerable improvement.

While further development of exciton based qubit systems will undoubtedly lead to impressive performance, much of the future work will be focused on extending the exciton system to the Λ system of Fig. 1(d) by doping a QD with a single electron. Moving to the spin based qubit will lead to longer coherence times with the promise of improved performance as aimed at quantum computing. The challenges here lie in developing further control of materials fabrication, which will be paired with more sophisticated schemes for optical excitation and coherent control. Full characterization of the density matrix will be essential for complete understanding of the system, and means for reliable readout and initialization will need to be experimentally investigated.

Advances in materials work combined with new optical probing and coherent optical control developments have enabled surprisingly fast development of optically driven quantum dot systems for quantum information processing. Innovative proposals by many groups have identified several approaches now to extend this work in a scalable manner.(18–20,24,36,37) Many critical questions remain, however, and further work is required in all areas.

ACKNOWLEDGMENTS

We thank the numerous coauthors of our papers in the references for their essential contributions and acknowledge with gratitude the financial support of ARDA/NSA/ARO, DARPA, ONR, NSF and FOCUS.

REFERENCES

1. A. Barenco, D. Deutsch, A. Ekert, and R. Jozsa, *Phys. Rev. Lett.* **74**, 4083 (1995).
2. D. Gammon, E. S. Snow, B. V. Shanabrook, D. S. Katzer, and D. Park, *Phys. Rev. Lett.* **76**, 3005 (1996).
3. D. Gammon, E. S. Snow, B. V. Shanabrook, D. S. Katzer, and D. Park, *Science* **273**, 87 (1996).
4. D. Gammon and D. G. Steel, *Phys. Today* **55**(10), 36 (2002).
5. F. Troiani, U. Hohenester, and E. Molinari, *Phys. Rev. B* **62**, 2263 (2000).
6. E. Biolatti, R. C. Iotti, P. Zanardi, and F. Rossi, *Phys. Rev. Lett.* **85**, 5647 (2000).
7. P. Chen, C. Piermarocchi, and L. J. Sham, *Phys. Rev. Lett.* **87**, 067401 (2001).
8. N. Bonadeo, G. Chen, D. Gammon, D. Park, D. S. Katzer, and D. G. Steel, *Phys. Rev. Lett.* **83**, 2267 (1998).
9. N. Bonadeo, J. Erland, D. Gammon, D. Park, D. S. Katzer, and D. G. Steel, *Science* **282**, 1473 (1998).
10. T. H. Stievater, X. Li, D. G. Steel, D. Gammon, D. Park, D. S. Katzer, C. Piermarocchi, and L. J. Sham, *Phys. Rev. Lett.* **87**, 133603 (2001).
11. H. Kamada, H. Gotoh, J. Temmyo, T. Takagahara, and H. Ando, *Phys. Rev. Lett.* **87**, 246401 (2001).

12. H. Htoon, T. Takagahara, D. Kulik, O. Baklenov, A. L. Holmes, Jr., and C. K. Shih, *Phys. Rev. Lett.* **88**, 087401 (2002).
13. A. Zrenner, E. Beham, S. Stufler, F. Findeis, M. Bichler, and G. Abstreiter, *Nature* **418**, 612 (2002).
14. G. Chen, N. H. Bonadeo, D. G. Steel, D. Gammon, D. S. Katzer, D. Park, and L. J. Sham, *Science* **289**, 1906 (2000).
15. G. Chen, T. H. Stievater, E. T. Batteh, X. Li, D. G. Steel, D. Gammon, D. S. Katzer, D. Park, and L. J. Sham, *Phys. Rev. Lett.* **88**, 117901 (2002).
16. X. Li, Y. Wu, D. G. Steel, D. Gammon, and L. J. Sham, Submitted.
17. X. Li, Y. Wu, D. G. Steel, D. Gammon, T. H. Stievater, D. S. Katzer, D. Park, C. Piermarocchi, and L. J. Sham, *Science* **301**, 809 (2003).
18. D. Loss and D. P. Dvincenzo, *Phys. Rev. A* **57**, 120 (1998).
19. C. Piermarocchi, P. Chen, L. J. Sham, and D. G. Steel, *Phys. Rev. Lett.* **89**, 167402 (2002).
20. E. Pazy, E. Biolatti, T. Calarco, I. Damico, P. Zanardi, F. Rossi, and P. Zoller, *Europhys. Lett.* **62**, 175 (2003); A. Imamoglu, E. Knill, L. Tian, and P. Zoller, *Phys. Rev. Lett.* **91**, 017402 (2003); P. Solinas, P. Zanardi, N. Zanghi, and F. Rossi, *Phys. Rev. B* **67**, 121307 (2003); F. Troiani, E. Molinari, and U. Hohenester, *Phys. Rev. Lett.* **90**, 206802 (2003); T. Calarco, A. Datta, P. Fedichev, E. Pazy, and P. Zoller, *Phys. Rev. A* **68**, 12310 (2003).
21. R. J. Warburton, C. Schaflein, D. Haft, F. Bickel, A. Lorke, K. Karrai, J. M. Garcia, W. Schoenfeld, and P. M. Petroff, *Nature* **405**, 926 (2001).
22. P. Chen, C. Piermarocchi, L. J. Sham, D. Gammon, and D. G. Steel, *Phys. Rev. B* **69**, 075320 (2004).
23. J. G. Tischler, A. S. Bracker, D. Gammon, and D. Park, *Phys. Rev. B* **66**, 81310R (2002).
24. D. P. DiVincenzo, in *Semiconductor Spintronics and Quantum Computation*, D. D. Awschalom, D. Loss, and N. Samarth (Eds), (Springer, Berlin, 2002), p. 221.
25. D. Birkedal, K. Leosson, and J. M. Hvam, *Phys. Rev. Lett.* **87**, 227401 (2001).
26. P. Borri, W. Langbein, S. Schneider, U. Woggon, R. L. Sellin, D. Ouyang, and D. Bimberg, *Phys. Rev. Lett.* **87**, 157401 (2001).
27. M. Bayer and A. Forchel, *Phys. Rev. B* **65**, 041308 (2002).
28. R. Hanson, B. Witkamp, L. M. K. Vandersypen, L. H. Willems van Beveren, J. M. Elzerman, and L. P. Kouwenhoven, *Phys. Rev. Lett.* **91**, 196802 (2003).
29. Ishikawa, T. Nishimura, S. Kohmoto, K. Asakawa, *Appl. Phys. Lett.* **76**, 167 (2000).
30. M. Bayer, P. Hawrylak, K. Hinzer, S. Fafard, M. Korkusinski, Z. R. Wasilewski, O. Stern, and A. Forchel, *Science* **291**, 451 (2001).
31. J. R. Guest, T. H. Stievater, G. Chen, E. T. Tabak, B. G. Orr, D. G. Steel, D. Gammon, and D. S. Katzer, *Science* **293**, 2224 (2001).
32. Q. Wu, D. Grober, D. Gammon, and D. S. Katzer, *Phys. Rev. Lett.* **83**, 2652 (1999).
33. K. Matsuda, T. Saiki, S. Nomura, M. Mihara, Y. Aoyagi, S. Nair, and T. Takagahara, *Phys. Rev. Lett.* **91**, 177401 (2003).
34. A. Lenihan, M. V. Gurudev Dutt, D. G. Steel, S. Ghosh, and P. K. Bhattacharya, *Phys. Rev. Lett.* **88**, 223601 (2002).
35. J. F. Poyatos, J. I. Cirac, and P. Zoller, *Phys. Rev. Lett.* **78**, 390 (1997).
36. A. Imamoglu, D. D. Awschalom, G. Burkard, D. P. Divincenzo, D. Loss, M. Sherwin, and A. Small, *Phys. Rev. Lett.* **83**, 4204 (1999).
37. T. A. Brun and H. Wang, Phys. Rev. A **61**, 032307 (2000).
38. A. Shabaev, A. L. Efros, D. Gammon, and I. A. Merkulov, *Phys. Rev. B* **68**, 201305 (2003).
39. W. Yao, R. Liu, and L. J. Sham, cond-mat/0312272.
40. K. Bergmann, H. Theuer, and B. W. Shore, *Rev. Mod. Phys.* **70**, 1003 (1998).

41. M. H. Levitt, *Prog. NMR Spect.* **18,** 61–122 (1986).
42. A. M. Wiener, *Prog. Quant. Electr.* **19,** 161–237 (1995).
43. C. Piermarocchi, P. Chen, Y. Dale, and L. J. Sham, *Phys. Rev. B* **65**, 075307 (2002).
44. D. Goswami, *Phys. Rep.* **374**, 385 (2003).

Implementing Qubits with Superconducting Integrated Circuits

Michel H. Devoret[1,4] and John M. Martinis[2,3]

Superconducting qubits are solid state electrical circuits fabricated using techniques borrowed from conventional integrated circuits. They are based on the Josephson tunnel junction, the only non-dissipative, strongly non-linear circuit element available at low temperature. In contrast to microscopic entities such as spins or atoms, they tend to be well coupled to other circuits, which make them appealling from the point of view of readout and gate implementation. Very recently, new designs of superconducting qubits based on multi-junction circuits have solved the problem of isolation from unwanted extrinsic electromagnetic perturbations. We discuss in this review how qubit decoherence is affected by the intrinsic noise of the junction and what can be done to improve it.

KEY WORDS: Quantum information; quantum computation; superconducting devices; Josephson tunnel junctions; integrated circuits.

PACS: 03.67.–a, 03.65.Yz, 85.25.–j, 85.35.Gv.

1. INTRODUCTION

1.1. The Problem of Implementing a Quantum Computer

The theory of information has been revolutionized by the discovery that quantum algorithms can run exponentially faster than their classical counterparts, and by the invention of quantum error-correction protocols.[(1)] These fundamental breakthroughs have lead scientists and engineers to imagine building entirely novel types of information processors. However, the construction of a computer exploiting quantum—rather than

[1]Applied Physics Department, Yale University, New Haven, CT 06520, USA.
[2]National Institute of Standards and Technology, Boulder, CO 80305, USA.
[3]Present address: Physics Department, University of California, Santa Barbara, CA 93106, USA.
[4]To whom correspondence should be addressed. E-mail: michel.devoret@yale.edu

classical—principles represents a formidable scientific and technological challenge. While quantum bits must be strongly inter-coupled by gates to perform quantum computation, they must at the same time be completely decoupled from external influences, except during the write, control and readout phases when information must flow freely in and out of the machine. This difficulty does not exist for the classical bits of an ordinary computer, which each follow strongly irreversible dynamics that damp the noise of the environment.

Most proposals for implementing a quantum computer have been based on qubits constructed from microscopic degrees of freedom: spin of either electrons or nuclei, transition dipoles of either atoms or ions in vacuum. These degrees of freedom are naturally very well isolated from their environment, and hence decohere very slowly. The main challenge of these implementations is enhancing the inter-qubit coupling to the level required for fast gate operations without introducing decoherence from parasitic environmental modes and noise.

In this review, we will discuss a radically different experimental approach based on "quantum integrated circuits." Here, qubits are constructed from *collective* electrodynamic modes of macroscopic electrical elements, rather than microscopic degrees of freedom. An advantage of this approach is that these qubits have intrinsically large electromagnetic cross-sections, which implies they may be easily coupled together in complex topologies via simple linear electrical elements like capacitors, inductors, and transmission lines. However, strong coupling also presents a related challenge: is it possible to isolate these electrodynamic qubits from ambient parasitic noise while retaining efficient communication channels for the write, control, and read operations? The main purpose of this article is to review the considerable progress that has been made in the past few years towards this goal, and to explain how new ideas about methodology and materials are likely to improve coherence to the threshold needed for quantum error correction.

1.2. Caveats

Before starting our discussion, we must warn the reader that this review is atypical in that it is neither historical nor exhaustive. Some important works have not been included or are only partially covered. On the other hand, the reader may feel we too frequently cite our own work, but we wanted to base our speculations on experiments whose details we fully understand. We have on purpose narrowed our focus: we adopt the point of view of an engineer trying to determine the best strategy for building a reliable machine given certain design criteria. This approach obviously runs the risk of presenting a biased and even incorrect account of recent scientific results, since the optimization of a complex system is always an intricate process with both hidden passageways and dead-ends. We hope nevertheless that the following sections will at least stimulate discussions on how to harness the physics of quantum integrated circuits into a mature quantum information processing technology.

2. BASIC FEATURES OF QUANTUM INTEGRATED CIRCUITS

2.1. Ultra-low Dissipation: Superconductivity

For an integrated circuit to behave quantum mechanically, the first requirement is the absence of dissipation. More specifically, all metallic parts need to be made out of a material that has zero resistance at the qubit operating temperature and at the qubit transition frequency. This is essential in order for electronic signals to be carried from one part of the chip to another without energy loss—a necessary (but not sufficient) condition for the preservation of quantum coherence. Low temperature superconductors such as aluminium or niobium are ideal for this task.[(2)] For this reason, quantum integrated circuit implementations have been nicknamed "superconducting qubits"[1].

2.2. Ultra-low Noise: Low Temperature

The degrees of freedom of the quantum integrated circuit must be cooled to temperatures where the typical energy kT of thermal fluctuations is much less that the energy quantum $\hbar\omega_{01}$ associated with the transition between the states $|\text{qubit} = 0\rangle$ and $|\text{qubit} = 1\rangle$. For reasons which will become clear in subsequent sections, this frequency for superconducting qubits is in the 5–20 GHz range and therefore, the operating temperature T must be around 20 mK (recall that 1 K corresponds to about 20 GHz). These temperatures may be readily obtained by cooling the chip with a dilution refrigerator. Perhaps more importantly though, the "electromagnetic temperature" of the wires of the control and readout ports connected to the chip must also be cooled to these low temperatures, which requires careful electromagnetic filtering. Note that electromagnetic

[1] In principle, other condensed phases of electrons, such as high-T_c superconductivity or the quantum Hall effect, both integer and fractional, are possible and would also lead to quantum integrated circuits of the general type discussed here. We do not pursue this subject further than this note, however, because dissipation in these new phases is, by far, not as well understood as in low-T_c superconductivity.

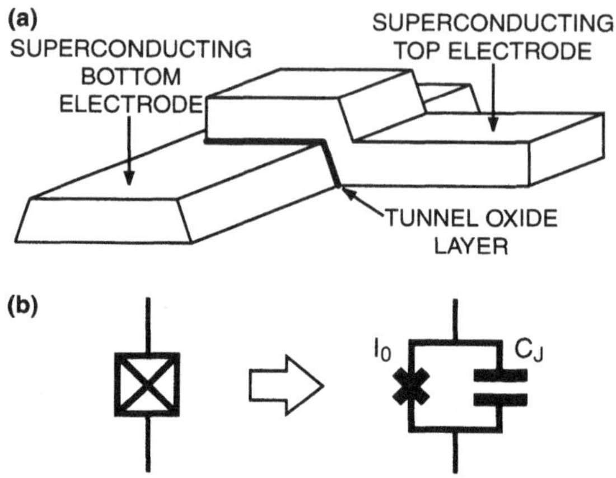

Fig. 1. (a) Josephson tunnel junction made with two superconducting thin films; (b) Schematic representation of a Josephson tunnel junction. The irreducible Josephson element is represented by a cross.

damping mechanisms are usually stronger at low temperatures than those originating from electron–phonon coupling. The techniques(3) and requirements(4) for ultra-low noise filtering have been known for about 20 years. From the requirements $kT \ll \hbar\omega_{01}$ and $\hbar\omega_{01} \ll \Delta$, where Δ is the energy gap of the superconducting material, one must use superconducting materials with a transition temperature greater than about 1 K.

2.3. Non-linear, Non-dissipative Elements: Tunnel Junctions

Quantum signal processing cannot be performed using only purely linear components. In quantum circuits, however, the non-linear elements must obey the additional requirement of being non-dissipative. Elements like PIN diodes or CMOS transistors are thus forbidden, even if they could be operated at ultra-low temperatures.

There is only one electronic element that is both non-linear and non-dissipative at arbitrarily low temperature: the superconducting tunnel junction[2] (also known as a Josephson tunnel junction(5)). As illustrated in Fig. 1, this circuit element consists of a sandwich of two superconducting thin films separated by an insulating layer that is thin enough (typically ~1 nm) to allow tunneling of discrete charges through the barrier. In later

[2]A very short superconducting weak link (see for instance Ref. 6) is a also a possible candidate, provided the Andreev levels would be sufficiently separated. Since we have too few experimental evidence for quantum effects involving this device, we do not discuss this otherwise important matter further.

sections we will describe how the tunneling of Cooper pairs creates an inductive path with strong non-linearity, thus creating energy levels suitable for a qubit. The tunnel barrier is typically fabricated from oxidation of the superconducting metal. This results in a reliable barrier since the oxidation process is self-terminating.[(7)] The materials properties of amorphous aluminum oxide, alumina, make it an attractive tunnel insulating layer. In part because of its well-behaved oxide, aluminum is the material from which good quality tunnel junctions are most easily fabricated, and it is often said that aluminium is to superconducting quantum circuits what silicon is to conventional MOSFET circuits. Although the Josephson effect is a subtle physical effect involving a combination of tunneling and superconductivity, the junction fabrication process is relatively straightforward.

2.4. Design and Fabrication of Quantum Integrated Circuits

Superconducting junctions and wires are fabricated using techniques borrowed from conventional integrated circuits[3]. Quantum circuits are typically made on silicon wafers using optical or electron-beam lithography and thin film deposition. They present themselves as a set of micron-size or sub-micron-size circuit elements (tunnel junctions, capacitors, and inductors) connected by wires or transmission lines. The size of the chip and elements are such that, to a large extent, the electrodynamics of the circuit can be analyzed using simple transmission line equations or even a lumped element approximation. Contact to the chip is made by wires bonded to mm-size metallic pads. The circuit can be designed using conventional layout and classical simulation programs.

Thus, many of the design concepts and tools of conventional semiconductor electronics can be directly applied to quantum circuits. Nevertheless, there are still important differences between conventional and quantum circuits at the conceptual level.

2.5. Integrated Circuits that Obey Macroscopic Quantum Mechanics

At the conceptual level, conventional and quantum circuits differ in that, in the former, the collective electronic degrees of freedom such as currents and voltages are classical variables, whereas in the latter, these degrees of freedom must be treated by quantum operators which do not necessarily commute. A more concrete way of presenting this rather

[3]It is worth mentioning that chips with tens of thousands of junctions have been successfully fabricated for the voltage standard and for the Josephson signal processors, which are only exploiting the speed of Josephson elements, not their macroscopic quantum properties.

abstract difference is to say that a typical electrical quantity, such as the charge on the plates of a capacitor, can be thought of as a simple number is conventional circuits, whereas in quantum circuits, the charge on the capacitor must be represented by a wave function giving the probability amplitude of all charge configurations. For example, the charge on the capacitor can be in a superposition of states where the charge is both positive and negative at the same time. Similarly the current in a loop might be flowing in two opposite directions at the same time. These situations have originally been nicknamed "macroscopic quantum coherence effects" by Tony Leggett[(8)] to emphasize that quantum integrated circuits are displaying phenomena involving the collective behavior of many particles, which are in contrast to the usual quantum effects associated with microscopic particles such as electrons, nuclei or molecules[4].

2.6. DiVicenzo Criteria

We conclude this section by briefly mentioning how quantum integrated circuits satisfy the so-called DiVicenzo criteria for the implementation of quantum computation.[(9)] The non-linearity of tunnel junctions is the key property ensuring that non-equidistant level subsystems can be implemented (criterion #1: qubit existence). As in many other implementations, initialization is made possible (criterion #2: qubit reset) by the use of low temperature. Absence of dissipation in superconductors is one of the key factors in the quantum coherence of the system (criterion #3: qubit coherence). Finally, gate operation and readout (criteria #4 and #5) are easily implemented here since electrical signals confined to and traveling along wires constitute very efficient coupling methods.

3. THE SIMPLEST QUANTUM CIRCUIT

3.1. Quantum LC Oscillator

We consider first the simplest example of a quantum integrated circuit, the LC oscillator. This circuit is shown in Fig. 2, and consists of an inductor L connected to a capacitor C, all metallic parts being superconducting. This simple circuit is the lumped-element version of a superconducting cavity or a transmission line resonator (for instance, the link between cavity resonators and LC circuits is elegantly discussed by

[4] These microscopic effects determine also the properties of materials, and explain phenomena such as superconductivity and the Josephson effect itself. Both classical and quantum circuits share this bottom layer of microscopic quantum mechanics.

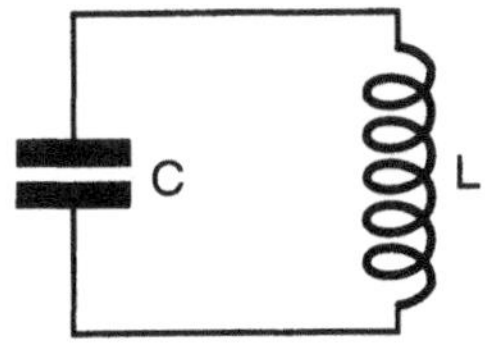

Fig. 2. Lumped element model for an electromagnetic resonator: LC oscillator.

Feynman[10]). The equations of motion of the LC circuit are those of an harmonic oscillator. It is convenient to take the position coordinate as being the flux Φ in the inductor, while the role of conjugate momentum is played by the charge Q on the capacitor playing the role of its conjugate momentum. The variables Φ and Q have to be treated as canonically conjugate quantum operators that obey $[\Phi, Q] = i\hbar$. The hamiltonian of the circuit is $H = (1/2)\Phi^2/L + (1/2)Q^2/C$, which can be equivalently written as $H = \hbar\omega_0(n + (1/2))$ where n is the number operator for photons in the resonator and $\omega_0 = 1/\sqrt{LC}$ is the resonance frequency of the oscillator. It is important to note that the parameters of the circuit hamiltonian are not fundamental constants of Nature. They are engineered quantities with a large range of possible values which can be modified easily by changing the dimensions of elements, a standard lithography operation. It is in this sense, in our opinion, that the system is unambiguously "macroscopic". The other important combination of the parameters L and C is the characteristic impedance $Z = \sqrt{L/C}$ of the circuit. When we combine this impedance with the residual resistance of the circuit and/or its radiation losses, both of which we can lump into a resistance R, we obtain the quality factor of the oscillation: $\mathcal{Q} = Z/R$. The theory of the harmonic oscillator shows that a quantum superposition of ground state and first excited state decays on a time scale given by $1/RC$. This last equality illustrates the general link between a classical measure of dissipation and the upper limit of the quantum coherence time.

3.2. Practical Considerations

In practice, the circuit shown in Fig. 2 may be fabricated using planar components with lateral dimensions around $10\,\mu$m, giving values of L and C approximately 0.1 nH and 1 pF, respectively, and yielding $\omega_0/2\pi \simeq 16\,$GHz and $Z_0 = 10\,\Omega$. If we use aluminium, a good BCS superconductor with transition temperature of 1.1 K and a gap $\Delta/e \simeq 200\,\mu V$, dissipation from the breaking of Cooper pairs will begin at frequencies greater than $2\Delta/h \simeq 100\,$GHz. The residual resistivity of a BCS superconductor decreases exponentially with the inverse of temperature and linearly with frequency, as shown by the Mattis-Bardeen (MB) formula $\rho(\omega) \sim \rho_0(\hbar\omega/k_B T)\exp(-\Delta/k_B T)$,[11] where ρ_0 is the resistivity of the metal in the normal state (we are treating here the case of the so-called "dirty" superconductor,[12] which is well adapted to thin film systems). According to MB, the intrinsic losses of the superconductor at the temperature and frequency (typically 20 mK and 20 GHz) associated with qubit dynamics can be safely neglected. However, we must warn the reader that the intrisinsic losses in the superconducting material do not exhaust, by far, sources of dissipation, even if very high quality factors have been demonstrated in superconducting cavity experiments.[13]

3.3. Matching to the Vacuum Impedance: A Useful Feature, not a Bug

Although the intrisinsic dissipation of superconducting circuits can be made very small, losses are in general governed by the coupling of the circuit with the electromagnetic environment that is present in the forms of write, control and readout lines. These lines (which we also refer to as ports) have a characteristic propagation impedance $Z_c \simeq 50\,\Omega$, which is constrained to be a fraction of the impedance of the vacuum $Z_{vac} = 377\,\Omega$. It is thus easy to see that our LC circuit, with a characteristic impedance of $Z_0 = 10\,\Omega$, tends to be rather well impedance-matched to any pair of leads. This circumstance occurs very frequently in circuits, and almost never in microscopic systems such as atoms which interact very weakly with electromagnetic radiation[5]. Matching to Z_{vac} is a useful feature because it allows strong coupling for writing, reading, and logic operations. As we mentioned earlier, the challenge with quantum circuits is to isolate them from parasitic degrees of freedom. The major task of this review is to explain how this has been achieved so far and what level of isolation is attainable.

3.4. The Consequences of being Macroscopic

While our example shows that quantum circuits can be mass-produced by standard micro-fabrication techniques and that their parameters can be easily engineered to reach some optimal condition, it also points out evident drawbacks of being "macroscopic" for qubits.

[5]The impedance of an atom can be crudely seen as being given by the impedance quantum $R_K = h/e^2$. We live in a universe where the ratio $Z_{vac}/2R_K$, also known as the fine structure constant 1/137.0, is a small number.

The engineered quantities L and C can be written as

$$\begin{aligned} L &= L^{\text{stat}} + \Delta L(t), \\ C &= C^{\text{stat}} + \Delta C(t). \end{aligned} \tag{1}$$

(a) The first term on the right-hand side denotes the static part of the parameter. It has **statistical variations**: unlike atoms whose transition frequencies in isolation are so reproducible that they are the basis of atomic clocks, circuits will always be subject to parameter variations from one fabrication batch to another. Thus prior to any operation using the circuit, the transition frequencies and coupling strength will have to be determined by "diagnostic" sequences and then taken into account in the algorithms.

(b) The second term on the right-hand side denotes the time-dependent fluctuations of the parameter. It describes **noise** due to residual material defects moving in the material of the substrate or in the material of the circuit elements themselves. This noise can affect for instance the dielectric constant of a capacitor. The low frequency components of the noise will make the resonance frequency wobble and contribute to the dephasing of the oscillation. Furthermore, the frequency component of the noise at the transition frequency of the resonator will induce transitions between states and will therefore contribute to the energy relaxation.

Let us stress that statistical variations and noise are not problems affecting superconducting qubit parameters only. For instance when several atoms or ions are put together in microcavities for gate operation, patch potential effects will lead to expressions similar in form to Eq. (1) for the parameters of the hamiltonian, even if the isolated single qubit parameters are fluctuation-free.

3.5. The Need for Non-linear Elements

Not all aspects of quantum information processing using quantum integrated circuits can be discussed within the framework of the LC circuit, however. It lacks an important ingredient: non-linearity. In the harmonic oscillator, all transitions between neighbouring states are degenerate as a result of the parabolic shape of the potential. In order to have a qubit, the transition frequency between states $|\text{qubit}=0\rangle$ and $|\text{qubit}=1\rangle$ must be sufficiently different from the transition between higher-lying eigenstates, in particular 1 and 2. Indeed, the maximum number of 1-qubit operations that can be performed coherently scales as $\mathcal{Q}_{01} \left|\omega_{01} - \omega_{12}\right| / \omega_{01}$ where $\mathcal{Q}_{01}$ is the quality factor of the $0 \rightarrow 1$ transition. Josephson tunnel junctions are crucial for quantum circuits since they bring a strongly non-parabolic inductive potential energy.

4. THE JOSEPHSON NON-LINEAR INDUCTANCE

At low temperatures, and at the low voltages and low frequencies corresponding to quantum information manipulation, the Josephson tunnel junction behaves as a pure non-linear inductance (Josephson element) in parallel with the capacitance corresponding to the parallel plate capacitor formed by the two overlapping films of the junction (Fig. 1b). This minimal, yet precise model, allows arbitrary complex quantum circuits to be analysed by a quantum version of conventional circuit theory. Even though the tunnel barrier is a layer of order ten atoms thick, the value of the Josephson non-linear inductance is very robust against static disorder, just like an ordinary inductance—such as the one considered in Sec. 3—is very insensitive to the position of each atom in the wire. We refer to[14] for a detailed discussion of this point.

4.1. Constitutive Equation

Let us recall that a linear inductor, like any electrical element, can be fully characterized by its constitutive equation. Introducing a generalization of the ordinary magnetic flux, which is only defined for a loop, we define the **branch flux of an electric element** by $\Phi(t) = \int_{-\infty}^{t} V(t_1)\,dt_1$, where $V(t)$ is the space integral of the electric field along a current line inside the element. In this language, the current $I(t)$ flowing through the inductor is proportional to its branch flux $\Phi(t)$:

$$I(t) = \frac{1}{L}\Phi(t). \tag{2}$$

Note that the generalized flux $\Phi(t)$ can be defined for any electric element with two leads (dipole element), and in particular for the Josephson junction, even though it does not resemble a coil. The Josephson element behaves inductively, as its branch flux–current relationship[5] is

$$I(t) = I_0 \sin[2\pi\,\Phi(t)/\Phi_0]. \tag{3}$$

This inductive behavior is the manifestation, at the level of collective electrical variables, of the inertia of Cooper pairs tunneling across the insulator (kinetic inductance). The discreteness of Cooper pair tunneling causes the periodic flux dependence of the current, with a period given by a universal quantum constant Φ_0, the superconducting flux quantum $h/2e$. The junction parameter I_0 is called the critical current of the tunnel element. It scales proportionally to the area of the tunnel layer and diminishes exponentially with the tunnel layer thickness. Note that the constitutive relation Eq. (3) expresses in only one equation the two Josephson relations.[5] This compact formulation is made possible by the introduction of the branch flux (see Fig. 3).

The purely sinusoidal form of the constitutive relation Eq. (3) can be traced to the perturbative nature of Cooper pair tunneling in a tunnel junction. Higher harmonics can appear if the tunnel layer becomes very thin, though their presence would not fundamentally change the discussion presented in this review. The quantity $2\pi\,\Phi(t)/\Phi_0 = \delta$ is called the gauge-invariant phase difference accross the junction (often abridged into "phase"). It is important to realize that at the level of the constitutive relation of the Josephson element, this variable is nothing else than an electromagnetic flux in dimensionless units. In general, we have

$$\theta = \delta \bmod 2\pi,$$

where θ is the phase difference between the two superconducting condensates on both sides of the junction. This last relation expresses how the superconducting ground state and electromagnetism are tied together.

4.2. Other Forms of the Parameter Describing the Josephson Non-linear Inductance

The Josephson element is also often described by two other parameters, each of which carry exactly the same information as the critical current. The first one is the Josephson effective inductance $L_{J0} = \varphi_0/I_0$, where $\varphi_0 = \Phi_0/2\pi$ is the reduced flux quantum. The name of this other form becomes obvious if we expand the sine function in Eq. (3) in powers of Φ around $\Phi = 0$. Keeping the leading term, we have $I = \Phi/L_{J0}$. Note that the junction behaves for small signals almost as a point-like kinetic

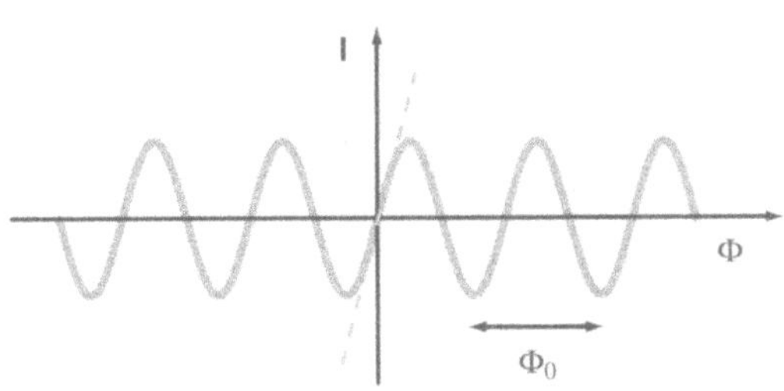

Fig. 3. Sinusoidal current-flux relationship of a Josephson tunnel junction, the simplest non-linear, non-dissipative electrical element (solid line). Dashed line represents current–flux relationship for a linear inductance equal to the junction effective inductance.

inductance: a 100 nm × 100 nm area junction will have a typical inductance of 100 nH, whereas the same inductance is only obtained magnetically with a loop of about 1 cm in diameter. More generally, it is convenient to define the phase-dependent Josephson inductance

$$L_J(\delta) = \left(\frac{\partial I}{\partial \Phi}\right)^{-1} = \frac{L_{J0}}{\cos\delta}.$$

Note that the Josephson inductance not only depends on δ, it can actually become infinite or negative! Thus, under the proper conditions, the Josephson element can become a switch and even an active circuit element, as we will see below.

The other useful parameter is the Josephson energy $E_J = \varphi_0 I_0$. If we compute the energy stored in the junction $E(t) = \int_{-\infty}^{t} I(t_1) V(t_1)\, dt_1$, we find $E(t) = -E_J \cos[2\pi\, \Phi(t)/\Phi_0]$. In contrast with the parabolic dependence on flux of the energy of an inductance, the potential associated with a Josephson element has the shape of a cosine washboard. The total height of the corrugation of the washboard is $2E_J$.

4.3. Tuning the Josephson Element

A direct application of the non-linear inductance of the Josephson element is obtained by splitting a junction and its leads into two equal junctions, such that the resulting loop has an inductance much smaller the Josephson inductance. The two smaller junctions in parallel then behave as an effective junction[15] whose Josephson energy varies with Φ_{ext}, the magnetic flux externally imposed through the loop

$$E_J(\Phi_{ext}) = E_J \cos(\pi\, \Phi_{ext}/\Phi_0)\,. \tag{4}$$

Here, E_J the total Josephson energy of the two junctions. The Josephson energy can also be modulated by applying a magnetic field in the plane parallel to the tunnel layer.

5. THE QUANTUM ISOLATED JOSEPHSON JUNCTION

5.1. Form of the Hamiltonian

If we leave the leads of a Josephson junction unconnected, we obtain the simplest example of a non-linear electrical resonator. In order to analyze its quantum dynamics, we apply the prescriptions of quantum circuit theory briefly summarized in Appendix 1. Choosing a representation privileging the branch variables of the Josephson element, the momentum corresponds to the charge $Q = 2eN$ having tunneled through the element and the canonically conjugate position is the flux $\Phi = \varphi_0 \theta$ associated with the superconducting phase difference across the tunnel layer. Here, N and θ are treated as operators that obey $[\theta, N] = i$. It is important to note that the operator N has integer eigenvalues whereas the phase θ is an operator corresponding to the position of a point on the unit circle (an angle modulo 2π).

By eliminating the branch charge of the capacitor, the hamiltonian reduces to

$$H = E_{CJ}(N - Q_r/2e)^2 - E_J \cos\theta \tag{5}$$

where $E_{CJ} = \frac{(2e)^2}{2C_J}$ is the Coulomb charging energy corresponding to one Cooper pair on the junction capacitance C_J and where Q_r is the residual offset charge on the capacitor.

One may wonder how the constant Q_r got into the hamiltonian, since no such term appeared in the corresponding LC circuit in Sec. 3. The continuous charge Q_r is equal to the charge that pre-existed on the capacitor when it was wired with the inductor. Such offset charge is not some nit-picking theoretical construct. Its physical origin is a slight difference in work function between the two electrodes of the capacitor and/or an excess of charged impurities in the vicinity of one of the capacitor plates relative to the other. The value of Q_r is in practice very large compared to the Cooper pair charge $2e$, and since the hamiltonian (5) is invariant under the transformation $N \rightarrow N \pm 1$, its value can be considered completely random.

Such residual offset charge also exists in the LC circuit. However, we did not include it in our description of Sec. 3 since a time-independent Q_r does not appear in the dynamical behavior of the circuit: it can be removed from the hamiltonian by performing a trivial canonical transformation leaving the form of the hamiltonian unchanged.

It is not possible, however, to remove this constant from the junction hamiltonian (5) because the potential is not quadratic in θ. The parameter Q_r plays a role here similar to the vector potential appearing in the hamiltonian of an electron in a magnetic field.

5.2. Fluctuations of the Parameters of the Hamiltonian

The hamiltonian 5 thus depends thus on three parameters which, following our discussion of the LC oscillator, we write as

$$
\begin{aligned}
Q_\mathrm{r} &= Q_\mathrm{r}^{\mathrm{stat}} + \Delta Q_\mathrm{r}(t),\\
E_\mathrm{C} &= E_\mathrm{C}^{\mathrm{stat}} + \Delta E_\mathrm{C}(t),\\
E_\mathrm{J} &= E_\mathrm{J}^{\mathrm{stat}} + \Delta E_\mathrm{J}(t)
\end{aligned}
\tag{6}
$$

in order to distinguish the static variation resulting from fabrication of the circuit from the time-dependent fluctuations. While $Q_\mathrm{r}^{\mathrm{stat}}$ can be considered fully random (see above discussion), $E_\mathrm{C}^{\mathrm{stat}}$ and $E_\mathrm{J}^{\mathrm{stat}}$ can generally be adjusted by construction to a precision better than 20%. The relative fluctuations $\Delta Q_\mathrm{r}(t)/2e$ and $\Delta E_\mathrm{J}(t)/E_\mathrm{J}$ are found to have a $1/f$ power spectral density with a typical standard deviations at 1 Hz roughly of order $10^{-3}\,\mathrm{Hz}^{-1/2}$ and $10^{-5}\,\mathrm{Hz}^{-1/2}$, respectively, for a junction with a typical area of $0.01\,\mu\mathrm{m}^2$.[16] The noise appears to be produced by independent two-level fluctuators.[17] The relative fluctuations $\Delta E_\mathrm{C}(t)/E_\mathrm{C}$ are much less known, but the behavior of some glassy insulators at low temperatures might lead us to expect also a $1/f$ power spectral density, but probably with a weaker intensity than those of $\Delta E_\mathrm{J}(t)/E_\mathrm{J}$. We refer to the three noise terms in Eq. (6) as offset charge, dielectric and critical current noises, respectively.

6. WHY THREE BASIC TYPES OF JOSEPHSON QUBITS?

The first-order problem in realizing a Josephson qubit is to suppress as much as possible the detrimental effect of the fluctuations of Q_r, while retaining the non-linearity of the circuit. There are three main stategies for solving this problem and they lead to three fundamental basic type of qubits involving only one Josephson element.

6.1. The Cooper Pair Box

The simplest circuit is called the "Cooper pair box" and was first described theoretically, albeit in a slightly different version than presented here, by Büttiker.[18] It was first realized experimentally by the Saclay group in 1997.[19] Quantum dynamics in the time domain were first seen by the NEC group in 1999.[20]

In the Cooper pair box, the deviations of the residual offset charge Q_r are compensated by biasing the Josephson tunnel junction with a voltage source U in series with a "gate" capacitor C_g (see Fig. 4a). One can easily show that the hamiltonian of the Cooper pair box is

$$H = E_\mathrm{C}\left(N - N_\mathrm{g}\right)^2 - E_\mathrm{J}\cos\theta. \tag{7}$$

Here $E_C = (2e)^2/(2(C_J + C_g))$ is the charging energy of the island of the box and $N_g = Q_r + C_g U/2e$. Note that this hamiltonian has the same form as hamiltonian (5). Often N_g is simply written as $C_g U/2e$ since U at the chip level will anyhow deviate substantially from the generator value at high-temperature due to stray emf's in the low-temperature cryogenic wiring.

In Fig. 5, we show the potential in the θ representation as well as the first few energy levels for $E_J/E_C = 1$ and $N_g = 0$. As shown in Appendix 2, the Cooper pair box eigenenergies and eigenfunctions can be calculated with special functions known with arbitrary precision, and in Fig. 6 we plot the first few eigenenergies as a function of N_g for $E_J/E_C = 0.1$ and $E_J/E_C = 1$. Thus, the Cooper box is to quantum circuit physics what the hydrogen atom is to atomic physics. We can modify the spectrum with the action of two externally controllable electrodynamic parameters: N_g, which is directly proportional to U, and E_J, which can be varied by applying a field through the junction or by using a split junction and applying a flux through the loop, as discussed in Sec. 3. These parameters bear some resemblance to the Stark and Zeeman fields in atomic physics. For the box, however much smaller values of the fields are required to change the spectrum entirely.

We now limit ourselves to the two lowest levels of the box. Near the degeneracy point $N_g = 1/2$ where the electrostatic energy of the two charge states $|N=0\rangle$ and $|N=1\rangle$ are equal, we get the reduced hamiltonian[19,21]

$$H_{\text{qubit}} = -E_z(\sigma_Z + X_{\text{control}}\sigma_X), \tag{8}$$

where, in the limit $E_J/E_C \ll 1$, $E_z = E_J/2$ and $X_{\text{control}} = 2(E_C/E_J)\left((1/2) - N_g\right)$. In Eq. (8), σ_Z and σ_X refer to the Pauli spin operators. Note that the X-direction is chosen along the charge operator, the variable of the box we can naturally couple to.

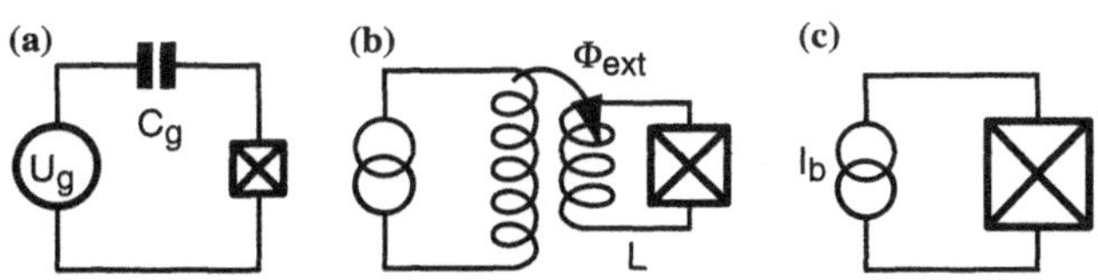

Fig. 4. The three basic superconducting qubits. (a) Cooper pair box (prototypal charge qubit); (b) RF-SQUID (prototypal flux qubit); and (c) current-biased junction (prototypal phase qubit). The charge qubit and the flux qubit requires small junctions fabricated with e-beam lithography while the phase qubit can be fabricated with conventional optical lithography.

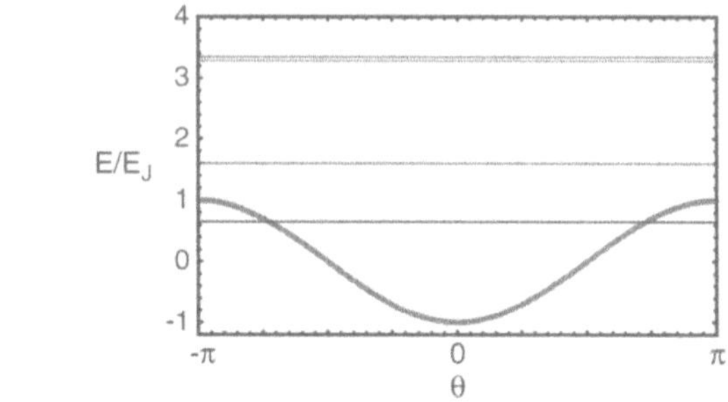

Fig. 5. Potential landscape for the phase in a Cooper pair box (thick solid line). The first few levels for $E_J/E_C = 1$ and $N_g = 1/2$ are indicated by thin horizontal solid lines.

If we plot the energy of the eigenstates of hamiltonian (8) as a function of the control parameter X_{control}, we obtain the universal level repulsion diagram shown in Fig. 7. Note that the minimum energy

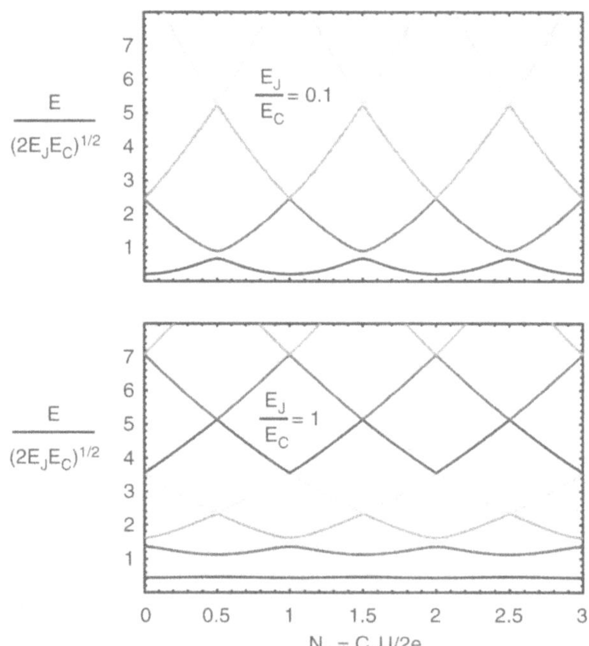

Fig. 6. Energy levels of the Cooper pair box as a function of N_g, for two values of E_J/E_C. As E_J/E_C increases, the sensitivity of the box to variations of offset charge diminishes, but so does the non-linearity. However, the non-linearity is the slowest function of E_J/E_C and a compromise advantageous for coherence can be found.

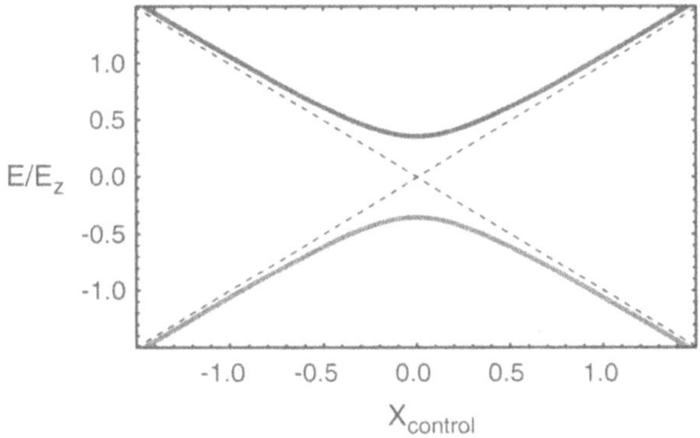

Fig. 7. Universal level anticrossing found both for the Cooper pair box and the RF-SQUID at their "sweet spot".

splitting is given by E_J. Comparing Eq. (8) with the spin hamiltonian in NMR, we see that E_J plays the role of the Zeeman field while the electrostatic energy plays the role of the transverse field. Indeed we can send on the control port corresponding to U time-varying voltage signals in the form of NMR-type pulses and prepare arbitrary superpositions of states.[22]

The expression 8 shows that at the "sweet spot" $X_{\text{control}} = 0$, i.e., the degeneracy point $N_g = 1/2$, the qubit transition frequency is to first order insentive to the offset charge noise ΔQ_r. We will discuss in Sec. 6.2 how an extension of the Cooper pair box circuit can display quantum coherence properties on long time scales by using this property.

In general, circuits derived from the Cooper pair box have been nicknamed "charge qubits". One should not think, however, that in charge qubits, quantum information is *encoded* with charge. Both the charge N and phase θ are quantum variables and they are both uncertain for a generic quantum state. Charge in "charge qubits" should be understood as refering to the "controlled variable", i.e., the qubit variable that couples to the control line we use to write or manipulate quantum information. In the following, for better comparison between the three qubits, we will be faithful to the convention used in Eq. (8), namely that σ_X represents the *controlled variable*.

6.2. The RF-SQUID

The second circuit—the so-called RF-SQUID[23]—can be considered in several ways the dual of the Cooper pair box (see Fig. 4b). It employs

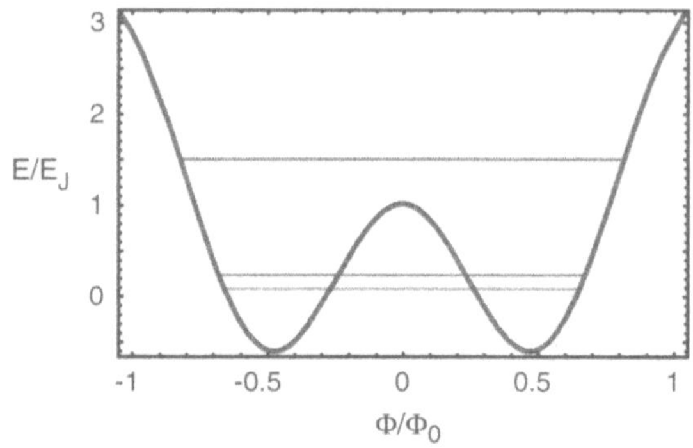

Fig. 8. Schematic potential energy landcape for the RF-SQUID.

a superconducting transformer rather than a gate capacitor to adjust the hamiltonian. The two sides of the junction with capacitance C_J are connected by a superconducting loop with inductance L. An external flux Φ_{ext} is imposed through the loop by an auxiliary coil. Using the methods of Appendix 1, we obtain the hamiltonian[8]

$$H = \frac{q^2}{2C_J} + \frac{\phi^2}{2L} - E_J \cos\left[\frac{2e}{\hbar}(\phi - \Phi_{ext})\right]. \tag{9}$$

We are taking here as degrees of freedom the integral ϕ of the voltage across the inductance L, i.e., the flux through the superconducting loop, and its conjugate variable, the charge q on the capacitance C_J; they obey $[\phi, q] = i\hbar$. Note that in this representation, the phase θ, corresponding to the branch flux across the Josephson element, has been eliminated. Note also that the flux ϕ, in contrast to the phase θ, takes its values on a line and not on a circle. Likewise, its conjugate variable q, the charge on the capacitance, has continuous eigenvalues and not integer ones like N. Note that we now have three adjustable energy scales: E_J, $E_{CJ} = (2e)^2/2C_J$ and $E_L = \Phi_0^2/2L$.

The potential in the flux representation is schematically shown in Fig. 8 together with the first few levels, which have been seen experimentally for the first time by the SUNY group.[24] Here, no analytical expressions exist for the eigenvalues and the eigenfunctions of the problem, which has two aspect ratios: E_J/E_{CJ} and $\lambda = L_J/L - 1$.

Whereas in the Cooper box the potential is cosine-shaped and has only one well since the variable θ is 2π-periodic, we have now in general a parabolic potential with a cosine corrugation. The idea here for curing the detrimental effect of the offset charge fluctuations is very different than in the box. First of all Q_r^{stat} has been neutralized by shunting the two metallic electrodes of the junction by the superconducting wire of the loop. Then, the ratio E_J/E_{CJ} is chosen to be much larger than unity. This tends to increase the relative strength of quantum fluctuations of q, making offset charge fluctuations ΔQ_r small in comparison. The resulting loss in the non-linearity of the first levels is compensated by taking λ close to zero and by flux-biasing the device at the half-flux quantum value $\Phi_{ext} = \Phi_0/2$. Under these conditions, the potential has two degenerate wells separated by a shallow barrier with height $E_B = (3\lambda^2/2)E_J$. This corresponds to the degeneracy value $N_g = 1/2$ in the Cooper box, with the inductance energy in place of the capacitance energy. At $\Phi_{ext} = \Phi_0/2$, the two lowest energy levels are then the symmetric and antisymmetric combinations of the two wavefunctions localized in each well, and the energy splitting between the two states can be seen as the tunnel splitting associated with the quantum motion through the potential barrier between the two wells, bearing close resemblance to the dynamics of the ammonia molecule. This splitting E_S depends exponentially on the barrier height, which itself depends strongly on E_J. We have $E_S = \eta\sqrt{E_B E_{CJ}} \exp\left(-\xi\sqrt{E_B/E_{CJ}}\right)$ where the numbers η and ξ have to be determined numerically in most practical cases. The non-linearity of the first levels results thus from a subtle cancellation between two inductances: the superconducting loop inductance L and the junction effective inductance $-L_{J0}$ which is opposed to L near $\Phi_{ext} = \Phi_0/2$. However, as we move away from the degeneracy point $\Phi_{ext} = \Phi_0/2$, the splitting $2E_\Phi$ between the first two energy levels varies linearly with the applied flux $E_\Phi = \zeta(\Phi_0^2/2L)(N_\Phi - 1/2)$. Here the parameter $N_\Phi = \Phi_{ext}/\Phi_0$, also called the flux frustration, plays the role of the reduced gate charge N_g. The coefficient ζ has also to be determined numerically. We are therefore again, in the vicinity of the flux degeneracy point $\Phi_{ext} = \Phi_0/2$ and for $E_J/E_{CJ} \gg 1$, in presence of the universal level repulsion behavior (see Fig. 7) and the qubit hamiltonian is again given by

$$H_{qubit} = -E_z\left(\sigma_Z + X_{control}\sigma_X\right), \tag{10}$$

where now $E_z = E_S/2$ and $X_{control} = 2(E_\Phi/E_S)((1/2) - N_\Phi)$. The qubits derived from this basic circuit[(25,33)] have been nicknamed "flux qubits". Again, quantum information is not directly represented here by the flux ϕ, which is as uncertain for a general qubit state as the charge q on the capacitor plates of the junction. The flux ϕ is the system variable to which we couple when we write or control information in the qubit, which is done by sending current pulses on the primary of the RF-SQUID transformer, thereby modulating N_Φ, which itself determines the strength of the pseudo-field in the X-direction in the hamiltonian 10. Note that the parameters E_S, E_Φ, and N_Φ are all influenced to some degree by the critical current noise, the dielectric noise and the charge noise. Another independent noise can also be present, the noise of the flux in the loop, which is not found in the box and which will affect only N_Φ. Experiments on DC-SQUIDS[(15)] have shown that this noise, in adequate conditions, can be as low as $10^{-8}(h/2e)/\mathrm{Hz}^{-1/2}$ at a few kHz. However, experimental results on flux qubits (see below) seem to indicate that larger apparent flux fluctuations are present, either as a result of flux trapping or critical current fluctuations in junctions implementing inductances.

6.3. Current-biased Junction

The third basic quantum circuit biases the junction with a fixed DC-current source (Fig. 7c). Like the flux qubit, this circuit is also insensitive to the effect of offset charge and reduces the effect of charge fluctuations by using large ratios of E_J/E_{CJ}. A large non-linearity in the Josephson inductance is obtained by biasing the junction at a current I very close to the critical current. A current bias source can be understood as arising from a loop inductance with $L \to \infty$ biased by a flux $\Phi \to \infty$ such that $I = \Phi/L$. The Hamiltonian is given by

$$H = E_{CJ}\,p^2 - I\varphi_0\delta - I_0\varphi_0 \cos\delta, \tag{11}$$

where the gauge invariant phase difference operator δ is, apart from the scale factor φ_0, precisely the branch flux across C_J. Its conjugate variable is the charge $2ep$ on that capacitance, a continuous operator. We have thus $[\delta, p] = i$. The variable δ, like the variable ϕ of the RF-SQUID, takes its value on the whole real axis and its relation with the phase θ is $\delta \bmod 2\pi = \theta$ as in our classical analysis of Sec. 4.

The potential in the δ representation is shown in Fig. 9. It has the shape of a tilted washboard, with the tilt given by the ratio I/I_0. When I approaches I_0, the phase is $\delta \approx \pi/2$, and in its vicinity, the potential is very well approximated by the cubic form

$$U(\delta) = \varphi_0(I_0 - I)(\delta - \pi/2) - \frac{I_0\varphi_0}{6}(\delta - \pi/2)^3, \tag{12}$$

Note that its shape depends critically on the difference $I_0 - I$. For $I \lesssim I_0$, there is a well with a barrier height $\Delta U = (2\sqrt{2}/3)I_0\varphi_0(1 - I/I_0)^{3/2}$ and the classical oscillation frequency at the bottom of the well (so-called plasma oscillation) is given by

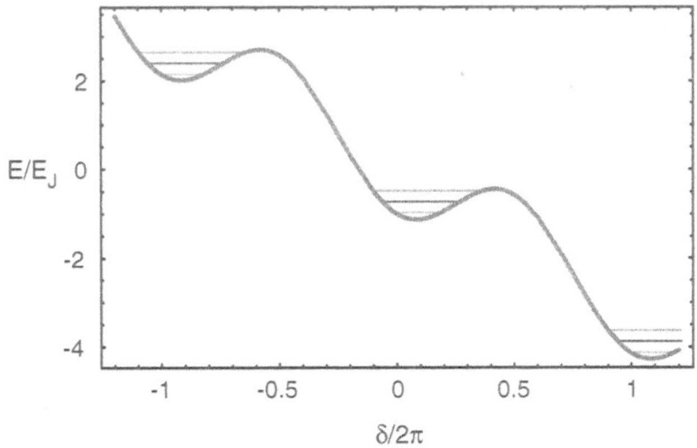

Fig. 9. Tilted washboard potential of the current-biased Josephson junction.

$$\begin{aligned}\omega_p &= \frac{1}{\sqrt{L_J(I)C_J}} \\ &= \frac{1}{\sqrt{L_{J0}C_J}}\left[1-(I/I_0)^2\right]^{1/4}.\end{aligned}$$

Quantum-mechanically, energy levels are found in the well (see Fig. 11)[3] with non-degenerate spacings. The first two levels can be used for qubit states,[26] and have a transition frequency $\omega_{01} \simeq 0.95\omega_p$.

A feature of this qubit circuit is built-in readout, a property missing from the two previous cases. It is based on the possibility that states in the cubic potential can tunnel through the cubic potential barrier into the continuum outside the barrier. Because the tunneling rate increases by a factor of approximately 500 each time we go from one energy level to the next, the population of the $|1\rangle$ qubit state can be reliably measured by sending a probe signal inducing a transition from the 1 state to a higher energy state with large tunneling probability. After tunneling, the particle representing the phase accelerates down the washboard, a convenient self-amplification process leading to a voltage as large as $2\Delta/e$ across the junction. Therefore, a finite voltage $V \neq 0$ suddenly appearing across the junction just after the probe signal implies that the qubit was in state $|1\rangle$, whereas $V = 0$ implies that the qubit was in state $|0\rangle$.

In practice, like in the two previous cases, the transition frequency $\omega_{01}/2\pi$ falls in the 5–20 GHz range. This frequency is only determined by material properties of the barrier, since the product $C_J\ L_J$ does not depend on junction area. The number of levels in the well is typically $\Delta U/\hbar\omega_p \approx 4$.

Setting the bias current at a value I and calling ΔI the variations of the difference $I - I_0$ (originating either in variations of I or I_0), the qubit Hamiltonian is given by

$$H_{\text{qubit}} = \hbar\omega_{01}\sigma_Z + \sqrt{\frac{\hbar}{2\omega_{01}C_J}}\Delta I(\sigma_X + \chi\sigma_Z), \tag{13}$$

where $\chi = \sqrt{\hbar\omega_{01}/3\Delta U} \simeq 1/4$ for typical operating parameters. In contrast with the flux and phase qubit circuits, the current-biased Josephson junction does not have a bias point where the $0 \rightarrow 1$ transition frequency has a local minimum. The hamiltonian cannot be cast into the NMR-type form of Eq. (8). However, a sinusoidal current signal $\Delta I(t) \sim \sin\omega_{01}t$ can still produce σ_X rotations, whereas a low-frequency signal produces σ_Z operations.[27]

In analogy with the preceding circuits, qubits derived from this circuit and/or having the same phase potential shape and qubit properties have been nicknamed "phase qubits" since the controlled variable is the phase (the X pseudo-spin direction in hamiltonian (13)).

6.4. Tunability versus Sensitivity to Noise in Control Parameters

The reduced two-level hamiltonians Eqs. (8), (10) and (13) have been tested thoroughly and are now well-established. They contain the very important parametric dependence of the coefficient of σ_X, which can be viewed on one hand as how much the qubit can be tuned by an external control parameter, and on the other hand as how much it can be dephased by uncontrolled variations in that parameter. It is often important to realize that even if the control parameter has a very stable value at the level of room-temperature electronics, the noise in the electrical components relaying its value at the qubit level might be inducing detrimental fluctuations. An example is the flux through a superconducting loop, which in principle could be set very precisely by a stable current in a coil, and which in practice often fluctuates because of trapped flux motion in the wire of the loop or in nearby superconducting films. Note that, on the other hand, the two-level hamiltonian does not contain all the non-linear properties of the qubit, and how they conflict with its intrinsic noise, a problem which we discuss in the next Sec. 6.5.

6.5. Non-linearity versus Sensitivity to Intrinsic Noise

The three basic quantum circuit types discussed above illustrate a general tendency of Josephson qubits. If we try to make the level structure very non-linear, i.e. $|\omega_{01} - \omega_{12}| \gg \omega_{01}$, we necessarily expose the system sensitively to at least one type of intrinsic noise. The flux qubit is contructed to reach a very large non-linearity, but is also maximally exposed, relatively speaking, to critical current noise and flux noise. On the other hand, the phase qubit starts with a relatively small non-linearity and acquires it at the expense of a precise tuning of the difference between the bias current and the critical current, and therefore exposes itself also to the noise in the latter. The Cooper box, finally, acquires non-linearity at the expense of its sensitivity to offset charge noise. The search for the optimal qubit circuit involves therefore a detailed knowledge of the relative intensities of the various sources of noise, and their variations with all the construction parameters of the qubit, and in particular—this point is crucial—the properties of the materials involved in the tunnel junction fabrication. Such in-depth knowledge does not yet exist at the time of this writing and one can only make educated guesses.

The qubit optimization problem is also further complicated by the necessity to readout quantum information, which we address just after reviewing the relationships between the intensity of noise and the decay rates of quantum information.

7. QUBIT RELAXATION AND DECOHERENCE

A generic quantum state of a qubit can be represented as a unit vector $\vec{S}$ pointing on a sphere—the so-called Bloch sphere. One distinguishes two broad classes of errors. The first one corresponds to the tip of the Bloch vector diffusing in the latitude direction, i.e., along the arc joining the two poles of the sphere to or away from the north pole. This process is called energy relaxation or state-mixing. The second class corresponds to the tip of the Bloch vector diffusing in the longitude direction, i.e., perpendicularly to the line joining the two poles. This process is called dephasing or decoherence.

In Appendix 3, we define precisely the relaxation and decoherence rates and show that they are directly proportional to the power spectral densities of the noises entering in the parameters of the hamiltonian of the qubit. More precisely, we find that the decoherence rate is proportional to the total spectral density of the quasi-zero-frequency noise in the qubit Larmor frequency. The relaxation rate, on the other hand, is proportional to the total spectral density, at the qubit Larmor frequency, of the noise in the field perpendicular to the eigenaxis of the qubit.

In principle, the expressions for the relaxation and decoherence rate could lead to a ranking of the various qubit circuits: from their reduced spin hamiltonian, one can find with what coefficient each basic noise source contributes to the various spectral densities entering in the rates. In the same manner, one could optimize the various qubit parameters to make them insensitive to noise, as much as possible. However, before discussing this question further, we must realize that the readout itself can provide substantial additional noise sources for the qubit. Therefore, the design of a qubit circuit that maximizes the number of coherent gate operations is a subtle optimization problem which must treat in parallel both the intrinsic noises of the qubit and the back-action noise of the readout.

8. READOUT OF SUPERCONDUCTING QUBITS

8.1. Formulation of the Readout Problem

We have examined so far the various basic circuits for qubit implementation and their associated methods to write and manipulate quantum information. Another important task quantum circuits must perform is the readout of that information. As we mentioned earlier, the difficulty of the readout problem is to open a coupling channel to the qubit for extracting information without at the same time submitting it to both dissipation and noise.

Ideally, the readout part of the circuit—referred to in the following simply as "readout"—should include both a switch, which defines an "OFF" and an "ON" phase, and a state measurement device. During the OFF phase, where reset and gate operations take place, the measurement device should be completely decoupled from the qubit degrees of freedom. During the ON phase, the measurement device should be maximally coupled to a qubit variable that distinguishes the 0 and the 1 state. However, this condition is not sufficient. The back-action of the measurement device during the ON phase should be weak enough not to relax the qubit.[28]

The readout can be characterized by 4 parameters. The first one describes the sensitivity of the measuring device while the next two describe its back-action, factoring in the quality of the switch (see Appendix 3 for the definitions of the rates Γ):

(i) the measurement time τ_m defined as the time taken by the measuring device to reach a signal-to-noise ratio of 1 in the determination of the state.

(ii) the energy relaxation rate Γ_1^{ON} of the qubit in the ON state.

(iii) the coherence decay rate Γ_2^{OFF} of the qubit information in the OFF state.

(iv) the dead time t_d needed to reset both the measuring device and qubit after a measurement. They are usually perturbed by the energy expenditure associated with producing a signal strong enough for external detection.

Simultaneously minimizing these parameters to improve readout performance cannot be done without running into conflicts. An important quantity to optimize is the readout fidelity. By construction, at the end of the ON phase, the readout should have reached one of two classical states: 0_c and 1_c, the outcomes of the measurement process. The latter can be described by two probabilities: the probability $p_{00_c}(p_{11_c})$ that starting from the qubit state $|0\rangle$ ($|1\rangle$) the measurement yields $0_c(1_c)$. The readout fidelity (or discriminating power) is defined as $F = p_{00c} + p_{11_c} - 1$. For a measuring device with a signal-to-noise ratio increasing like the square of measurement duration τ, we would have, if back-action could be neglected, $F = \mathrm{erf}\left(2^{-1/2}\tau/\tau_m\right)$.

8.2. Requirements and General Strategies

The fidelity and speed of the readout, usually not discussed in the context of quantum algorithms because they enter marginally in the evaluation of their complexity, are actually key to experiments studying the coherence properties of qubits and gates. A very fast and sensitive readout will gather at a rapid pace information on the imperfections and drifts of qubit parameters, thereby allowing the experimenter to design fabrication strategies to fight them during the construction or even correct them in real time.

We are thus mostly interested in "single-shot" readouts,[28] for which F is of order unity, as opposed to schemes in which a weak measurement is performed continuously.[29] If $F \ll 1$, of order F^{-2} identical preparation and readout cycles need to be performed to access the state of the qubit. The condition for "single-shot" operation is

$$\Gamma_1^{\mathrm{ON}}\tau_m < 1.$$

The speed of the readout, determined both by τ_m and t_d, should be sufficiently fast to allow a complete characterization of all the properties of the qubit before any drift in parameters occurs. With sufficient speed, the automatic correction of these drits in real time using feedback will be possible.

Rapidly pulsing the readout to full ON and OFF states is done with a fast, strongly non-linear element, which is provided by one or more auxiliary Josephson junctions. Decoupling the qubit from the readout in the OFF phase requires balancing the circuit in the manner of a Wheatstone bridge, with the readout input variable and the qubit variable corresponding to two orthogonal electrical degrees of freedom. Finally, to be as complete as possible even in presence of small asymmetries, the decoupling also requires an impedance mismatch between the qubit and the dissipative degrees of freedom of the readout. In Sec. 8.3, we discuss how these general ideas have been implemented in second generation quantum circuits. The examples we have chosen all involve a readout circuit which is built-in the qubit itself to provide maximal coupling during the ON phase, as well as a decoupling scheme which has proven effective for obtaining long decoherence times.

8.3. Phase Qubit: Tunneling Readout with a DC-SQUID On-chip Amplifier.

The simplest example of a readout is provided by a system derived from the phase qubit (see Fig. 10). In the phase qubit, the levels in the cubic potential are metastable and decay in the continuum, with level $n+1$ having roughly a decay rate Γ_{n+1} that is 500 times faster than the decay Γ_n of level n. This strong level number dependence of the decay rate leads naturally to the following readout scheme: when readout needs to be performed, a microwave pulse at the transition frequency ω_{12} (or better at ω_{13}) transfers the eventual population of level 1 into level 2, the latter decaying rapidly into the continuum, where it subsequently loses energy by friction and falls into the bottom state of the next corrugation of the potential (because the qubit junction is actually in a superconducting loop of large but finite inductance, the bottom of this next corrugation is in fact the absolute minimum of the potential and the particle representing the system can stay an infinitely long time there). Thus, at the end of the readout pulse, the sytem has either decayed out of the cubic well (readout state 1_c) if the qubit was in the $|1\rangle$ state or remained in the cubic well (readout state 0_c) if the qubit was in the $|0\rangle$ state. The DC-SQUID amplifier is sensitive enough to detect the change in flux accompanying the exit of the cubic well, but the problem is to avoid sending the back-action noise of its stabilizing resistor into the qubit circuit. The solution to this problem involves balancing the SQUID loop in such a way, that for readout state 0_c, the small signal gain of the SQUID is zero, whereas for readout state 1_c, the small signal gain is non-zero.[17] This signal dependent gain is obtained by having two junctions in one arm of the SQUID whose total Josephson inductance equals that of the unique junction in the other arm.

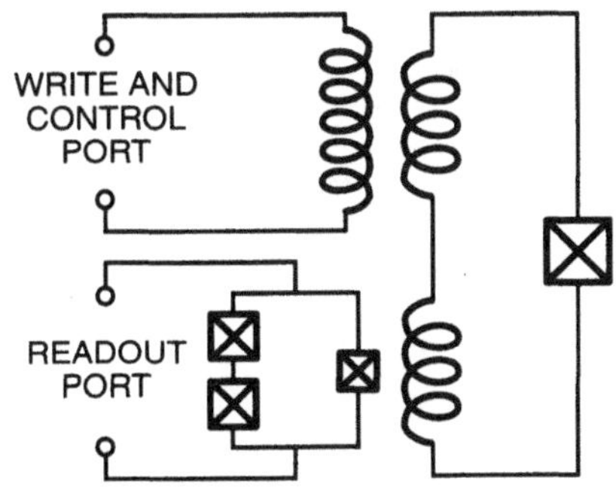

Fig. 10. Phase qubit implemented with a Josephson junction in a high-inductance superconducting loop biased with a flux sufficiently large that the phase across the junction sees a potential analogous to that found for the current-biased junction. The readout part of the circuit is an asymmetric hysteretic SQUID which is completely decoupled from the qubit in the OFF phase. Isolation of the qubit both from the readout and control port is obtained through impedance mismatch of transformers.

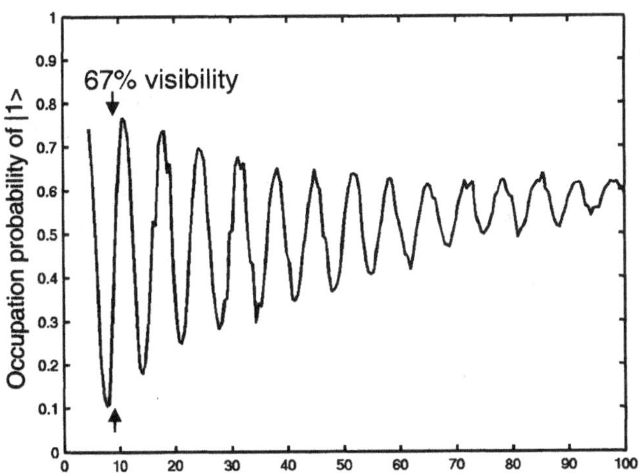

Fig. 11. Rabi oscillations observed for the qubit of Fig. 10.

Finally, a large impedance mismatch between the SQUID and the qubit is obtained by a transformer. The fidelity of such readout is remarkable: 95% has been demonstrated. In Fig. 11, we show the result of a measurement of Rabi oscillations with such qubit+readout.

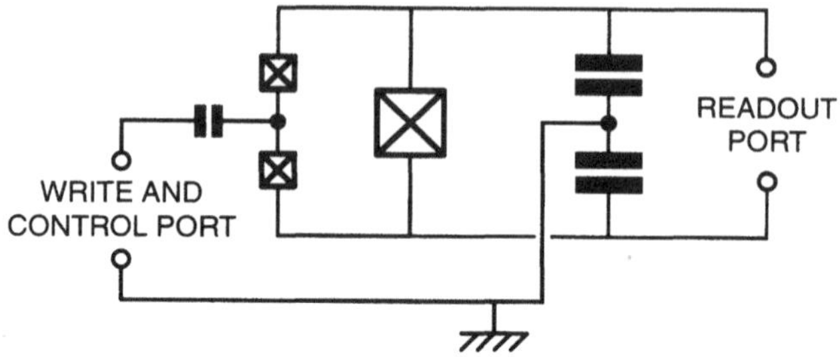

Fig. 12. "Quantronium" circuit consisting of a Cooper-pair box with a non-linear inductive readout. A Wheatstone bridge configuration decouples qubit and readout variables when readout is OFF. Impedance mismatch isolation is also provided by additional capacitance in parallel with readout junction.

8.4. Cooper-pair Box with Non-linear Inductive Readout: The "Quantronium" Circuit

The Cooper-pair box needs to be operated at its "sweet spot" (degeneracy point) where the transition frequency is to first order insensitive to offset charge fluctuations. The "Quantronium" circuit presented in Fig. 12 is a 3-junction bridge configuration with two small junctions defining a Cooper box island, and thus a charge-like qubit which is coupled capacitively to the write and control port (high-impedance port). There is also a large third junction, which provides a non-linear inductive coupling to the read port. When the read port current I is zero, and the flux through the qubit loop is zero, noise coming from the read port is decoupled from the qubit, provided that the two small junctions are identical both in critical current and capacitance. When I is non-zero, the junction bridge is out of balance and the state of the qubit influences the effective non-linear inductance seen from the read port. A further protection of the impedance mismatch type is obtained by a shunt capacitor across the large junction: at the resonance frequency of the non-linear resonator formed by the large junction and the external capacitance C, the differential mode of the circuit involved in the readout presents an impedance of the order of an ohm, a substantial decoupling from the $50\,\Omega$ transmission line carrying information to the amplifier stage. The readout protocol involves a DC pulse(22,30) or an RF pulse(31) stimulation of the readout mode. The response is bimodal, each mode corresponding to a state of the qubit. Although the theoretical fidelity of the DC readout can attain 95%, only a maximum of 40% has been obtained so far. The cause of this discrepancy is still under investigation.

In Fig. 13 we show the result of a Ramsey fringe experiment demonstrating that the coherence quality factor of the quantronium can reach

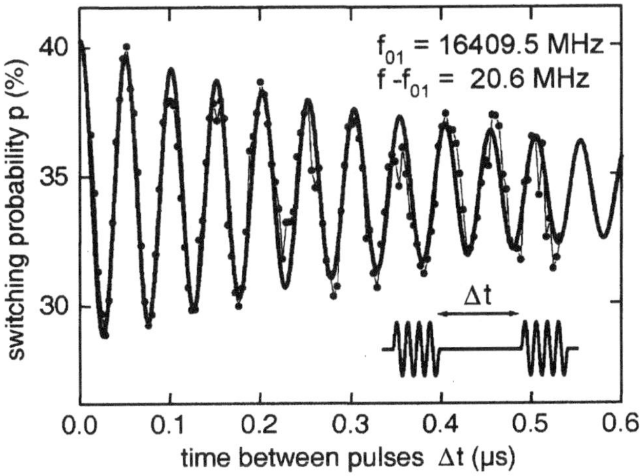

Fig. 13. Measurement of Ramsey fringes for the Quantronium. Two $\pi/2$ pulses separated by a variable delay are applied to the qubit before measurement. The frequency of the pulse is slightly detuned from the transition frequency to provide a stroboscopic measurement of the Larmor precession of the qubit.

25,000 at the sweet spot.[22] By studying the degradation of the qubit absorption line and of the Ramsey fringes as one moves away from the sweet spot, it has been possible to show that the residual decoherence is limited by offset charge noise and by flux noise.[32] In principle, the influence of these noises could be further reduced by a better optimization of the qubit design and parameters. In particular, the operation of the box can tolerate ratios of E_J/E_C around 4 where the sensitivity to offset charge is exponentially reduced and where the non-linearity is still of order 15%. The quantronium circuit has so far the best coherence quality factor. We believe this is due to the fact that critical current noise, one dominant intrinsic source of noise, affects this qubit far less than the others, relatively speaking, as can be deduced from the qubit hamiltonians of Sec. 6.

8.5. 3-Junction Flux Qubit with Built-in Readout

Figure 14 shows a third example of built-in readout, this time for a flux-like qubit. The qubit by itself involves three junctions in a loop, the larger two of the junctions playing the role of the loop inductance in the basic RF-SQUID.[33] The advantage of this configuration is to reduce the sensitivity of the qubit to external flux variations. The readout part of the circuit involves two other junctions forming a hysteretic DC-SQUID whose offset flux depends on the qubit flux state. The critical current of this DC-SQUID has been probed by a DC pulse, but an

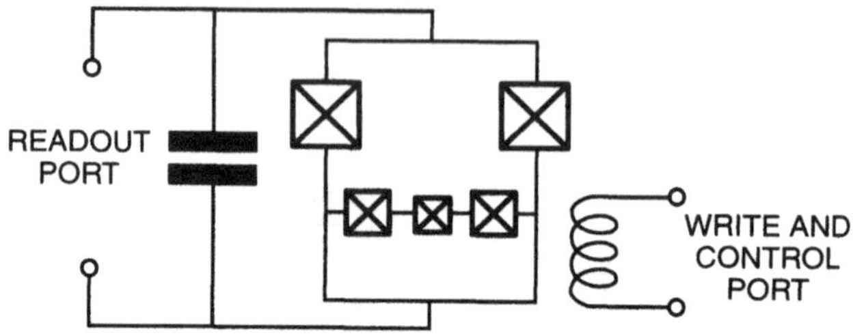

Fig. 14. Three-junction flux qubit with a non-linear inductive readout. The medium-size junctions play the role of an inductor. Bridge configuration for nulling out back-action of readout is also employed here, as well as impedance mismatch provided by additional capacitance.

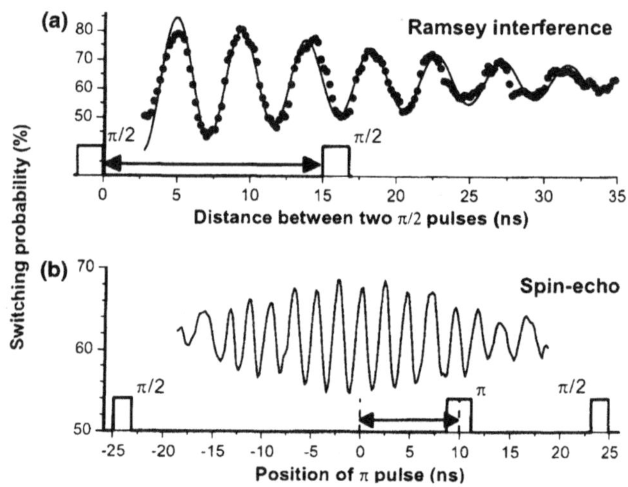

Fig. 15. Ramsey fringes obtained for qubit of Fig. 14.

RF pulse could be applied as in another flux readout. Similarly to the two previous cases, the readout states 1_c and 0_c, which here correspond to the DC-SQUID having switched or not, map very well the qubit states $|1\rangle$ and $|0\rangle$, with a fidelity better than 60%. Here also, a bridge technique orthogonalizes the readout mode, which is the common mode of the DC-SQUID, and the qubit mode, which is coupled to the loop of the DC-SQUID. External capacitors provide additional protection through impedance mismatch. Figure 15 shows Ramsey oscillations obtained with this system.

8.6. Too much On-chip Dissipation is Problematic: Do not Stir up the Dirt

All the circuits above include an on-chip amplification scheme producing high-level signals which can be read directly by high-temperature low-noise electronics. In the second and third examples, these signals lead to non-equilibrium quasi-particle excitations being produced in the near vicinity of the qubit junctions. An elegant experiment has recently demonstrated that the presence of these excitations increases the offset charge noise.[34] More generally, one can legitimately worry that large energy dissipation on the chip itself will lead to an increase of the noises discussed in Sec. 5.2. A broad class of new readout schemes addresses this question.[31,35,36] They are based on a purely dispersive measurement of a qubit susceptibility (capacitive or inductive). A probe signal is sent to the qubit. The signal is coupled to a qubit variable whose average value is identical in the two qubit states (for instance, in the capacitive susceptibility, the variable is the island charge in the charge qubit at the degeneracy point). However, the susceptibility, which is the derivative of the qubit variable with respect to the probe, differs from one qubit state to the other. The resulting state-dependent phase shift of the reflected signal is thus amplified by a linear low temperature amplifier and finally discriminated at high temperature against an adequately chosen threshold. In addition to being very thrifty in terms of energy being dissipated on chip, these new schemes also provide a further natural decoupling action: when the probe signal is off, the back-action of the amplifier is also completely shut off. Finally, the interrogation of the qubit in a frequency band excluding zero facilitates the design of very efficient filters.

9. COUPLING SUPERCONDUCTING QUBITS

A priori, three types of coupling scheme can be envisioned:

(a) In the first type, the transition frequency of the qubits are all equal and the coupling between any pair is switched on using one or several junctions as non-linear elements.[37,38]
(b) In the second type, the couplings are fixed, but the transition frequencies of a pair of qubits, originally detuned, are brought on resonance when the coupling between them needs to be turned on.[39–41]
(c) In the third type, which bears close resemblance to the methods used in NMR,[1] the couplings and the resonance frequencies of the qubits remain fixed, the qubits being always detuned. Being off-diagonal, the coupling elements have negligible action on the qubits. However, when a strong micro-wave field is applied to the target and control

qubits their resonant frequency with an appropriate amplitude, they become in “speaking terms” for the exchange of energy quanta and gate action can take place.[42]

So far only scheme (b) has been tested experimentally.

The advantage of schemes (b) and (c) is that they work with purely passive reactive elements like capacitors and inductors which should remain very stable as a function of time and which also should present very little high-frequency noise. In a way, we must design quantum integrated circuits in the manner that vacuum tube radios were designed in the 1950s: only six tubes were used for a complete heterodyne radio set, including the power supply. Nowadays several hundreds of transistors are used in a radio or any hi-fi system. In that ancient era of classical electronics, linear elements like capacitors, inductors or resistors were “free” because they were relatively reliable whereas tubes could break down easily. We have to follow a similar path in quantum integrated circuit, the reliability issues having become noise minimization issues.

10. CAN COHERENCE BE IMPROVED WITH BETTER MATERIALS?

Up to now, we have discussed how, given the power spectral densities of the noises ΔQ_{r}, ΔE_{C} and ΔE_{J}, we could design a qubit equipped with control, readout and coupling circuits. It is worthwhile to ask at this point if we could improve the material properties to improve the coherence of the qubit, assuming all other problems like noise in the control channels and the back-action of the readout have been solved. A model put forward by one of us (JMM) and collaborators shed some light on the direction one would follow to answer this question. The $1/f$ spectrum of the materials noises suggests that they all originate from 2-level fluctuators in the amorphous alumina tunnel layer of the junction itself, or its close vicinity. The substrate or the surface of the superconducting films are also suspect in the case of ΔQ_{r} and ΔE_{C} but their influence would be relatively weaker and we ignore them for simplicity. These two-level systems are supposed to be randomly distributed positional degrees of freedom ξ_i with effective spin-1/2 properties, for instance an impurity atom tunneling between two adjacent potential well. Each two-level system is in principle characterized by three parameters: the energy splitting $\hbar\omega_i$, and the two coefficients α_i and β_i of the Pauli matrix representation of $\xi_i = \alpha_i \sigma_{iz} + \beta_i \sigma_{ix}$ (z is here by definition the eigen-energy axis). The random nature of the problem leads us to suppose that α_i and β_i are both Gaussian random variables with the same standard deviation ρ_i.

By carrying a charge, the thermal and quantum motion of ξ_i can contribute to $\Delta Q_r = \sum_i q_i \xi_i$ and $\Delta E_C = \sum_i c_i \frac{\beta_i^2}{\omega_i} \sigma_{iz}$. Likewise, by modifying the transmission of a tunneling channel in its vicinity, the motion of ξ_i can contribute to $\Delta E_J = \sum_i g_i \xi_i$. We can further suppose that the quality of the material of the junction is simply characterized by a few numbers. The essential one is the density ν of the transition frequencies ω_i in frequency space and in real space, assuming a ω^{-1} distribution (this is necessary to explain the $1/f$ behavior) and a uniform spatial distribution on the surface of the junction. Recent experiments indicate that the parameter ν is of order $10^5\,\mu\text{m}^{-2}$ per decade. Then, assuming a value for ρ_i independent of frequency, only one coefficient is needed per noise, namely, the average modulation efficiency of each fluctuator. Such analysis provides a common language for describing various experiments probing the dependence of decoherence on the material of the junction. Once the influence of the junction fabrication parameters (oxydation pressure and temperature, impurity contents, and so on) on these noise intensities will be known, it will be possible to devise optimized fabrication procedures, in the same way perhaps as the $1/f$ noise in C-MOS transistors has been reduced by careful material studies.

11. CONCLUDING REMARKS AND PERSPECTIVES

The logical thread through this review of superconducting qubits has been the question "What is the best qubit design?". Because some crucial experimental data is still missing, we unfortunately, at present, cannot conclude by giving a definitive answer to this complex optimization problem.

Yet, a lot has already been achieved, and superconducting qubits are becoming serious competitors of trapped ions and atoms. The following properties of quantum circuits have been demonstrated:

(a) Coherence quality factors $Q_\varphi = T_\varphi \omega_{01}$ can attain at least 2×10^4.
(b) Readout and reset fidelity can be greater than 95%.
(c) All states on the Bloch sphere can be addressed.
(d) Spin echo techniques can null out low frequency drift of offset charges.
(e) Two qubits can be coupled and RF pulses can implement gate operation.
(f) A qubit can be fabricated using only optical lithography techniques.

The major problem we are facing is that these various results have not been obtained at the same time IN THE SAME CIRCUIT, although succesful design elements in one have often been incorporated into the next generation of others. The complete optimization of the single qubit+readout has not been achieved yet. However, we have presented in this review the elements of a systematic methodology resolving the various conflicts that are generated by all the different requirements. Our opinion is that, once noise sources are better characterized, an appropriate combination of all the known circuit design strategies for improving coherence, as well as the understanding of optimal tunnel layer growth conditions for lowering the intrinsic noise of Josephson junctions, should lead us to reach the 1-qubit and 2-qubit coherence levels needed for error correction.(45) Along the way, good medium term targets to test overall progress on the simultaneous fronts of qubit coherence, readout and gate coupling are the measurement of Bell 's inequality violation or the implementation of the Deutsch–Josza algorithm, both of which requiring the simultaneous satisfaction of properties (a)–(e).

ACKNOWLEDGMENTS

The authors have greatly benefited from discussions with I. Chuang, D. Esteve, S. Girvin, S. Lloyd, H. Mooij, R. Schoelkopf, I. Siddiqi, C. Urbina and D. Vion. They would like also to thank the participants of the Les Houches Summer School on Quantum Information Processing and Entanglement held in 2003 for useful exchanges. Finally, funding from ARDA/ARO and the Keck Fundation is gratefully acknowledged.

APPENDIX 1. QUANTUM CIRCUIT THEORY

The problem we are addressing in this section is, given a superconducting circuit made up of capacitors, inductors and Josephson junctions, how to systematically write its quantum hamiltonian, the generating function from which the quantum dynamics of the circuit can be obtained. This problem has been considered first by Yurke and Denker(46) in a seminal paper and analyzed in further details by Devoret.(47) We will only summarize here the results needed for this review.

The circuit is given as a set of branches, which can be capacitors, inductors or Josephson tunnel elements, connected at nodes. Several independent paths formed by a succession of branches can be found between nodes. The circuit can therefore contain one or several loops. It is important to note that a circuit has not one hamiltonian but many, each one depending on a particular representation. We are describing here one particular type of representation, which is usually well adapted to circuits containing Josephson junctions. Like in classical circuit theory, a set of independent current and voltages has to be found for a particular representation. We start by associating to each branch b of the circuit, the current i_{b} flowing through it and the voltage v_{b} across it (a convention has to be made first on the direction of the branches). Kirchhoff's laws impose relations among branch variables and some of them are redundant. The following procedure is used to eliminate redundant branches: one node of the circuit is first chosen as ground. Then from the ground, a loop-free set of branches called spanning tree is selected. The rule behind the selection of the spanning tree is the following: each node of the circuit must be linked to the ground by one and only one path belonging to the tree. In general, inductors (linear or non-linear) are preferred as branches of the tree but this is not necessary. Once the spanning tree is chosen (note that we still have many possibilities for this tree), we can associate to each node a "node voltage" v_{n} which is the algebraic sum of the voltages along the branches between ground and the node. The conjugate "node current" i_{n} is the algebraic sum of all currents flowing to the node through capacitors ONLY. The dynamical variables appearing in the hamiltonian of the circuit are the node fluxes and node charges defined as

$$\phi_{\mathrm{n}} = \int_{-\infty}^{t} v\,(t_1)\,\mathrm{d}t_1,$$

$$q_{\mathrm{n}} = \int_{-\infty}^{t} i\,(t_1)\,\mathrm{d}t_1.$$

Using Kirchhoff's laws, it is possible to express the flux and the charge of each branch as a linear combination of all the node fluxes and charges, respectively. In this inversion procedure, important physical parameters appear: the magnetic fluxes through the loops imposed by external static magnetic fields and the polarization charges imposed by charge bias sources.

If we now sum the energies of all branches of the circuit expressed in terms of node flux and charges, we will obtain the hamiltonian of the circuit corresponding to the representation associated with the particular spanning tree. In this hamiltonian, capacitor energies behave like kinetic terms while the inductor energies behave as potential terms. The hamiltonian of the LC circuit written in Sec. 2 is an elementary example of this procedure.

Once the hamiltonian is obtained it is easy get its quantum version by replacing all the node fluxes and charges by their quantum operator equivalent. The flux and charge of a node have a commutator given by $i\hbar$, like the position and momentum of a particle

$$\phi \rightarrow \hat{\phi},$$
$$q \rightarrow \hat{q},$$
$$\left[\hat{\phi}, \hat{q}\right] = i\hbar.$$

One can also show that the flux and charge operators corresponding to a branch share the same commutation relation. Note that for the special case of the Josephson element, the phase $\hat{\theta}$ and Cooper pair number $\hat{N}$, which are its dimensionless electric variables, have the property

$$\left[\hat{\theta}, \hat{N}\right] = i.$$

In the so-called charge basis, we have

$$\hat{N} = \sum_N N \left|N\right\rangle \left\langle N\right|,$$
$$\cos\hat{\theta} = \tfrac{1}{2}\sum_N \left(\left|N\right\rangle \left\langle N+1\right| + \left|N+\right\rangle \left\langle N\right|\right)$$

while in the so-called phase basis, we have

$$\hat{N} = \left|\theta\right\rangle \frac{\partial}{i\partial} \left\langle \theta\right|.$$

Note that since the Cooper pair number $\hat{N}$ is an operator with integer eigenvalues, its conjugate variable $\hat{\theta}$, has eigenvalues behaving like angles, i.e., they are defined only modulo 2π.

In this review, outside this appendix, we have dropped the hat on operators for simplicity.

APPENDIX 2. EIGENENERGIES AND EIGENFUNCTIONS OF THE COOPER PAIR BOX

From Appendix 1, it easy to see that the hamiltonian of the Cooper pair box leads to the Schrodinger equation

$$\left[E_{\mathrm{C}}\left(\frac{\partial}{i\,\partial\theta} - N_{\mathrm{g}}\right)^2 - E_{\mathrm{J}}\cos\theta\right]\Psi_k(\theta) = E_k\Psi_k(\theta).$$

The functions $\Psi_k(\theta)\,\mathrm{e}^{-\mathrm{i}N_\mathrm{g}}$ and energies E_k are solutions of the Mathieu equation and can be found with arbitrary precision for all values of the parameters N_g and $E_\mathrm{J}/E_\mathrm{C}$.[48] For instance, using the program Mathematica, we find

$$E_k = E_\mathrm{C}\mathcal{M}_\mathrm{A}\left[k+1-(k+1)\,\mathrm{mod}\,2+2N_\mathrm{g}(-1)^k, -2E_\mathrm{J}/E_\mathrm{C}\right],$$

$$\Psi_k(\theta) = \frac{\mathrm{e}^{\mathrm{i}N_\mathrm{g}\theta}}{\sqrt{2\pi}}\left\{\mathcal{M}_\mathrm{C}\left[\frac{4E_k}{E_\mathrm{C}}, \frac{-2E_\mathrm{J}}{E_\mathrm{C}}, \frac{\theta}{2}\right]+i(-1)^{k+1}\mathcal{M}_\mathrm{S}\left[\frac{4E_k}{E_\mathrm{C}}, \frac{-2E_\mathrm{J}}{E_\mathrm{C}}, \frac{\theta}{2}\right]\right\},$$

where $\mathcal{M}_\mathrm{A}(r,q)=$ `MathieuCharacteristicA[r,q]`,
$\mathcal{M}_\mathrm{C}(a,q,z)=$ `MathieuC[a,q,z]`,
$\mathcal{M}_\mathrm{S}(a,q,z)=$ `MathieuS[a,q,z]`.

APPENDIX 3. RELAXATION AND DECOHERENCE RATES FOR A QUBIT

Definition of the Rates

We start by introducing the spin eigenreference frame $\hat{z}$, $\hat{x}$ and $\hat{y}$ consisting of the unit vector along the eigenaxis and the associated orthogonal unit vectors ($\hat{x}$ is in the XZ plane). For instance, for the Cooper pair box, we find that $\hat{z}=\cos\alpha\hat{Z}+\sin\alpha\hat{X}$, with $\tan\alpha=2E_\mathrm{C}\left(N_\mathrm{g}-1/2\right)/E_\mathrm{J}$, while $\hat{x}=-\sin\alpha\hat{Z}+\cos\alpha\hat{X}$.

Starting with $\vec{S}$ pointing along $\hat{x}$ at time $t=0$, the dynamics of the Bloch vector in absence of relaxation or decoherence is

$$\vec{S}_0(t)=\cos(\omega_{01})\hat{x}+\sin(\omega_{01})\hat{y}$$

In presence of relaxation and decoherence, the Bloch vector will deviate from $\vec{S}_0(t)$ and will reach eventually the equilibrium value $S_z^{\mathrm{eq}}\hat{z}$, where $S_z^{\mathrm{eq}}=\tanh\,(\hbar\omega_{01}/2k_\mathrm{B}T)$.

We define the relaxation and decoherence rates as

$$\Gamma_1 = \lim_{t\to\infty}\frac{\ln\left\langle S_z(t)-S_z^{\mathrm{eq}}\right\rangle}{t},$$

$$\Gamma_\phi = \lim_{t\to\infty}\frac{\ln\left[\frac{\left\langle\vec{S}(t).\vec{S}_0(t)\right\rangle}{\left|\vec{S}(t)-S_z^{\mathrm{eq}}\hat{z}\right|}\right]}{t}.$$

Note that these rates have both a useful and rigorous meaning only if the evolution of the components of the average Bloch vector follows, after a negligibly short settling time, an exponential decay. The Γ_1 and Γ_ϕ rates are related to the NMR spin relaxation times T_1 and T_2[49] by

$$T_1 = \Gamma_1^{-1},$$
$$T_2 = \left(\Gamma_\phi + \Gamma_1/2\right)^{-1}.$$

The T_2 time can be seen as the net decay time of quantum information, including the influence of both relaxation and dephasing processes. In our discussion of superconducting qubits, we must separate the contribution of the two type of processes since their physical origin is in general very different and cannot rely on the T_2 time alone.

Expressions for the Rates

The relaxation process can be seen as resulting from unwanted transitions between the two eigenstate of the qubit induced by fluctuations in the effective fields along the x and y axes. Introducing the power spectral density of this field, one can demonstrate from Fermi's Golden Rule that, for perturbative fluctuations,

$$\Gamma_1 = \frac{S_x(\omega_{01}) + S_y(\omega_{01})}{\hbar^2}.$$

Taking the case of the Cooper-pair box as an example, we find that $S_y(\omega_{01}) = 0$ and that

$$S_x(\omega) = \int_{-\infty}^{+\infty} dt\, e^{i\omega t} \langle A(t)A(0)\rangle + \langle B(t)B(0)\rangle,$$

where

$$A(t) = \frac{\Delta E_J(t) E_{el}}{2\sqrt{E_J^2 + E_{el}^2}},$$
$$B(t) = \frac{E_J \Delta E_{el}(t)}{2\sqrt{E_J^2 + E_{el}^2}},$$
$$E_{el} = 2E_C(N_g - 1/2).$$

Since the fluctuations $\Delta E_{el}(t)$ can be related to the impedance of the environment of the box,[19,21,50] an order of magnitude estimate of the relaxation rate can be performed, and is in rough agreement with observations.[22,51]

The decoherence process, on the other hand, is induced by fluctuations in the effective field along the eigenaxis z. If these fluctuations are Gaussian, with a white noise spectral density up to frequencies of order several Γ_ϕ (which is often not the case because of the presence of 1/f noise) we have

$$\Gamma_\phi = \frac{S_z(\omega \simeq 0)}{\hbar^2}.$$

In presence of a low frequency noise with an 1/f behavior, the formula is more complicated.[52] If the environment producing the low frequency noise consists of many degrees of freedom, each of which is very weakly coupled to the qubit, then one is in presence of classical dephasing which, if slow enough, can in principle be fought using echo techniques. If, one the other hand, only a few degrees of freedom like magnetic spins or glassy two-level systems are dominating the low frequency dynamics, dephasing is quantum and not correctable, unless the transition frequencies of these few perturbing degrees of freedom is itself very stable.

REFERENCES

1. M. A. Nielsen and I. L. Chuang, *Quantum Computation and Quantum Information* (Cambridge, 2000).
2. M. Tinkham, *Introduction to Superconductivity* (Krieger, Malabar, 1985).
3. J. M. Martinis, M. H. Devoret, J. Clarke, *Phys. Rev. Lett.* **55**, 1543 (1985); M. H. Devoret, J. M. Martinis, J. Clarke, *Phys. Rev. Lett.* **55**, 1908 (1985); J. M. Martinis, M. H. Devoret and J. Clarke, *Phys. Rev.* **35**, 4682 (1987).
4. J. M. Martinis and M. Nahum, *Phys Rev.* B **48**, 18316 (1993).
5. B. D. Josephson, in *Superconductivity*, R. D. Parks (ed.) (Marcel Dekker, New York, 1969).
6. K. K. Likharev, *Dynamics of Josephson Junctions and Circuits* (Gordon and Breach, New York, 1986).
7. I. Giaever, *Phys. Rev. Lett.* **5**, 147, 464 (1960).
8. A. O. Caldeira and A. J. Leggett, *Ann. Phys. (N.Y.)* **149**, 347 (1983); A. J. Leggett, *J. Phys. CM* **14**, R415 (2002).
9. D. P. DiVincenzo, arXiv:quant-ph/0002077.
10. R. P. Feynman, *Lectures on Physics*, Vol. 2, Chap. 23, (Addison-Wesley, Reading, 1964).
11. D. C. Mattis and J. Bardeen, *Phys. Rev.* **111**, 412 (1958).
12. P. G. de Gennes, *Superconductivity of Metals and Alloys* (Benjamin, New York, 1966).
13. J. M. Raimond, M. Brune, and S. Haroche, *Rev. Mod. Phys.* **73**, 565 (2001).
14. J. M. Martinis and K. Osborne, in *Quantum Information and Entanglement*, eds. J. M. Raimond, D. Esteve, and J. Dalibard, Les Houches Summer School Series, arXiv:cond-mat/0402430.
15. J. Clarke, *Proc. IEEE* **77**, 1208 (1989).
16. D. J. Van Harlingen, B. L. T. Plourde, T. L. Robertson, P. A. Reichardt, and John Clarke, preprint.

17. R. W. Simmonds, K. M. Lang, D. A. Hite, D. P. Pappas, and J. M. Martinis, accepted for publication in *Phys. Rev. Lett.*
18. M. Büttiker, *Phys. Rev.* B **36**, 3548 (1987).
19. V. Bouchiat, D. Vion, P. Joyez, D. Esteve, M. H. Devoret, *Physica Scripta T* **76**, 165 (1998).
20. Y. Nakamura , Yu. A. Pashkin, and J. S. Tsai, *Nature* **398**, 786 (1999).
21. Yu. Makhlin, G. Schön, and A. Shnirman, *Rev. Mod. Phys.* **73**, 357 (2001).
22. D. Vion, A. Aassime, A. Cottet, P. Joyez, H. Pothier, C. Urbina, D. Esteve, and M. H. Devoret, *Science* **296**, 286 (2002).
23. A. Barone and G. Paternò, *Physics and Applications of the Josephson Effect* (Wiley, New York, 1992).
24. S. Han, R. Rouse, and J. E. Lukens, *Phys. Rev. Lett.* **84**, 1300 (2000); J. R. Friedman, V. Patel, W. Chen, S. K. Tolpygo, and J. E. Lukens, *Nature* **406**, 43 (2000).
25. J. E. Mooij, T. P. Orlando, L. Levitov, Lin Tian, C. H. van der Wal, and S. Lloyd, *Science* **285**, 1036 (1999); C. H. van der Wal, A. C. J. ter Haar, F. K. Wilhem, R. N. Schouten, C. Harmans, T. P. Orlando, S. Loyd, and J. E. Mooij, *Science* **290**, 773 (2000).
26. J. M. Martinis, S. Nam, J. Aumentado, and C. Urbina, *Phys. Rev. Lett.* **89**, 117901 (2002).
27. M. Steffen, J. Martinis, and I. L. Chuang, *PRB* **68**, 224518 (2003).
28. M. H. Devoret and R. J. Schoelkopf, *Nature* **406**, 1039 (2002).
29. A. N. Korotkov and D. V. Averin, *Phys. Rev. B* **64**, 165310 (2001).
30. A. Cottet, D. Vion, A. Aassime, P. Joyez, D. Esteve, and M. H. Devoret, *Physica C* **367**, 197 (2002).
31. I. Siddiqi, R. Vijay, F. Pierre, C. M. Wilson, M. Metcalfe, C. Rigetti, L. Frunzio, and M. H. Devoret, cond-mat/0312623, submitted to *Phys. Rev. Lett.*
32. D. Vion, A. Aassime, A. Cottet, P. Joyez, H. Pothier, C. Urbina, D. Esteve, and M. H. Devoret, *Fortschritte der Physik* **51**, 462 (2003).
33. I. Chiorescu, Y. Nakamura, C. J. P. M. Harmans, and J. E. Mooij, *Science* **299**, 1869 (2003).
34. J. Mannik and J. E. Lukens, *Phys. Rev. Lett.* **92**, 057004 (2004).
35. A. Wallraff, D. I. Schuster, A. Blais, L. Frunzio, R.-S. Huang, J. Majer, S. Kumar, S. M. Girvin, R. J. Schoelkopf, cond-mat/0408367, accepted by *Nature*.
36. A. Lupascu, C. J. M. Verwijs, R. N. Schouten, C. J. P. M. Harmans, J. E. Mooij, cond-mat/0311510, submitted to *Phys. Rev. Lett.*
37. Variable electrostatic transformer: controllable coupling of two charge qubits, D.V. Averin, C. Bruder, *Phys. Rev. Lett.* **91**, 057003 (2003).
38. A. Blais, A. Maassen van den Brink, A. M. Zagoskin, *Phys. Rev. Lett.* **90**, 127901 (2003)
39. A. Pashkin Yu, T. Yamamoto, O. Astafiev, Y. Nakamura, D. V. Averin, and J. S. Tsai, *Nature* **421** (2003).
40. J. B. Majer, *Superconducting Quantum Circuits*, PhD Thesis, TU Delft, (2002); J. B. Majer, F. G. Paauw, A. C. J. ter Haar, C. P. J. Harmans, and J. E. Mooij, arXiv:cond-mat/0308192.
41. A. J. Berkley, H. Xu, R. C. Ramos, M. A. Gubrud, F. W. Strauch, P. R. Johnson, J. R. Anderson, A. J. Dragt, C. J. Lobb, and F. C. Wellstood, *Science* **300**, 1548 (2003).
42. C. Rigetti and M. Devoret, unpublished.
43. Y. Nakamura, A. Pashkin Yu, and J. S. Tsai, *Phys. Rev. Lett.* **88**, 047901 (2002).
44. D. Vion, A. Aassime, A. Cottet, P. Joyez, H. Pothier, C. Urbina, D. Esteve, and M. H. Devoret, *Forts. der Physik* **51**, 462 (2003); E. Collin, G. Ithier, A. Aassime, P. Joyez, D. Vion and D. Esteve, submitted.
45. J. Preskill, *J. Proc. R. Soc. London Ser. A* **454**, 385 (1998).

46. B. Yurke and J. S. Denker, *Phys. Rev. A* **29**, 1419 (1984).
47. M. H. Devoret in *"Quantum Fluctuations"* , S. Reynaud, E. Giacobino, J. Zinn-Justin (eds.) (Elsevier, Amsterdam, 1996), p. 351.
48. A. Cottet, *Implementation of a quantum bit in a superconducting circuit,* PhD Thesis, Université Paris 6, 2002.
49. A. Abragam, *Principles of Nuclear Magnetic Resonance* (Oxford University Press, Oxford, 1985).
50. R. J. Schoelkopf, A. A. Clerk, S. M. Girvin, K. W. Lehnert, and M. H. Devoret. arXiv:cond-mat/0210247.
51. K. W. Lehnert, K. Bladh, L. F. Spietz, D. Gunnarsson, D. I. Schuster, P. Delsing, and R. J. Schoelkopf, *Phys. Rev.Lett.* **90**, 027002 (2002).
52. J. M. Martinis, S. Nam, J. Aumentado, K. M. Lang, and C. Urbina, *Phys. Rev. B* **67**, 462 (2003).

Towards Scalable Linear-Optical Quantum Computers

J. P. Dowling,[1,5] J. D. Franson,[2] H. Lee,[1,4] and G. J. Milburn[3]

Scalable quantum computation with linear optics was considered to be impossible due to the lack of efficient two-qubit logic gates, despite the ease of implementation of one-qubit gates. Two-qubit gates necessarily need a non-linear interaction between the two photons, and the efficiency of this non-linear interaction is typically very small in bulk materials. However, it has recently been shown that this barrier can be circumvented with effective non-linearities produced by projective measurements, and with this work linear-optical quantum computing becomes a new avenue towards scalable quantum computation. We review several issues concerning the principles and requirements of this scheme.

KEY WORDS: Linear optics, Logic gates, Single photon, Quantum memory, Quantum repeater.

PACS: 03.67.Lx, 03.67.Pp, 42.50.Dv, 42.65.Lm.

1. PRINCIPLES

There are three key principles in the Knill, Laflamme, and Milburn (KLM) proposal[(1)] for efficient and scalable linear-optical quantum information processing (QIP):

(1) Conditional non-linear gates for two photon states.
(2) Teleportation to achieve high efficiency.
(3) Error correction to achieve scalability.

[1]Quantum Computing Technologies Group, Section 367, Jet Propulsion Laboratory, MS 126-347 and California Institute of Technology, 4800 Oak Grove Drive, CA 91109, USA.

[2]Applied Physics Laboratory, Johns Hopkins University, Laurel, MD 20723, USA.

[3]Centre for Quantum Computer Technology, University of Queensland, QLD 4072, Australia.

[4]To whom correspondence should be addressed.

[5]Department of Physics, Louisiana State University, Baton Rouge, LA 70803, USA.

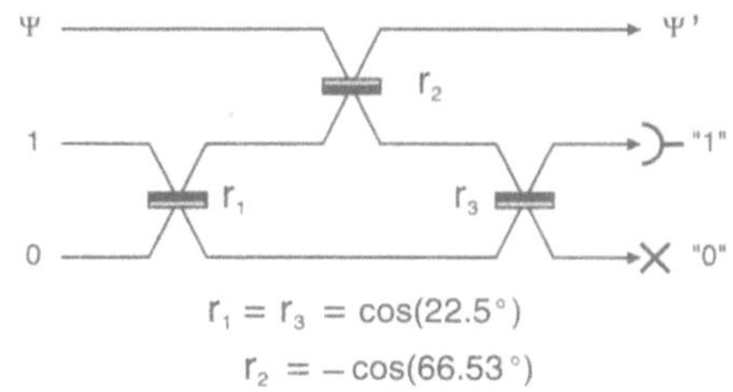

Fig. 1. The conditional non-linear sign (NS) gate.

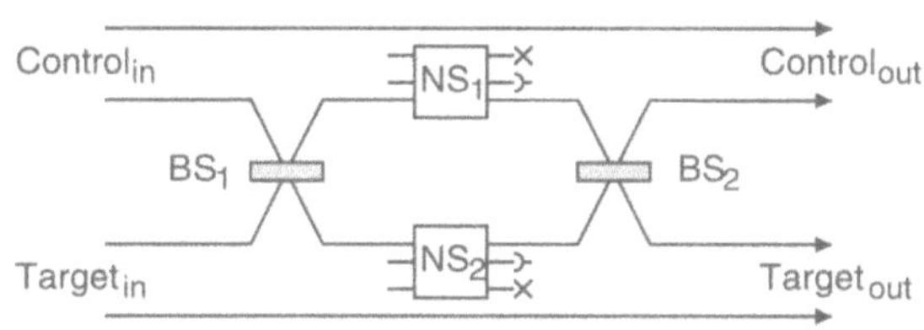

Fig. 2. Controlled-σ_z gate with dual-rail logic and two NS gates.

Conditional non-linear gates are based on the non-unitary state change due to measurement. The gate works with some probability, but correct functioning is heralded by the measurement result. We seek to implement a non-linear transformation, the so-called non-linear sign (NS) gate, on an arbitrary two-photon state of a single mode field:

$$|\psi\rangle = \alpha_0|0\rangle_1 + \alpha_1|1\rangle_1 + \alpha_2|2\rangle_1 \rightarrow |\psi'\rangle = \alpha_0|0\rangle_1 + \alpha_1|1\rangle_1 - \alpha_2|2\rangle_1. \quad (1)$$

This is done using the linear-optical network shown in Fig. 1. The signal state is first combined with two ancilla modes, one in a single-photon state and one in the vacuum. At the end of the optical processing photon counting is done on the ancilla modes. If the number of photons is unchanged from the input, the desired transformed state exits the signal mode port. This will happen with a probability of 0.25.

In order to use this result to implement QIP we code the logical states as physical qubits using one photon in one of two modes: $|0\rangle_L = |1\rangle_1 \otimes |0\rangle_2$, $|1\rangle_L = |0\rangle_1 \otimes |1\rangle_2$. Single-qubit gates are then implemented by a beam splitter. A two qubit gate, the conditional sign-flip gate, can then be implemented using the Hong-Ou-Mandel (HOM) interference effect to first convert two single modes, each with one photon, into an appropriate entangled two-photon state (Fig. 2). Such a gate uses two NS gates and thus succeeds with probability of 0.125. A general formalism for calculating the effective photon non-linearities generated by such conditional measurement schemes in linear optics can be found in Ref. 2

We have also shown that probabilistic quantum logic operations can be performed using polarization-encoded qubits,(3) as illustrated by the controlled-NOT gate shown in Fig. 3. This device consists of two polarizing beam splitters and two polarization-sensitive detectors, along with a pair of entangled ancilla photons.

The correct controlled-NOT logic operation will have been performed whenever one and only one photon is detected in each of the two detectors, which occurs with a probability of 0.25. Feed-forward control(4) must also be applied, depending on what polarization states were measured. From an experimental perspective, this approach has the advantage of being relatively simple and insensitive to phase drifts.

A sequence of such probabilistic gates is of course not scalable. However, the Gottesman and Chuang protocol for implementing gates via teleportation(5) can be used to fix this. Implementation of the gate then reduces to preparing the appropriate entangled-state resource. That can be done off-line by using conditional gates, and only when success is achieved is the teleportation gate completed. Using a resource with n photons in $2n$ modes decreases the gate failure probability as n^{-1} or even as n^{-2}.(6) Gates can thus be implemented efficiently. The controlled-σ_z (CZ) gate is now achieved with success probability $n^2/(n+1)^2$, denoted as $CZ_{n^2/(n+1)^2}$.

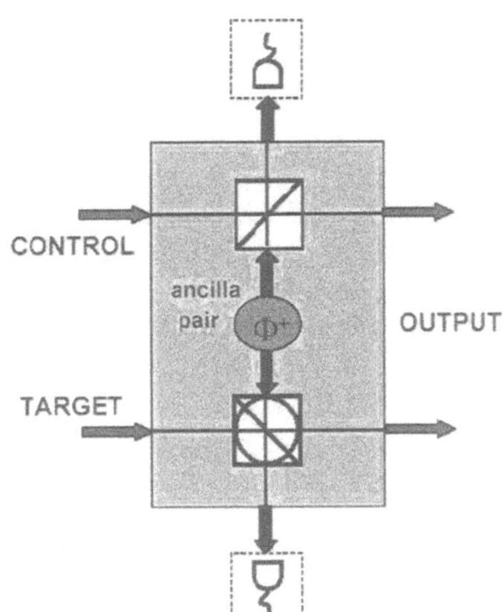

Fig. 3. Implementation of a probabilistic controlled-NOT gate using polarization-encoded qubits.

When a teleportation gate fails it does so by making a measurement of an incoming qubit. This is always heralded and can be fixed using detected-measurement codes. This enables the scheme to be scalable (i.e. fault-tolerant) provided the intrinsic error probability is <0.5,[1] but at the expense of very complicated, multi-mode, entangled, resource states for teleportation. The other major source of error is photon loss. In principle, this can also be corrected using teleportation gates. Scalability requires that loss probability per gate be <0.01.[7] Not detecting a photon is equivalent to loss. While there are proposals for single-photon detectors that have quantum efficiency higher than 99%, using cycling transitions in atomic vapors,[8,9] currently the best known value is around 93%.[10]

Three experiments have implemented conditional two-qubit gates: Pittman et al.;[11] O'Brien et al.;[12] and Sanaka et al.[13] The first experiment is based on the Pittman and Franson's polarization-encoded scheme.[11] The second is based on a simplification of the KLM-NS gate that requires only two photons.[12] The last is a full four photon version of KLM.[13] However, all experiments so far only work in the coincidence basis. This means that successful implementations correspond to two- or four-fold coincidence counts. However, no light leaves the device as all photons are detected.

Experimental results obtained from a CNOT gate in Ref. 11 are shown in Fig. 4. Here, a single ancilla photon was used, which restricts the operation of the device to the case in which a single photon is detected in each output port (the so-called coincidence basis). This was a three-photon experiment in which two of the single photons were obtained using parametric down-conversion while the third photon was obtained by attenuating the pump laser beam. Optical fibres were used instead of free-space components in order to reduce errors due to mode mismatch. The fidelity of the output qubits was limited in this case by the degree of indistinguishability of the three photons. Experimental demonstrations of several other simple quantum logic gates have also been performed, including a quantum parity check[14] and a quantum encoder.[15]

2. REQUIREMENTS

There are four major technical requirements that must be met:

(1) Single photon sources
(2) Number-discriminating photon detectors
(3) Feed-forward control and quantum memory
(4) Design and implementation of the very complex quantum circuit architectures.

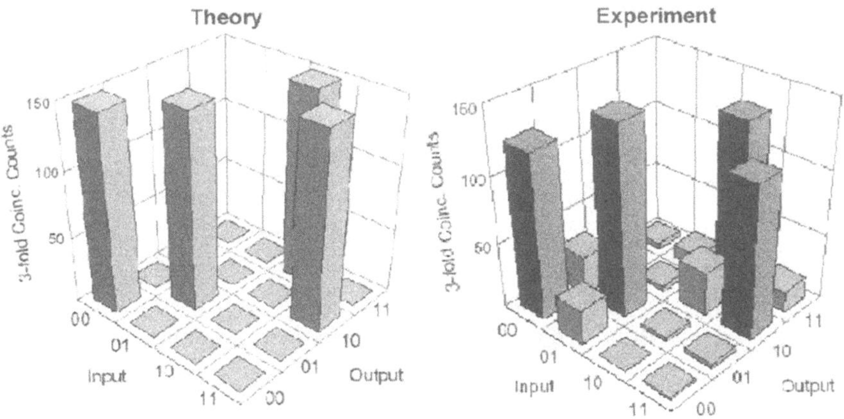

Fig. 4. Implementation of a probabilistic CNOT gate using polarization-encoded qubits.[11]

The required ideal single-photon sources are transform-limited pulses with one and only one photon per pulse. In practice, this means that one must be able to exhibit HOM interference between photons from different pulses. The required single-photon detectors must be able to detect a single photon with efficiency >0.99 and discriminate between 0, 1, and 2 photon counts.[16] Although two-photon interference visibilities >99% have been achieved,[11] it may be necessary to achieve even higher visibilities for quantum computing applications.

One approach to implementing such a single-photon source is illustrated in Fig. 5. A pulsed laser beam generates pairs of photons in a parametric down-conversion crystal. Detection of one member of a pair signals the presence of the other member of the pair, which is then switched into an optical storage loop. The single photon can then be switched out of the storage loop when needed.[18] Although this kind of approach cannot produce photons at arbitrary times, it can produce them at periodic time intervals that could be synchronized with the cycle time of a quantum computer. A prototype experiment of this kind demonstrated the ability to store and retrieve single photons in this way, but its performance at the time was limited by losses in the optical switch.

The ability to switch a single photon into an optical storage loop and then retrieve it when needed can also be used to implement a quantum memory for single photons. This application is more demanding than the single-photon source described above, since the polarization state of the photons must be maintained in order to preserve the value of the qubits. A prototype experiment of this kind has also been performed, where the primary limitation was once again the losses in the optical switch.[19]

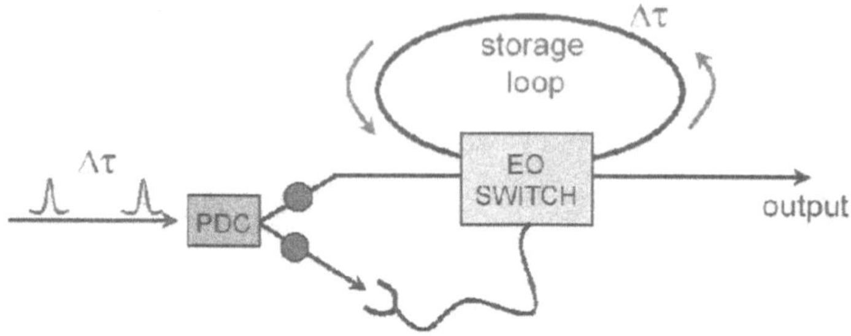

Fig. 5. Single-photon source using parametric down-conversion and an optical storage loop. Similar storage techniques can also be used to implement a quantum memory device for single photons.

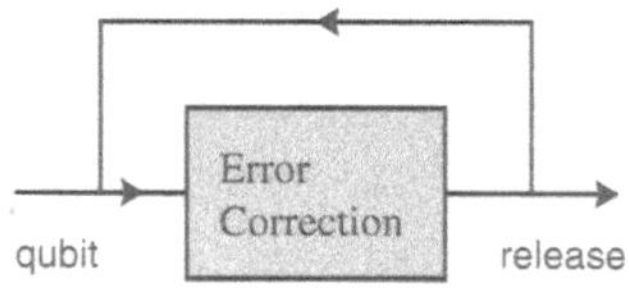

Fig. 6. A cyclic quantum memory based on quantum error correction.

Furthermore, the ability to perform quantum logic operations using linear elements raises the possibility of using quantum error correction techniques to extend the coherent storage time of the quantum memory described above, see Fig. 6. The primary source of error is expected to be photon loss, which can be corrected using a simple four-qubit encoding scheme(20) as illustrated in Fig. 7. Provided that the errors in the logic gates and storage loops are sufficiently small, techniques of this kind can be used to store photonic qubits for an indefinitely long period of time.

An essential component in this kind of quantum memory is the single-photon quantum non-demolition (QND) measurement device.(21) Again, a simple way to perform a single-photon QND measurement is provided by quantum teleportation techniques. If the input state is in a arbitrary superposition of zero and one photon, with a fixed polarization, the detector coincidence in a Bell-state measurement signals the present of a single photon in the input and output states.(22)

Similar techniques can also be used to compensate for the photon loss in optical fibre transmission lines, which would allow the development of a quantum repeater. A quantum repeater is a device for achieving remote, shared entanglement by using quantum purification and swapping protocols.(23) A simple protocol for optical quantum repeaters based on linear optical elements and an entangled-photon source has been developed.(24) On the other hand, utilizing quantum error correction, one can relay an

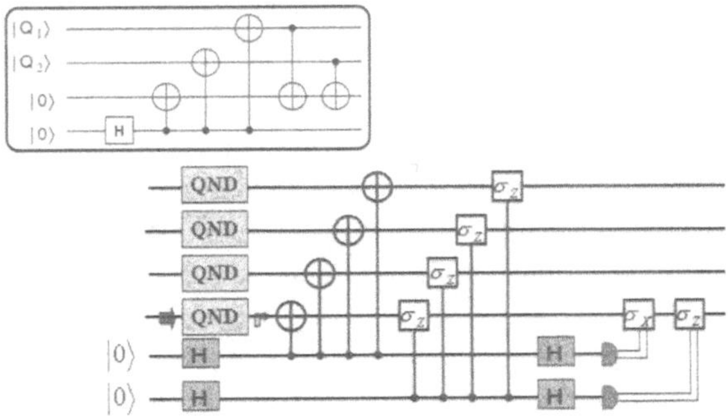

Fig. 7. Quantum error-correction code that recovers photon loss using two ancilla photons. The QND box represents a single-photon quantum non-demolition measurement device. The inset shows the two-to-four qubit encoding.

unknown quantum state with high fidelity down a quantum channel. This relay device we call a quantum transponder, and it has direct applications to quantum repeater and memory applications.

The fact that the LOQC architecture employs a non-deterministic protocol implies that large number of ancilla photons are required to make the scheme nearly deterministic. The implementation of quantum circuits then becomes more demanding with these increased number of resources. For example, implementing the error-correction circuit in Fig. 7 at >80% success probability requires more than 300 ancilla photons for each two-qubit gate and the same number of single-photon detections at the extremely high quantum efficiency.[20] Using a concatenation scheme to manage a single two-qubit gate with 95% success probability requires about 300 successful $CZ_{9/16}$ gate operations.[1] Therefore, development of techniques for efficient simplification of quantum circuits is an important task, as is the ability to fabricate these circuits on an opto-electonic chip.

On the other hand, QIP schemes implemented in unconventional ways, such as cluster state quantum computing,[25] may be utilized to reduce the number of resources. It has recently been shown that the required number of optical elements and resources can be reduced by an order of magnitude from the original KLM scheme by using multi photon-linked states,[26] and even more reduction has been found by using the cluster-state approach.[27] Furthermore, it may be possible to avoid the need for large number of nested interferometers and entangled ancillæ by using hybrid approaches that combine linear optics techniques with some amount of non-linearity.[28]

3. CONCLUSION

Quantum mechanics enables some exponentially more efficient algorithms than those that can be implemented on a classical computer. This discovery has led to the explosive growth of the field of quantum computation. Many physical systems have been suggested for building a quantum computer, but the final architecture is still to be determined. These systems include ion traps, non-linear optical systems, quantum dots, and superconducting circuits, among others.[(29)] In linear optical quantum computing the desired non-linearities come from projective measurements. Projective measurements over some part of the quantum system simply project the rest of the system into a desired quantum state. Additional photons, known as ancillæ, are mixed with the inputs to the logic devices using beam splitters, while single-photon detectors are used to make measurements on the ancilla photons after the interaction. The non-linear nature of single-photon detection and the quantum measurement process then project out the desired logical output. Therefore, although logic operations are inherently non-linear, our approach uses only simple linear optical elements, such as beam splitters and phase shifters. Building a quantum computer will be a major challenge for a future quantum technology; requiring the ability to manipulate quantum-entangled states of large number of sub components. Systematic development of each component of preparation, control, and measurement will facilitate the task of building a quantum computer.

ACKNOWLEDGMENTS

Part of this work was carried out at the Jet Propulsion Laboratory, California Institute of Technology, under a contract with the National Aeronautics and Space Administration. Authors wish to acknowledge support from the National Security Agency, the Advanced Research and Development Activity, the Defense Advanced Research Projects Agency, the National Reconnaissance Office, the Office of Naval Research, the Army Research Office, the IR&D funding, and the NASA Intelligent Systems Program.

REFERENCES

1. E. Knill, R. Laflamme, and G. J. Milburn, *Nature* **409**, 46 (2001).
2. G. G. Lapaire, P. Kok, J. P. Dowling, and J. E. Sipe, *Phys. Rev. A* **68**, 042314 (2003).

3. T. B. Pittman, B. C. Jacobs, and J. D. Franson, *Phys. Rev. A* **64**, 062311 (2001).
4. T. B. Pittman, B. C. Jacobs, and J. D. Franson, *Phys. Rev. A* **66**, 052305 (2002).
5. D. Gottesman and I. L. Chuang, *Nature* **402**, 390 (1999).
6. J. D. Franson *et al.*, *Phys. Rev. Lett.* **89**, 137901 (2002).
7. E. Knill, R. Laflamme, and G. J. Milburn, quant-ph/0006120 (2000).
8. D. F. V. James and P. G. Kwiat, *Phys. Rev. Lett.* **89**, 183601 (2002).
9. A. Imamoglu, *Phys. Rev. Lett.* **89**, 163602 (2002).
10. E. Waks, *et al.*, *IEEE J. Sel. Top. Quant. Ele.* **9**, 1502 (2003).
11. T. B. Pittman, M. J. Fitch, B. C. Jacobs, and J. D. Franson, *Phys. Rev. A* **68**, 032316 (2003).
12. J. L. O'Brien *et al.*, *Nature* **426**, 264 (2003).
13. K. Sanaka *et al.*, *Phys. Rev. Lett.* **92**, 017902 (2004).
14. T. B. Pittman, B. C. Jacobs, and J. D. Franson, *Phys. Rev. Lett.* **88**, 257902 (2002).
15. T. B. Pittman, B. C. Jacobs, and J. D. Franson, *Phys. Rev. Lett. A* **69**, 042306 (2004).
16. D. Achilles *et al.* (in press) *J. Mod. Opt.* quant-ph/0310183 (2003); H. Lee *et al.*, *J. Mod. Opt.*, quant-ph/0310161 (2003).
17. T. B. Pittman and J. D. Franson, *Phys. Rev. Lett.* **90**, 240401 (2003).
18. T. B. Pittman, B. C. Jacobs, and J. D. Franson, *Phys. Rev. A* **66**, 042303 (2002).
19. T. B. Pittman and J. D. Franson, *Phys. Rev. A* **66**, 062302 (2002).
20. R. M. Gingrich *et al.*, *Phys. Rev. Lett.* **91**, 217901 (2003).
21. G. Nogues *et al.*, *Nature* **400**, 239 (1999).
22. P. Kok, H. Lee, and J. P. Dowling, *Phys. Rev. A* **66**, 063814 (2002).
23. H.-J. Briegel, W. Dür, J. I. Cirac, and P. Zoller, *Phys. Rev. Lett.* **81**, 5932 (1998).
24. P. Kok, C. P. Williams, and J. P. Dowling, *Phys. Rev. A* **68**, 022301 (2003).
25. R. Raussendorf and H.-J. Briegel, *Phys. Rev. Lett.* **86**, 5188 (2001).
26. N. Yoran and B. Reznik, *Phys. Rev. Lett.* **91**, 037903 (2003).
27. M. Nielsen, quant-ph/0402005 (2004).
28. J. D. Franson, T. B. Pittman, and B. C. Jacobs, quant-ph/0401133 (2004).
29. J. P. Dowling and G. J. Milburn, *Phil. Trans. R. Soc. Lond. A* **361**, 1655 (2003).

Photonic Technologies for Quantum Information Processing

Prem Kumar,[1,7] Paul Kwiat,[2] Alan Migdall,[3] Sae Woo Nam,[4] Jelena Vuckovic,[5] and Franco N. C. Wong[6]

The last several years have seen tremendous progress toward practical optical quantum information processing, including the development of single- and entangled-photon sources and high-efficiency photon counting detectors, covering a range of wavelengths. We review some of the recent progress in the development of these photonic technologies.

KEY WORDS: Quantum dot; entanglement; down-conversion; single-photon detector.

PACS: 03.67.–a, 42.50.Dv, 42.65.Lm, 78.67.Hc, 85.60.Gz.

1. INTRODUCTION

It is now generally realized that fundamentally quantum-mechanical phenomena can enable significant, and in some cases, tremendous, improvement for a variety of tasks important to emergent technologies. Building on decades of successes in the experimental demonstration of such fundamental phenomena, it is not surprising that photonics is playing a

[1]Departments of Electrical and Computer Engineering, and Physics and Astronomy, Northwestern University, Evanston, Illinois 60208-3118, USA. E-mail: kumarp@northwestern.edu
[2]Department of Physics, University of Illinois, Urbana-Champaign, Illinois 61801-3080, USA.
[3]Optical Technology Div., NIST, Gaithersburg, Maryland 20899-8441, USA.
[4]Quantum Electrical Metrology Division, NIST, Boulder, Colorado 80305-3328, USA.
[5]Department of Electrical Engineering, Stanford University, Stanford, California 94305, USA.
[6]Research Laboratory of Electronics, MIT, Cambridge, Massachusetts 02139, USA.
[7]To whom correspondence should be addressed. E-mail: kumarp@northwestern.edu

preeminent role in this nascent endeavor. Many of the objectives of quantum information processing are inherently suited to optics (e.g., quantum cryptography[1] and optical metrology[2]), while others may have a strong optical component (e.g., distributed quantum computing[3]). In addition, it is now known that, at least in principle, one can realize *scalable* linear optics quantum computing (LOQC).[4] For these applications to attain their full potential, various photonic technologies are needed, including high fidelity sources of single and entangled photons, and high efficiency photon-counting detectors, both at visible and telecommunication wavelengths. Much progress has been made on the development of these, though they are still not up to the demanding requirements of LOQC. Nevertheless, even at their present stage they have direct application to initial experiments. Moreover, they may find use in various "adjacent" technologies, such as biomedical and astronomical imaging, and low-power classical telecommunications. Here we describe a number of the leading schemes for implementing approximations of sources of single photons on-demand and entangled photons, followed by a review of methods for detecting individual photons.

2. SINGLE-PHOTON SOURCES

Photon-based quantum cryptography, communication, and computation schemes have increased the need for light sources that produce individual photons. Ideally a single-photon source would produce completely characterized single photons on demand. When surveying attempts to create such sources, however, it is important to realize that there never has been and will never be such an ideal source. All of the currently available sources fall significantly short of this ideal. While other factors (such as rate, robustness, and complexity) certainly do matter, two of the most important parameters for quantifying how close a "single-photon source" approaches the ideal are the fraction of the time the device delivers light in response to a request, and the fraction of time that that light is just a single photon.

In general single-photon sources fall into two categories—isolated quantum systems, or two-photon emitters. The first type relies on the fact that a single isolated quantum system can emit only one photon each time it is excited. The trick here is obtaining efficient excitation, output collection, and good isolation of individual systems. The second type uses light sources that emit two photons at a time. Here the detection of one photon indicates the existence of the second photon. That knowledge allows the second photon to be manipulated and delivered to where it is needed.

2.1. Quantum Dot Single-Photon Sources

A quantum dot is essentially an artificial atom that is easily isolated, so it is an obvious choice as the basis of a single-photon source. Single photons on-demand have been generated by a combination of pulsed excitation of a single self-assembled semiconductor quantum dot and spectral filtering.(5) When such a quantum dot is excited, either with a short (e.g., 3 ps) laser pulse, or with an electrical pulse,(6) electron–hole pairs are created. For laser excitation, this can occur either within the dot itself, when the laser frequency is tuned to a resonant transition between confined states of the dot, or in the surrounding semiconductor matrix, when the laser frequency is tuned above the semiconductor band gap. In the latter case, carriers diffuse toward the dot, where they relax to the lowest confined states. Created carriers recombine in a radiative cascade, leading to the generation of several photons for each laser pulse; all of these photons have slightly different frequencies, resulting from the Coulomb interaction among carriers. The last emitted photon for each pulse has a unique frequency, and can be spectrally isolated.

If the dots are grown in a *bulk* semiconductor material,(6) the out-coupling efficiency is poor, since the majority of emitted photons are lost in the semiconductor substrate. To increase the efficiency, an optical microcavity can be fabricated around a quantum dot. An additional advantage is that the duration of photon pulses emitted from semiconductor quantum dots is reduced, due to an enhancement of the spontaneous emission rate. This enhancement, also known as the Purcell factor, is proportional to the ratio of the mode quality factor to the mode volume. In addition, the spontaneous emission becomes directional; the photons emitted into the nicely shaped cavity mode can be more easily coupled into downstream optical components.

By embedding InGaAs/GaAs quantum dots inside micropost microcavities with quality (Q)-factors of around 1300 and Purcell factors around five, the properties of a single-photon source have been significantly improved;(7) see Fig. 1. The probability of generating two photons for the same laser pulse [estimated from the zero-time correlation parameter $g^{(2)}(0)$] can be as small as 2% compared to a Poisson-distributed source (i.e., an attenuated laser) of the same mean photon rate, the duration of single-photon pulses is below 200 ps, and the sources emit identical (indistinguishable) photons, as confirmed by two-photon interference in a Hong-Ou-Mandel type experiment.(7) Such sources have been employed to realize the BB84 QKD protocol, and to generate post-selected polarization-entangled photons.(8)

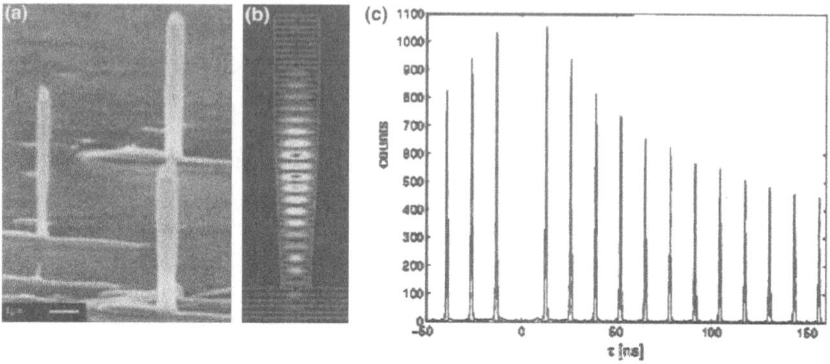

Fig. 1. (a) Scanning electron micrograph showing a fabricated array of GaAs/AlAs microposts (~0.3-μm diameters, 5-μm heights), with InAs/GaAs quantum dots embedded at the cavity center. (b) Electric field magnitude of the fundamental HE_{11} mode in a micropost microcavity with a realistic wall profile. (c) Photon correlation histogram for a single quantum dot embedded inside a micropost and on resonance with the cavity, under pulsed, resonant excitation. The histogram is generated using a Hanbury Brown and Twiss-type setup—the vanishing central peak (at $\tau = 0$) indicates a large suppression of two-photon pulses (to ~2% compared to a Poisson-distributed source, e.g., an attenuated laser, of the same intensity. The 13-ns peak-to-peak separation corresponds to the repetition period of excitation pulses.

These sources still face several great challenges, however. They require cryogenic cooling (<10 K), the output wavelengths are not yet readily tunable (present operation is around 900 nm), the out-coupling efficiency into a single-mode traveling wave is still rather low (<40%),[(9)] and excitation of quantum dots in microcavities presently requires *optical* pumping (electrical pumping would be more desirable and efforts in that direction are underway[(6)]). In the future, photonic-crystal microcavities may lead to much higher ratios of the quality factor to mode volumes, and therefore, much stronger cavity QED effects should be possible.[(10)] This would enable an increase in the efficiency and speed of the single-photon devices, and thus open the possibility for building integrated quantum information systems. The spontaneous emission lifetime could be reduced further to on the order of several picoseconds, which would allow the generation of single photons at a rate higher than 10 GHz. Moreover, the Purcell effect would also help in bringing the emitted photons closer to being Fourier-transform limited in bandwidth. Finally, photonic-crystal based cavities could even enable the realization of the strong coupling regime with a single quantum dot exciton, opening the possibility for the generation of completely indistinguishable single photons by coherent excitation schemes.

2.2. Other Single-Emitter Approaches

Other isolated quantum system approaches to producing single photons include isolated single fluorescence molecules[11] and isolated nitrogen vacancies in diamond.[12] Two significant deficiencies of these sources for many applications are that it is not easy to efficiently out-couple the photons, and that the spectral spread of the light is typically quite large (~120 nm), though widths as low as 12 nm have been seen in new results.[11] This spectral width is non-optimal for applications relying on two-photon interference effects, and also for quantum cryptographic applications (where one typically desires fairly narrow bandwidths to exclude background light).

More recently, single atoms[13] coupled to a high-finesse optical cavity have demonstrated features of single-photon operation. Despite the technological challenges, this approach does offer the large potential advantage that the photons are emitted preferentially into the cavity modes, which are easier to couple out of, with couplings of 40–70% already achieved. Also, the frequency of the photons is necessarily matched to a strong atomic transition, which may allow for efficient quantum *communication* using photons, while other quantum information processing tasks, such as memory or state readout, are carried out in the atomic system.[14,15]

2.3. Downconversion Single-Photon Sources

Another effort toward single-photon sources relies on producing photons in pairs, typically via the process of optical parametric down conversion (PDC).[16] The PDC process effectively takes an input photon from a pump beam and converts it into output pairs in a crystal possessing a $\chi^{(2)}$ nonlinearity. Thus the detection of one photon can be used to indicate (or herald) the existence of the second photon, which is available for further use. This second photon is, at low photon rates, left in an excellent approximation to a single-photon number state.[17] It has been demonstrated how these photons may then be converted into completely arbitrary quantum states with fidelities of 99.9%.[18] Recent efforts have focused on improving the collection of those pairs and improving the "single-photon accuracy," e.g., the value of $g^{(2)}$ (0).

The physics of the PDC process guarantees that the output pairs will possess certain energy and momentum constraints, so that under appropriate conditions the detected location of the herald photon tightly defines the location of its twin, a significant advantage over other single-photon schemes. There have been many mode-engineering efforts to improve this collection into a *single* mode,[19] but the current best collection efficiency is

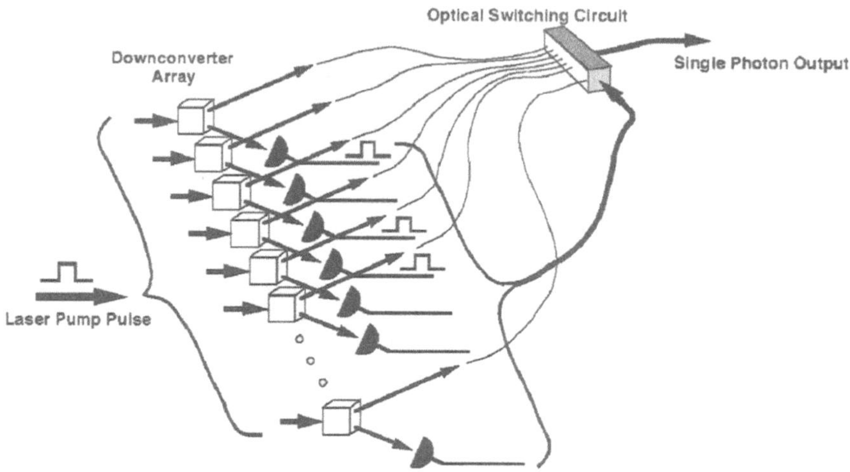

Fig. 2. Multiplexed PDC scheme to better approximate a source of single photons on demand. By operating an array of simultaneously pumped PDC sources at low photon production rates and optically switching the output of one of the PDC sources that *did* produce a photon to the single output channel, it is possible to increase the single-photon rate, while maintaining a low rate of unwanted multiphoton pulses.

still only 58%[8], including 15% optical-transmittance losses. (Contrast this to the required single-photon efficiency of over 99% for LOQC.)[(4)] One example of a method to improve this is to directly modify the spatial emission profile of the photon pairs (which are usually emitted along cones) so that the photons are emitted preferentially into "beacon"-like beams, which couple more naturally into single-mode optical fibers.[(20)] Another approach yet to be explored is the use of adaptive optics to tailor the output modes. It should be noted that not all quantum information processing applications require single-mode performance; for example, free-space quantum key distribution is likely to work nearly as well with a small number of modes.

Because the conversion of pump photons into pairs via PDC is a random process, these sources suffer from the same problem that afflicts faint laser sources—one cannot guarantee that one and only one photon pair is created at a time (i.e., $g^{(2)}(0) \neq 0$). Multiplexing and storage schemes have been proposed to deal with this. They both work by similar principles (one scheme is based on space multiplexing[(21)]—see Fig. 2—and the other is based on temporal multiplexing[(18,22)])—photons are created at relatively low rates where the probability of simultaneous multi-pair production is

[8]An 83% coupling efficiency has recently been reported. See quant-ph/0408093.

low; contingent on the detection of a herald photon, the twin is then "stored", to be emitted in a controlled fashion at some later desired time. The overall emission rate is reduced, but the rate of producing one and only one photon at regular intervals is improved.

3. ENTANGLED-PHOTON SOURCES

Entangled states are now known to be a critical resource for realizing many quantum information protocols, such as teleportation and quantum networking. An on-demand source of entangled photons would also greatly aid the realization of all-optical quantum computing.

3.1. Down-Conversion Schemes

At present, by far the most prevalent source of entangled photon pairs is parametric down conversion based on crystals with a $\chi^{(2)}$ non-linearity. As discussed above, it is precisely the temporal and spatial correlations between the photon pairs which make them very promising for the realization of an on-demand source of single photons. Much of the effort in studying these sources has been devoted to the generation of *polarization*-entangled photon pairs, an area which has seen tremendous growth—more than a million-fold improvement in the detected rates of polarization-entangled photons has been achieved in the past two decades (see Fig. 3).

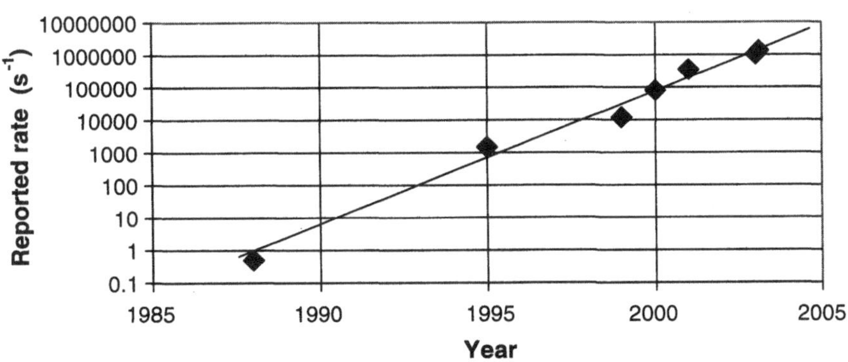

Fig. 3. The apparent "Moore's Law" for entanglement. Shown are the reported detection rates of (polarization-)entangled photon pairs (from down conversion), as a function of year. The solid line—drawn to guide the eye—indicates the ×100 gain every 5 years. The primary limiting factor has now become the lack of single-photon counting detectors with saturation rates above 10 MHz.

There are now several ways to realize polarization entanglement using the PDC process. One method uses a single nonlinear crystal, cut for "type-II" phase matching, and selecting out a particular pair of output directions.[23] Although initially these sources used large gas lasers for pumping, the recent availability of ultraviolet diode lasers has led to much more compact sources.[24] A potentially important disadvantage, in addition to the need to compensate the birefringent walk-off with this scheme, is that the entanglement is present only over a particular pair of modes (corresponding to the intersection of two cones). One method to eliminate this disadvantage is to pump the crystal from two different directions,[25,26] or to allow the PDC to occur in either of *two* crystals, the outputs of which are superposed directly[27,28] or using a beam splitter.[29] By proper alignment, nearly all of the output modes can display polarization entanglement, which moreover is completely tunable.[30] Nearly perfect entanglement (within statistical uncertainty) has been observed with such sources. Results with short-pulse pumps[28,29,31] are encouraging, but the quality of the entanglement is typically not as high, a problem that will need to be addressed for future applications.

One disadvantage of all of these techniques is that the output spectral bandwidth is still quite wide (typically 1–10 nm) for possible coupling to atomic states. Research is underway to circumvent this problem by placing the nonlinear crystals inside high finesse optical cavities, which significantly increases the probability of downconversion into a narrow spectral bandwidth.[14]

As discussed above, there are a number of approaches for improving the coupling efficiency into single spatial modes. Improving conversion efficiency by finding higher non-linearity bulk crystals is limited by the choice of available crystals (with BBO and $LiIO_3$ being two of the better ones). Engineering crystals by processes such as periodic poling[32] allows one to take advantage of crystals (e.g., Lithium Niobate) with somewhat higher nonlinearities. The conversion efficiency into a specific mode can be further enhanced by some 1–2 orders of magnitude by creating waveguides in these crystals.[33] Because the waveguide is small, possibly even single mode, it can be much easier to collect the output light. However, the net outcoupling efficiencies achieved to date (10–20%) still require substantial improvement. Finally, by using a buildup cavity to recycle the unconverted pump photons, the effective conversion efficiency may be increased (at the expense of a more complicated setup).[34]

Entanglement in non-polarization degrees of freedom, such as energy/time-bin[35] and orbital angular momentum,[36] has also been realized recently. These may present some advantages over the polarization case, e.g., they allow implementation of higher-order quantum structures, such as qu-trits (3-level systems), and timing entanglement is more robust for transmission through optical fibers.

One problem plaguing all of these sources is that the production of pairs is a random process. By using short pulsed pumps, it is possible to define the times when *no* photon pairs will be produced, but there is still no way to guarantee production of exactly one photon-pair during any given pulse. At least one theoretical scheme has been proposed to circumvent this problem,[37] but practical implementations have yet to be realized.

3.2. $\chi^{(3)}$-Nonlinearity Schemes

The difficulty of coupling the entangled photons into optical fibers has been overcome by directly producing them *inside* of the fiber, by exploiting the $\chi^{(3)}$ (Kerr) nonlinearity of the fiber itself.[38] By placing the pump wavelength close to the zero-dispersion wavelength of the fiber, the probability amplitude for inelastic four-photon scattering can be significantly enhanced. Two pump photons at frequency ω_{p} scatter through the Kerr nonlinearity to create simultaneous energy-time-entangled signal and idler photons at frequencies ω_{s} and ω_{i}, respectively, such that $2\omega_{\mathrm{p}} = \omega_{\mathrm{s}} + \omega_{\mathrm{i}}$. Because of the isotropic nature of the Kerr nonlinearity in fused-silica-glass fibers, the correlated scattered photons are predominantly co-polarized with the pump photons. Two such correlated down-conversion events from temporally multiplexed orthogonally polarized pumps can be configured to create polarization entanglement as well. In this way all four polarization-entangled Bell states have recently been prepared, violating Bell inequalities by up to ten standard deviations of measurement uncertainty.[39] One drawback is the existence of Raman scattering in standard optical fibers due to coupling of the pump photons with optical phonons in the fiber. However, for small pump-signal detunings the imaginary part of $\chi^{(3)}$ in standard fibers is small enough that a 10-fold higher probability of creating a correlated photon-pair in a suitable detection window can be obtained than the probability of two uncorrelated Raman-scattered photons in the same detection window.[40] Further work to quantify Raman scattering at the single-photon level is needed.

3.3. Quantum Dot Entangled-Photon Sources

A biexcitonic cascade from a semiconductor quantum dot might also allow the generation of polarization-entangled photon pairs on demand, since the selection rules should translate the anticorrelation of electron and hole spins in the biexcitonic state into polarization anticorrelation of photons.[41] However, this requires that the two decay paths from the biexcitonic state are indistinguishable; therefore, the effects such as dot anisotropy, strain, piezoelectric effects, and dephasing processes need to be minimized.[42] To accomplish this, one needs to optimize quantum dot growth conditions and employ novel high-Q photonic crystal microcavities, which would increase the radiative recombination rate over the dephasing rate.[43]

4. SINGLE-PHOTON DETECTORS

As noted in the introduction, photon-based quantum information processing applications require that single photons, or more generally, the photon number in a multiphoton state, be detected with efficiency approaching unity. To that end much progress has been made in recent years towards developing high efficiency, low noise, and high count-rate detectors, which can reliably distinguish the photon number in an incident quantum state.

4.1. Avalanche Devices

Detection of single photons with avalanche photodiodes[44] (APDs) biased above the breakdown voltage is convenient (no cryogenic temperatures are needed) and relatively efficient. When one or more photons are absorbed, the generated carriers that undergo avalanche gain may cause a detectable macroscopic breakdown of the diode p–n junction. APD photon counters suffer both from dark counts, where thermally generated charge carriers cause a detection event, and from after-pulses, where carriers from a previous avalanche cause subsequent detection events when the APD is reactivated.

The best counters at visible wavelengths have been made with silicon APDs. These work well because of both the material system's ability to provide very low-noise avalanche gain and the availability of silicon of nearly perfect quality. For example, the single-photon counting modules (SPCMs), made by Perkin-Elmer (SPCM-AQR-16), can have 50–70% quantum efficiency near 700-nm wavelength, dark-count rate < 25/s, and can count at rates up to 10–15 MHz.†[45] The dark-count rate is low enough for the SPCMs to be operated continuously except for a 50-ns avalanche quench time, although heating effects limit the CW counting rate to about 5 MHz. After-pulsing is less than 0.5%. The quantum efficiency of

†Certain trade names and company products are mentioned in the text in order to specify adequately the experimental procedure and equipment used. In no case does such identification imply recommendation or endorsement by the National Institute of Standards and Technology, nor does it imply that the products are necessarily the best available for the purpose.

the SPCMs drops at longer wavelengths (2% at 1 μm). Attempts to resolve multiple photons by splitting a multi-photon pulse into several time bins (e.g., with a storage loop) have been made, but they are limited by losses in the device switching photons into and out of the loop, and by the non-unity detector efficiencies.[(46)]

The Visible Light Photon Counter[(47)] (VLPC) and Solid State Photomultiplier[(48)] (SSPM) are modified Si devices which operate using a spatially localized avalanche from an impurity band to the conduction band. They possess high quantum efficiency (estimated to be ~95%) with low multiplication noise. The localized nature of the avalanche allows high efficiency photon-number discrimination,[(49)] which is not possible with conventional APDs. Using this capability, the non-classical nature of PDC has been investigated and violations of classical statistics demonstrated.[(50)] Unfortunately, these detectors require cooling to 6 K for optimal performance, and even then they display dark count rates in excess of $10^4\,s^{-1}$.

In the infrared, 1–1.6 μm, the best results to date have come from APDs having InGaAs as the absorption region that is separate from a multiplication layer of InP[(51)]; see Fig. 4. This has proven to be a

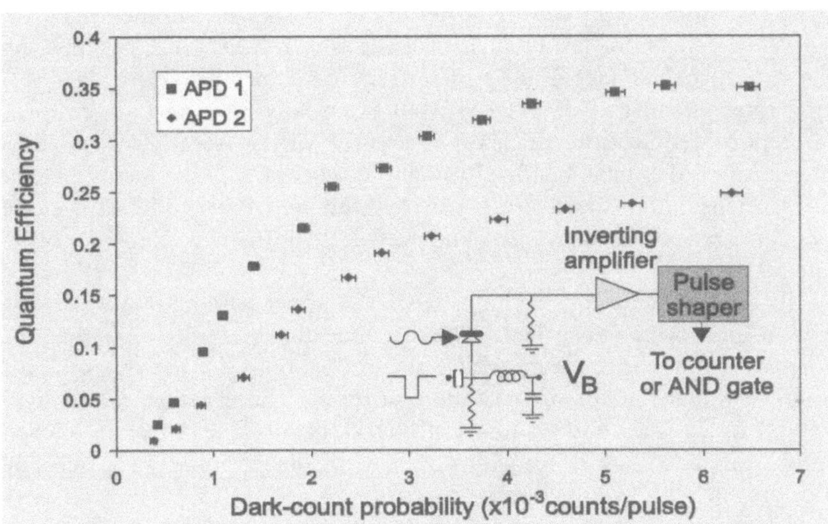

Fig. 4. Quantum efficiency versus dark-count probability for two InGaAs APDs operated in gated Geiger mode near 1537 nm wavelength. In the gated Geiger mode, the APD is biased below breakdown and a short electrical pulse (~1 ns), coincident with the incident light pulse containing the photon to be detected, brings it momentarily into the breakdown region. The inset shows a schematic of the electronic circuit used with the APDs (from Ref. 38).

better solution than germanium APDs.[52] To suppress the high dark count rate in these devices, at best thousands of times worse than in silicon APDs, cooled InGaAs/InP APDs are usually activated for only ~1–10 ns duration to coincide with the arrival of the photon to be detected. The reported quantum efficiencies are typically between 10–30%, and the APDs are usually operated at a count rate of 100 kHz in order to alleviate after-pulsing caused by carriers trapped between the InGaAs and InP layers.

4.2. Superconducting Devices

Superconducting devices offer the potential to achieve levels of performance that exceed those of conventional semiconductor APDs. Although there are many types of superconducting detectors, only three have been used to observe single optical photons: the transition-edge sensor[53] (TES), the superconducting tunnel junction[54] (STJ), and the superconducting single-photon detector (SSPD).[55] Both the TES and the STJ detectors have been able to detect single photons and count the number of photons absorbed by the detector. The TES detector uses the steep slope of the resistance as a function of temperature at the superconducting transition as a very sensitive thermometer. This thermometer is able to measure the temperature change in an absorber when one or more photons are absorbed (see Fig. 5). The TES detectors are slow, capable of count rates at most up to 100 kHz, but essentially have no dark counts.[53] The reported detection efficiency currently varies from 20% to 40% in the telecom to optical band, although significant improvements in detection efficiency and speed are being realized with better detector designs (e.g., anti-reflection coatings) and research into new superconducting materials.

In an STJ detector, excitations of the superconductor are generated when a photon is absorbed. The excited quasiparticles can create an enhanced tunneling current which is proportional to the energy of the photon (or the number of photons absorbed). These detectors are similar in speed to the TES and also have no dark counts. The detection efficiency demonstrated to date is roughly 40% for visible photons,[54] which could be improved with AR coatings.

The SSPD detectors are extremely fast detectors (~100-ps total pulse duration) that have single photon sensitivities.[55] In an SSPD, the detector is a narrow superconducting current path on a substrate. This path is current-biased at a point just below the superconducting critical current. A local hot spot is formed where a photon(s) is absorbed, locally destroying the superconductivity. This forces the current to flow around the hot spot

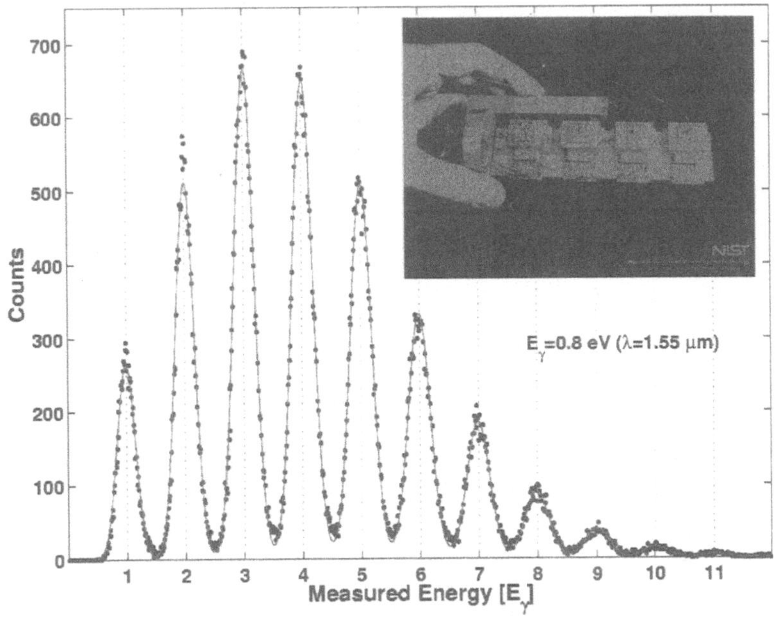

Fig. 5. Measured Poisson photon-number distribution of an attenuated, pulsed 1550-nm laser, repeatedly measured using a TES. The TES devices are made of superconducting tungsten and operated at a temperature of 100 mK. The horizontal axis is the pulse height of the photon absorption events in units of the energy of one 1550-nm photon, 0.8 eV (from Ref. 53). The inset shows a photograph of four fiber-coupled devices prepared to be cooled to 100 mK.

causing the current density around the hot-spot to exceed the critical current density. As a result, the device develops a resistance, causing a voltage to appear across the device. These detectors are single-photon-threshold devices and are not able to resolve the photon number in multiphoton pulses. Typical implementations use meandering paths to increase the sensitive area, which is otherwise very small due to the narrowness required for the conducting path. Much improvement in device fabrication and design is needed to improve the quantum efficiencies of these devices beyond the current values of ~20%; the detection efficiency is lower still, due to the area effect mentioned above.

4.3. Frequency Upconversion

Detection techniques based on frequency upconversion allow IR photons to be converted into the visible where single photon detection is more efficient and convenient. Frequency upconversion uses sum-frequency

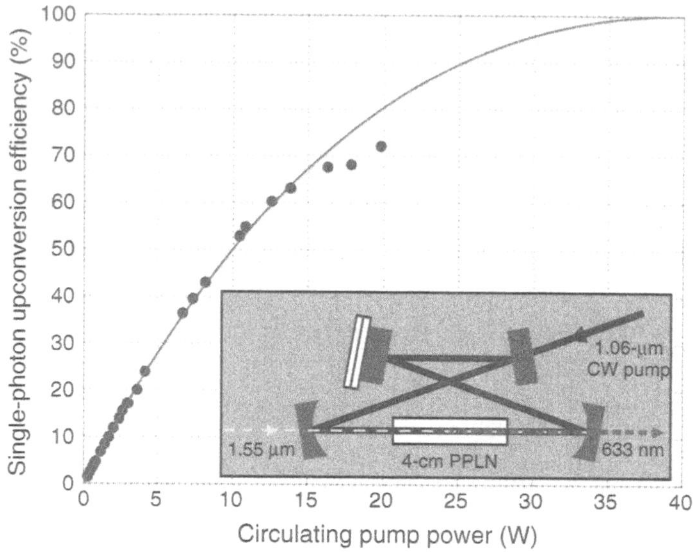

Fig. 6. CW single-photon upconversion efficiency versus circulating pump power in the pump enhancement ring cavity (inset). Solid line is a theoretical fit to data. At high pump powers lower than expected efficiencies are due to heating in PPLN that caused thermal instability in the ring cavity lock. See Ref. 56 for results with improved cavity lock.

generation in a non-linear optical crystal to mix a weak input signal at ω_{in} with a strong pump at ω_{p} to yield a higher-frequency output field at $\omega_{\text{out}} = \omega_{\text{in}} + \omega_{\text{p}}$. With sufficient pump power this upconversion can occur with near unity efficiency even for weak light fields at the single-photon level. For LOQC and quantum key distribution applications, telecommunication-wavelength photons at 1.55 μm can then be efficiently detected with low-noise, high quantum-efficiency Si APDs. Recently, upconversion of single photons from 1.55 to 0.63 μm in bulk periodically poled lithium niobate (PPLN) has been demonstrated with an efficiency of 90%,[(56)] limited only by the available continuous wave (CW) pump power at 1.06 μm, see Fig. 6. The bulk PPLN crystal is embedded inside a pump enhancement cavity that also imposes a well-defined spatial mode for the single-pass input photons. One approach to eliminate the need for a stabilized buildup cavity is to use a bright pulsed escort beam which is temporally mode-matched to the input photon. Such a system has enabled single-photon conversion efficiencies of ~80% and backgrounds less than 10^{-3} per pulse.[(57)]

The pump power requirement can be relaxed by using a waveguide PPLN crystal,[(58)] but the effect of waveguide losses must be addressed to achieve the required near-unity net upconversion efficiency. The next step is to demonstrate frequency upconversion of a quantum state,[59] i.e., high fidelity frequency translation of a single photon in an arbitrary quantum polarization state. This will allow a modular approach to developing LOQC technologies. For example, the photonic qubits and ancilla photons can be prepared at wavelengths with the most convenient and efficient methods, and then converted with near-unity efficiency to wavelengths that are optimal for photonic logic gates employing quantum interference. Similarly, tunable quantum frequency upconversion can be used to match the required wavelengths to the resonant transitions in various atomic systems, for applications such as quantum repeaters.[14] As another example, there have also been proposals[15] to couple the photons to an atomic vapor system—the excitation of a single atom can be made very probable by having many atoms, and that excitation can be read out with very high efficiency by using a cycling transition. Such schemes could potentially yield efficiencies in excess of 99.9%. However, there are critical noise issues which must still be addressed.

5. CONCLUSIONS

For reasons noted in the introduction, there is intense current interest in creating robust, high-precision sources and detectors of single photons. In the last year alone, two special issues have appeared in the literature focusing just on these topics.[60,61] Though tremendous progress has been achieved, more development is clearly necessary to bring these technologies to the level of operation needed for LOQC. Nevertheless, already they have shown promise, enabling the realization of simple quantum gates, and improved quantum key distribution protocols. We anticipate that further improvements over the next few years will continue to make optical qubits an attractive system, though it remains to be seen whether the extremely demanding LOQC requirements can be met.

ACKNOWLEDGMENTS

P. Kumar and F. Wong would like to acknowledge support of the MURI Center for Quantum Information Technology: Entanglement, Teleportation, and Quantum Memory (ARO program DAAD19-00-1-0177); P. Kwiat, A. Migdall, Sae Woo Nam, and J. Vuckovic would like to acknowledge support by the MURI Center for Photonic Quantum Information Systems (ARO/ARDA program DAAD19-03-1-0199). A. Migdall would also like to acknowledge DARPA/QuIST support.

REFERENCES

1. N. Gisin *et al.*, *Rev. Modern Phys.* **74**, 145 (2002).
2. A. Migdall, *Phys. Today* (January, 1999), 41.
3. S. J. van Enk *et al.*, *J. Mod. Opt.* **44**, 1727 (1997); J. I. Cirac, A. K. Ekert, S. F. Huelga, and C. Macchiavello, *Phys. Rev. A* **59**, 4249 (1999); H. Buhrman, R. Cleve, and W. van Dam, *SIAM J. Comput.* **30**, 1829 (2001).
4. E. Knill, R. Laflamme, and G. J. Milburn, *Nature* **409**, 46 (2001).
5. P. Michler *et al.*, *Science* **290**, 2282 (2000); C. Santori *et al.*, *Phys. Rev. Lett.* **86**, 1502 (2001).
6. Z. Yuan *et al.*, *Science* **295**, 102 (2002).
7. J. Vuckovic *et al.*, *Appl. Phys. Lett.* **82**, 3596 (2003); C. Santori *et al.*, *Nature* **419**, 594 (2002).
8. E. Waks *et al.*, *Nature* **420**, 762 (2002); D. Fattal *et al.*, *Phys. Rev. Lett.* 92, 037903 (2004).
9. M. Pelton *et al.*, *Phys. Rev. Lett.* **89**, 233602 (2002).
10. J. Vuckovic and Y. Yamamoto, *Appl. Phys. Lett.*, **82**, 2374 (2003).
11. L. Brunel, B. Lounis, P. Tamarat, and M. Orrit, *Phys. Rev. Lett.* **83**, 2722 (1999); T.-H. Lee et al., *Phys. Rev. Lett.* **85**, 100 (2004).
12. C. Kurtsiefer, S. Mayer, P. Zarda, and H. Weinfurter, *Phys. Rev. Lett.* **85**, 290 (2000); R. Brouri *et al.*, *Eur. Phys. J. D* 18, 191 (2002); A. Beveratos *et al.*, *Phys. Rev. Lett.* **89**, 187901 (2002).
13. A. Kuhn, M. Hennrich, and G. Rempe, *Phys. Rev. Lett.* **89**, 067901 (2002); J. McKeever *et al.*, *Science* **303**, 1992 (2004).
14. J. H. Shapiro, *New J. Phys.* **4**, 47.1 (2002).
15. A. Imamoglu, *Phys. Rev. Lett.* **89**, 163602 (2002); D. F. V. James and P. G. Kwiat, *Phys. Rev. Lett.* **89**, 183601 (2002).
16. D. C. Burnham and D. L. Weinberg, *Phys. Rev. Lett.* **25**, 84 (1970).
17. C. K. Hong and L. Mandel, *Phys. Rev. Lett.* **56**, 58 (1986).
18. N. Peters *et al.*, *Quant. Inform and Comput.* **3**, 503–517 (2003); E. Jeffrey, N. A. Peters, and P. G. Kwiat, *New J. Phys.* **6**, 100 (2004).
19. C. Kurtsiefer, M. Oberparleiter, and H. Weinfurter, *Phys. Rev. A* **64**, 023802 (2001); F. A. Bovino *et al.*, *Opt. Commun.* **227**, 343 (2003).
20. S. Takeuchi, *Opt. Lett.* **26**, 843 (2001).
21. A. L. Migdall, D. Branning, and S. Castelletto, *Phys. Rev. A* **66**, 053805 (2002).
22. T. B. Pittman, B. C. Jacobs, and J. D. Franson, *Phys. Rev. A* **66**, 42303 (2002); P. G. Kwiat *et al.*, *Proc. SPIE* **5161**, 87 (2004).
23. P. G. Kwiat *et al.*, *Phys. Rev. Lett.* **75**, 4337 (1995).
24. P. Trojek, Ch. Schmid, M. Bourennane and H. Weinfurter, *Opt. Exp.* **12**, 276 (2004).
25. D. Branning, W. Grice, R. Erdmann, and I. A. Walmsley, *Phys. Rev. A* **62**, 013814 (2000).
26. M. Fiorentino *et al.*, quant-ph/0309071; *Phys. Rev. A* **69**, 041801(R) (2004).
27. P. G. Kwiat *et al.*, *Phys. Rev. A* **60**, R773 (1999).
28. G. Bitton, W. P. Grice, J. Moreau, and L. Zhang *Phys. Rev. A* **65**, 063805 (2002).
29. Y. -H. Kim *et al.*, *Phys. Rev. A* **63**, 062301 (2001).
30. A. G. White, D. F. V. James, P. H. Eberhard, and P. G. Kwiat, *Phys. Rev. Lett.* **83**, 3103 (1999).
31. Y. Nambu *et al.*, *Phys. Rev. A* **66**, 033816 (2002); B.-S. Shi and A. Tomita, *Phys. Rev. A* **69**, 013803 (2004).
32. S. Tanzilli *et al.*, *Electron. Lett.* **37**, 26 (2001); C. E. Kuklewicz *et al.*, *Phys. Rev. A* **69**, 013807 (2004).

33. K. Banaszek, A. B. U'Ren, and I. A. Walmsley, *Opt. Letts.* **26**, 1367 (2001); K. Sanaka, K. Kawahara, and T. Kuga, *Phys. Rev. Lett.* **86**, 5620 (2001).
34. M. Oberparleiter and H. Weinfurter, Opt. Commun. **183**, 133 (2000).
35. I. Marcikic *et al.*, *Phys. Rev. A* **66**, 062308 (2002).
36. A. Mair, A. Vaziri, G. Weihs, and A. Zeilinger, *Nature* **412**, 312 (2001); N. K. Langford, quant-ph/0312072.
37. T. B. Pittman *et al.*, in *IEEE J. Selec. Top. Quant. Electron., special issue on "Quantum Internet Technologies"* (2003).
38. M. Fiorentino, P. L. Voss, J. E. Sharping, and P. Kumar, *IEEE Photonics Tech. Lett.* **14**, 983 (2002).
39. X. Li, P. Voss, J. E. Sharping, and P. Kumar, Quant. Electr. and Laser Science Conf., Baltimore, MD, June 1–6, 2003, paper QTuB4 in QELS'03 Technical Digest (Optical Society of America, Washington, D.C. 2003); *ibid*, quant-ph/ 0402191.
40. P. L. Voss and P. Kumar, *Opt. Lett.* **29**, 445 (2004); X. Li, J. Chen, P. Voss, J. Sharping, and P. Kumar, *Opt. Exp.* **12**, 3737 (2004).
41. O. Benson, C. Santori, M. Pelton, and Y. Yamamoto, *Phys. Rev. Lett.* **84**, 2513 (2000).
42. C. Santori *et al.*, *Phys. Rev. B* **66**, 045308 (2002).
43. J. Vuckovic and Y. Yamamoto, *Appl. Phys. Lett.* **82**, 2374 (2003).
44. W. G. Oldham, R. R. Samuelson, and P. Antognetti, *IEEE Trans. Electron. Dev.* **ED-19**, 1056 (1972).
45. http://optoelectronics.perkinelmer.com/
46. M. J. Fitch, B. C. Jacobs, T. B. Pittman, and J. D. Franson, *Phys. Rev. A* **68**, 043814 (2003); D. Achilles *et al.*, *Opt. Lett.* **28**, 2387 (2003); J. Rehacek *et al.*, quant-ph/0303032 (2003).
47. E. Waks, K. Inoue, E. Diamanti, and Y. Yamamoto, quant-ph/0308054 (2003).
48. P. G. Kwiat *et al.*, *Appl. Opt.* **33**, 1844 (1994).
49. J. Kim, S. Takeuchi, Y. Yamamoto, and H. H. Hogue, *Appl. Phys. Lett.* **74**, 902 (1999).
50. E. Waks *et al.*, quant-ph/0307162 (2003).
51. A. Lacaita, F. Zappa, S. Cova, and P. Lovati, *Appl. Opt.* **35**, 2986 (1996); G. Ribordy, J.-D. Gautier, H. Zbinden, and N. Gisin, *Appl. Opt.* **37**, 2272 (1998); P. A. Hiskett, G. S. Buller, A. Y. Loudon, J. M. Smith, Ivair Gontijo, Andrew C. Walker, Paul D. Townsend, and Michael J. Robertson, *Appl. Opt.* **39**, 6818 (2000); J. G. Rarity, T. E. Wall, K. D. Ridley, P. C. M. Owens, and P. R. Tapster, *Appl. Opt.* **39**, 6746 (2000); N. Namekata, Y. Makino, S. Inoue, *Opt. Lett.* **27**, 954 (2002); A. Tomita and K. Nakamura, *Opt. Lett.* **27**, 1827 (2002); D. S. Bethune, W. P. Risk, and G. W. Pabst, quant-ph/03111120 (2003); P. L. Voss, K. G. Köprülü, S.-K. Choi, S. Dugan, and P. Kumar, *J. Mod. Opt.* **51**, 1369 (2004).
52. A. Lacaita, P. A. Francese, F. Zappa, and S. Cova, *Appl. Opt.* **33**, 6902 (1994).
53. A. J. Miller, S. W. Nam, J. M. Martinis, and A. V. Sergienko, *Appl. Phys. Lett.* **83**, 791 (2003).
54. A. Peacock *et al.*, *J. Appl. Phys.* **81**, 7641 (1997).
55. G. N. Gol'tsman *et al.*, *Appl. Phys. Lett.* **79**, 705 (2001).
56. M. A. Albota and F. N. C. Wong, *Opt. Lett.* **29**, 1449 (2004).
57. A. P. VanDevender and P. G. Kwiat, *J. Mod. Opt.* **51**, 1433 (2004).
58. K. R. Parameswaran *et al.*, *Opt. Lett.* **27**, 179 (2002).
59. J. M. Huang and P. Kumar, *Phys. Rev. Lett.* **68**, 2153 (1992).
60. "Special Issue: Single-photon: detectors, applications, and measurement methods," A. Migdall and J. Dowling, eds., *J. Mod. Opt.* **51**, 1265–1557 (2004).
61. "Focus Issue: Single photons on demand," P. Grangier, B. Sanders, and J. Vuckovic, eds., *New J. Phys.* **6** (2004). http://www.iop.org/EJ/abstract/1367-2630/6/1/E04

Quantum Computer Development with Single Ion Implantation

A. Persaud,[1] S. J. Park,[1] J. A. Liddle,[1] I. W. Rangelow,[2] J. Bokor,[1,3] R. Keller,[1] F. I. Allen,[1] D. H. Schneider,[4] and T. Schenkel[1]

Spins of single donor atoms are attractive candidates for large scale quantum information processing in silicon. Formation of devices with a few qubits is crucial for validation of basic ideas and development of a scalable architecture. We describe our development of a single ion implantation technique for placement of single atoms into device structures. Collimated highly charged ion beams are aligned with a scanning probe microscope. Enhanced secondary electron emission due to high ion charge states (e.g., $^{31}P^{13+}$, or $^{126}Te^{33+}$) allows efficient detection of single ion impacts. Studies of electrical activation of low dose, low energy implants of ^{31}P in silicon show a drastic effect of dopant segregation to the SiO_2/Si interface, while Si_3N_4/Si retards ^{31}P segregation. We discuss resolution limiting factors in ion placement, and process challenges for integration of single atom arrays with control gates and single electron transistors.

KEY WORDS: Electron emission; single electron devices; coulomb blockade; ion doping; scanning probe; quantum computation.

PACS: 03.67. Lx, 34.50. Dy, 85.35. Gv, 73.23, 61.72, 86.40. py, 07.79.–v.

1. INTRODUCTION

Several promising proposals for implementation of large scale quantum information processing devices in solids are based on the manipulation

[1]E. O. Lawrence Berkeley National Laboratory, Berkeley, CA 94720, USA. E-mail: T_Schenkel@LBL.gov

[2]Institute of Microstructure Technologies and Analytics, University of Kassel, 34109 Kassel, Germany.

[3]Department of Electrical Engineering and Computer Science, University of California, Berkeley, CA 94720, USA.

[4]Lawrence Livermore National Laboratory, Livermore, CA 94550, USA.

of spins of electrons and nuclei of single dopant atoms.[1–3] The accurate placement of single atoms and their integration with control gates and readout structures (such as single electron transistors) represents a nanofabrication challenge that needs to be mastered in order to be able to test, validate and improve device architectures. Spins of electrons of phosphorous atoms are the most attractive candidates to date, since their decoherence time is quite long (~60 ms for isolated donors),[4] and their integration in a silicon matrix allows use of a highly developed semiconductor processing infrastructure.

In this article we describe our program on single atom array formation by single ion implantation, and the integration of single P-atom arrays with control gates and readout SETs.[5,6]

2. SINGLE ION IMPLANTATION

In single ion implantation, every ion impact has to be detected, and the position of each ion impact has to be controlled with high resolution and accuracy.[6–8] The ease or difficulty of detecting single ions depends on their energy and charge state. Low energy (< 20 keV), singly charged ions are difficult to detect, because emission of secondary electrons from inelastic collisions ("kinetic electron emission") is ineffective with secondary electron yields of one or less than one electron per ion.[9] Detection in the bulk of a detector is difficult because most of the ion's energy is dissipated in elastic collisions and not in the formation of electron hole pairs. This is especially so for low energy heavy ions (e.g., Te at 20 keV), where only about 10% of the ion energy is transferred into measurable electronic excitations in the solid. Using high charge state ions allows efficient detection of single ions, because secondary electron emission is proportional to the ion charge state, $Y_e \sim 1.5 \times q$, and tens to hundreds of electrons can be emitted from single ions.[10] Secondary electron emission from high charge state ions is rather insensitive to the ions kinetic energy for ion energies above 2 keV. Secondary electron emission yields increase for very low impact energies (below 2 keV) due to increased time for above surface charge relaxation.[11]

Next to detection, the second challenge in single ion implantation is to ensure accurate alignment and high resolution in ion placement. In conventional focused ion beam systems, beams of Ga^+ ions are focused to a spot size of ~ 10 nm (full width half maximum of a beam with approximately Gaussian shape). This is enabled by the high brightness of the liquid metal ion source used to form Ga^+ beams.[12] While progress is being made in source development, the brightness of highly charged ion sources still remains orders of magnitude lower than that of liquid metal ion guns.[13] Besides beam resolution, i.e., spot size, the accuracy of alignment is crucial for placement of ions into desired positions. Alignment is complicated by the fact that the ion beam must not be used for imaging of the region of interest since this would be accompanied by undesired implantation.

A schematic of our single ion implantation setup is shown in Fig. 1. Beams of highly charged dopant ions are extracted from our electron beam ion trap/source.[14] Typical ion beam energies are 1–10 kV $\times q$, where q is the charge of the selected ion species. Specific ion species are selected by momentum analysis in a double focusing bending magnet. The beam is first focused by a series of electrostatic lenses and is then collimated by a set of apertures. Ions transmit the nanometer scale apertures at a rate of a few ions per second.[15] When a highly charged dopant ion reaches a wafer surface, the potential energy that corresponds to its high charge state is released. Once the ion reaches a critical distance above the

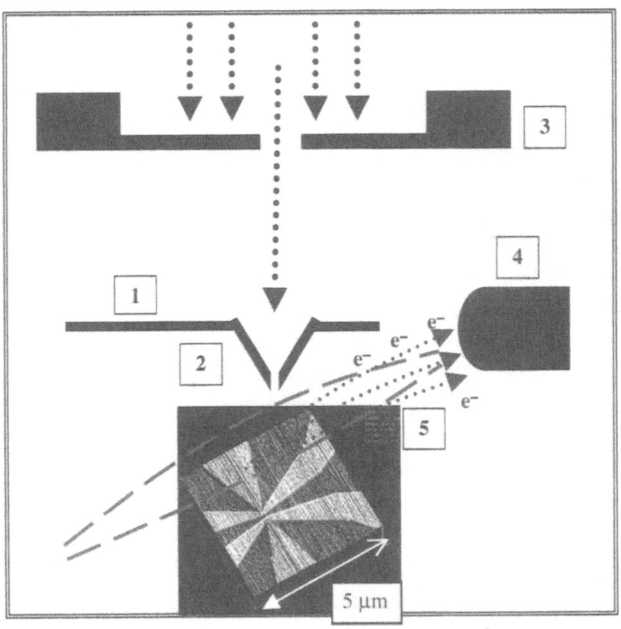

Fig. 1. Schematic illustration of the Single Ion Implantation setup with SFM alignment. Piezo-resistive cantilever with hollow tip (1) and small aperture (2), mounted on a pre-collimation membrane (3), and secondary electron detector (4). The sample is an SET pair shown here in a scanning force microscope image (5, planar view). Dashed lines indicate magnetic field lines for guidance of secondary electrons to the detector.

surface, it can capture electrons from the solid into excited Rydberg states, while vacancies in inner shells remain empty. This process, also referred to as formation of a transient "hollow atom", is very effective and the ion actually neutralizes completely before it impinges on the surface. Rapid di-electronic processes mediate the relaxation to the ground state. The ion travels only for about 10 fs in the solid during dissipation of the potential energy.[16] Secondary electron emission yields depend on the specific material. For $^{31}P^{13+}$ on silicon with a native oxide, about 20 electrons are emitted by each ion. It is important to note that these electrons are emitted within less than 100 fs. Most of the secondary electrons have energies of ~5–20 eV. The beam collimating aperture is formed in the tip of a scanning force microscope, and the tip is placed in close proximity ($<5\,\mu m$) to the sample surface to minimize beam divergence. In order to maximize the extraction efficiency of secondary electrons into the detector, we formed a magnetic guide field with an angle of 10° with respect to the surface. Secondary electrons spiral along the magnetic field lines to the surface of a plastic scintillator. The scintillator is biased at high voltage (10–18 kV). Electrons are stopped in the scintillator and a fraction of their energy is converted into electronic excitations which relax radiatively. A typical conversion efficiency is 100 photons per 10 keV electron. Photons are emitted with a decay time of about 2 ns, and are guided through a Teflon coated quartz light guide to a photo multiplier tube (PMT). The PMT detects photons from the scintillator (central wavelength 425 nm) with a quantum efficiency of 15–20%. Multiple photons arrive as a burst and the resulting avalanche of electrons produces a pulse with a height proportional to the number of secondary electrons collected from a single ion impact. The detection of multiple secondary electrons in a burst is crucial since this allows us to discriminate spurious background in the PMT from the pulse height spectrum. Single electron or single photon background events are well separated from true single ion impact events. Examples of pulse height distributions are shown in Fig. 2 for Te^{33+} and P^{13+} ions with kinetic energies of $3\,kV \times q$. Tellurium, a double donor in silicon, is a candidate for implementation of spin dependent charge measurements.[17] Atoms of both phosphorous and tellurium were injected into the highly charged ion source by evaporation of solid stock in ovens. The detection efficiencies for single ions of P^{13+} and Te^{33+} are 85% and 95%, respectively. The collection efficiency for secondary electrons, and guidance of photons to the PMT has been optimized to achieve 100% detection efficiency for Xe^{40+}. In these measurements some background from high energy events in the scintillator was present. This background was reduced by reduction of the scintillator voltage, and allowed us to detect Xe^{32+} ions with 85% efficiency and zero background in the main peak region

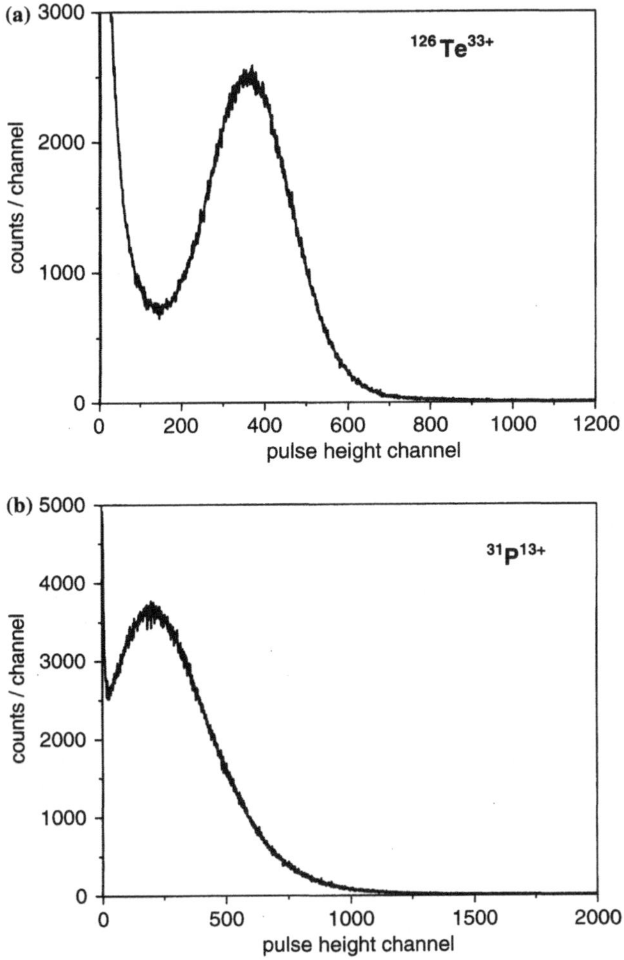

Fig. 2. Pulse height distributions from detection of secondary electrons bursts following the impact of $^{126}Te^{33+}$ (a) and $^{31}P^{13+}$ ions (b). The detection efficiencies were 95% and 85% respectively.

over periods of hours. Here, ions were not decelerated and their kinetic energies corresponded to a 5 kV extraction potential. Detection of very low energy heavy ions was achieved by deceleration at the target, and we detected 7 keV Xe^{32+} ions also with 85% efficiency.[6] With further optimization of the secondary electron collection and reduction of losses in the light pipe from the scintillator to the PMT we anticipate to be able to detect low energy P^{13+} ions with 100% efficiency integrated with scanning probe alignment.

The aperture that defines the beam spot is formed in the tip of a scanning force microscope (SFM). The SFM is based on piezoelectric sensors with a Wheatstone bridge circuit formed in silicon cantilevers.[18] The SFM is needed to allow accurate alignment of the ion beam with marker structures on the samples. An alignment accuracy of ~5 nm is highly desirable for effective device development. The probe tip is a hollow pyramid into which a large hole (~500 nm diameter) is drilled with a FIB (Fig. 3). The SFM cantilever is mounted on a silicon nitrite membrane, and when the hole is drilled in the tip, a second hole is also pierced through the membrane. These two holes are aligned and the membrane provides additional beam collimation. The large hole in the tip is then reduced in diameter by monitored, *in situ* thin film deposition.[19] Holes with diameters as small as 4.3 nm have been formed this way. Additionally, "sensing poles", i.e., poles with radii of curvature of less than 50 nm to ensure high imaging resolution have been formed by electron beam assisted deposition of

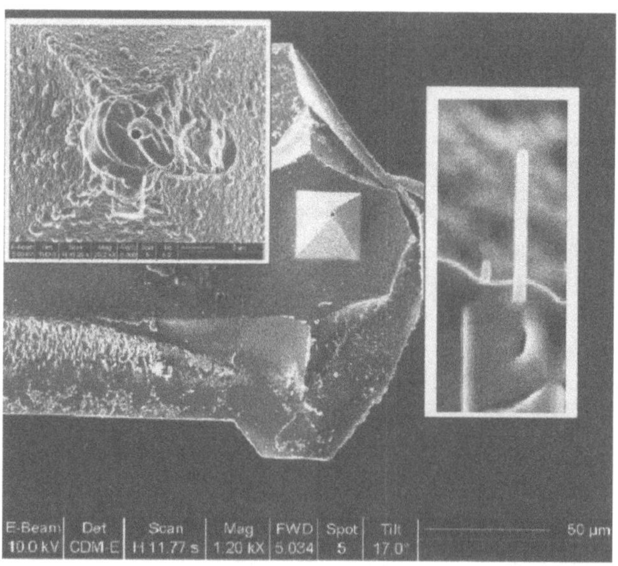

Fig. 3. SFM cantilever with hollow pyramid tip. A hole was drilled in a single step both through the tip and the membrane on which the cantilever is mounted. The hole in the membrane is visible due to the tilt angle of 17°. The insert on the top left shows a hole that went through several cycles of drilling and closing by electron beam assisted Pt deposition, and redrilling. The insert on the right shows a 90 nm wide "sensing pole" grown next to a hole with reduced diameter.

Pt. The sensors of the piezoresistive SFM allow imaging of the surface topography in conventional contact mode, and also in tapping mode with rapid thermal actuators.[20] The latter is important for increased imaging and alignment speed. To date, our SFM with sensors have been operated in vacuum and integration with the implant beam is in progress. Studies of tip lifetime and charge exchange of ions during transmission of the apertures indicate that the tips are robust for days of exposure to ion beams with nA intensities.[5]

3. ACTIVATION OF LOW DOSE, LOW ENERGY P IMPLANTS IN SILICON

Ion implantation is accompanied by damage to the host crystal as incident ions transfer energy to target atoms in elastic collisions. Atoms are displaced from their lattice positions, forming vacancies and interstitials. A fraction of these defects anneals during the collision cascade that accompanies the slowdown process of the incident ions. Implanted ions come to rest mostly on interstitial positions in the lattice. Following implantation, arrays of P atoms have to be annealed in order to incorporate the P atoms into the silicon lattice, and to heal implantation damage to the silicon crystal. Incorporation into the lattice ensures that the bound excess electrons (at low temperature) exhibit the quasi hydrogenic wave function that makes them so attractive for quantum information processing. A qubit spacing of 20 nm corresponds to $\sim 10^{11}$ P-atoms/cm^2, representing a regime of ultra low dose in ion implantation. Diffusion during activation annealing of dopants in silicon is a well studied problem, but most studies have focused on the high dose ($> 10^{13}$ cm^{-2}) regime relevant for conventional silicon devices.[21] For low energy implants with a shallow range, the interface to the silicon crystal plays a crucial role during annealing. Phosphorous diffusion is mediated by silicon interstitials. The SiO_2/Si interface injects interstitials during annealing, and this leads to the segregation of P atoms to the interface. The consequence of this segregation effect is that P atoms are not electrically active, since they are bound at the interface, and that any pre-formed atom array is completely dissolved. Even a "perfect" SiO_2/Si interface can absorb a dose of at least 10^{12} P/cm^2.[22] The Si_3N_4/Si interface shows a complementary effect, since it injects vacancies during annealing. Vacancies retard P diffusion in Si. We have studied the activation of low dose P implants in Si and use spreading resistance analysis (SRA) as a method for characterization of carrier concentrations. SRA has exquisite sensitivity (down to at least 10^{12} carriers/cm^3), but the depth resolution is limited to about 5 nm, and the accuracy of absolute carrier levels is also limited (error of

~25%). However, SRA does allow the qualification of trends, where depth profiling with secondary ion mass spectrometry (SIMS) is in turn limited by a detection efficiency for ^{31}P in Si of about $10^{16}\,cm^{-3}$. In Fig. 4, we show SRA depth profiles of carrier concentrations for silicon (100) samples (5 Ω cm, n-type), implanted with a phosphorus dose of 10^{11} P/cm^2. Samples were annealed for 10 s at 1000 °C in a dry nitrogen atmosphere. The implant energies were 15, 30 and 60 keV. We compare samples with

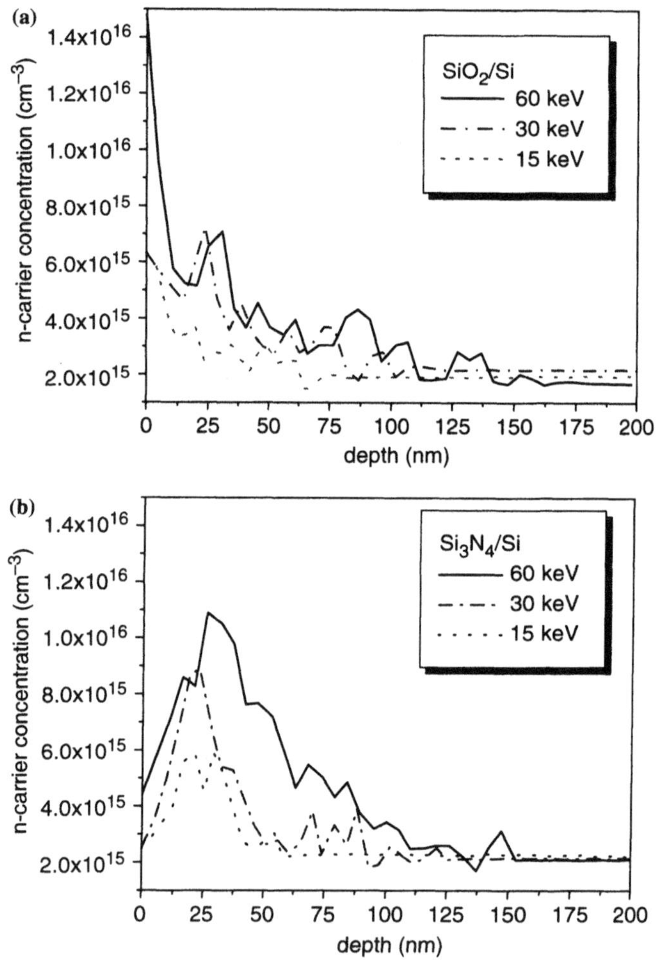

Fig. 4. SRA depth profiles of carrier concentration for silicon samples with SiO_2 (a) and Si_3N_4 (b) layers. The implant dose was 10^{1131}P/cm^2, and implant energies were 60, 30 and 15 keV.

a 5 nm thick SiO_2 layer and a 7 nm thick Si_3N_4 layer. For the oxide, a large fraction of the dose is driven to the interface, while the nitride retards dopant movement. The fraction of activated dopants increases with the implant energy from 25% to 75% for both interfaces. Residual oxygen from water in the ambient of the annealing chamber might contribute to some dopant deactivation also for the samples with silicon nitride interface. Increasing the number of available vacancies during annealing by co-implantation with Si^+ increases the activation probability,(5) and experiments are underway to combine co-implantation with Si_3N_4/Si interfaces to achieve full electrical activation of low energy, low dose implants.

This effect of enhanced dopant segregation rules out the use of SiO_2 interfaces close to P atom arrays. An additional reason to abandon SiO_2 (and Si_3N_4) is that electrical defects in amorphous dielectrics are sources of random polarization that preclude large scale spin qubit integration. We are currently developing a dielectric free, metal silicide based process for epitaxial integration of P atom arrays.

Small scale test devices are important for proof of principle demonstrations. While SiO_2 is incompatible with phosphorus, it is not with antimony. Antimony diffusion is mediated by vacancies,(21) and retarded by interstitials, making formation of Sb atom devices with SiO_2/Si interfaces an attractive test bed for single atom electronics.

4. SINGLE ELECTRON TRANSISTORS IN SILICON

Single electron transistors (SET) are sensitive electrometers that make them prime candidates for implementation of single spin measurements in spin dependent charge measurements.(17) We have formed SET pairs in SOI in order to develop two atom devices for single charge transfer experiments. An example of an SET pair is shown in Fig. 5, together with a typical $I - V$ curve of source-drain current and voltage. Coulomb blockade effects have been studied in silicon nanostructures for many years.(23) Patterning of SOI with electron beam lithography allows reliable formation of silicon nanowires down to 10 nm line width.(5,6) Mechanisms for formation of effective tunnel barriers are controversial. While deliberate tunnel barriers can be formed, charging effects also result from defects and dopant fluctuations. In our fabrication process, silicon nanowires are defined using the negative resist hydrogen silsesquioxane (HSQ) together with an organic resist in a bi-layer electron beam lithography process.(6) Nanowire size reduction by oxidation was not applied, due to direct lithographic access to features around 10–15 nm. The SET with a 14 nm wide wire had a capacitance of 10 aF and a charging energy of about 10 meV. SOI is a very attractive material for formation of single atom test devices

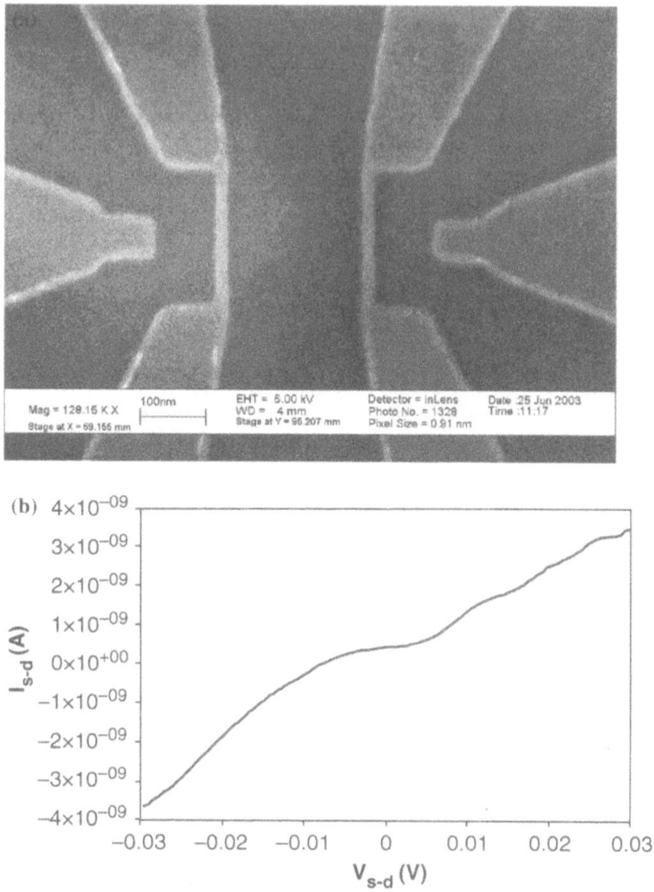

Fig. 5. Scanning electron micrograph of an SET pair in SOI (a) and typical Coulomb blockade signature in the $I - V$ curve of a device with 14 nm wide silicon wire. The temperature was 4.2 K, and the gate voltage 20 mV.

since charging energies are large enough to allow testing at liquid helium temperatures. SET pair structures with a few Sb dopants are promising for single electron transfer tests, but large scale integration requires the development of a dielectric free architecture.

5. OUTLOOK

The effective ion placement resolution that can be achieved with single ion implantation is limited by the following contributions: (1) Range straggling from statistical energy loss processes leads to a distribution of implantation depths. Range straggling is proportional to the implant energy, and implantation with very low energies (< 1 keV) can achieve placement accuracies of a few nm. For heavier ions, implantation into a given mean depth is accompanied by smaller range straggling. E.g., the range straggling for implantation into a depth of 20 nm is 9 nm for 13 keV P, but only 6 nm for 25 keV Sb ions. (2) The beam spot size can be controlled with the aperture size in the SFM tip, and ~5 nm wide apertures have been formed. Ion transport through nanotubes (carbon, or boron nitride based) would reduce the aperture size to 1–2 nm, and experiments are underway to test the efficiency of nanotube ion guides. (3) Alignment to markers can be achieved with scanning force microscopy to an imaging resolution of 1–5 nm. (4) Diffusion during annealing can completely dissolve atom arrays. Most dopants diffuse through distinct and specific mechanisms.[21] Phosphorous diffuses in silicon through interstitials, while antimony diffuses through vacancies. P diffusion is retarded by silicon nitride (which injects vacancies), while antimony diffusion is retarded by silicon dioxide (which injects interstitials). Choice of appropriate dopant interface combinations can reduce dopant movement during annealing to a few nm. Taking these contributions into account we conclude that an overall placement accuracy of 5–10 nm is possible with our current single ion implantation approach, and this enables the formation of prototype devices for spin qubits. A placement resolution below 5 nm will be possible if low energy ions can be transported through nanotubes efficiently. While single atom placement by scanning tunneling microscope (STM) based hydrogen lithography has atomic resolution in single atom placement, issues of diffusion and segregation are dependent on device integration and remain to be resolved also for STM based arrays.[24,25] In contrast to STM, single ion implantation does not require specific surface chemistries, so that ions can be placed into any matrix, and ions across the periodic table can be implanted. It is also perceivable to form devices where the strengths of both techniques are utilized.

Integration of several thousand qubits is necessary to realize the full potential of quantum computation. The development of tools for single atom array formation, process integration and formation of device infrastructure (control gates and SETs) has to take this goal into account at every level, while also focusing on the demonstration of basic steps in single and few qubit experiments.

ACKNOWLEDGMENTS

We thank the staff of LBNL's National Center for Electron Microscopy (NCEM) and the U. C. Berkeley Microlab for their support. This work was supported by the National Security Agency and Advanced Research and Development Activity under Army Research Office contract number MOD707501, and by the U.S. Department of Energy under contract No. DE-AC03-76SF00098. Work at LLNL was performed under the auspices of the U.S. DOE under contract No. W-7405-ENG-48.

REFERENCES

1. B. E. Kane, *Nature* **393**, 133 (1998).
2. A. J. Skinner, M. E. Davenport, and B. E. Kane, *Phys. Rev. Lett.* **90**, 087901 (2003).
3. R. Vrijen, E. Yablonovitch, K. Wang, H. W. Jiang, A. Balandin, V. Roychowdhury, T. Mor, and D. DiVincenzo, *Phys. Rev. A* **62**, 12306 (2000).
4. A. M. Tyryshkin, S. A. Lyon, A. V. Astashkin, and A. M. Raitsimring, *Phys. Rev. B* **68**, 193207 (2003).
5. T. Schenkel, A. Persaud, S. J. Park, J. Nilsson, J. Bokor, J. A. Liddle, R. Keller, D. H. Schneider, D. W. Cheng, and D. E. Humphries, *J. Appl. Phys.* **94**, 7017 (2003).
6. S.-J. Park, A. Persaud, J. A. Liddle, J. Nilsson, J. Bokor, D. H. Schneider, I. Rangelow, and T. Schenkel, *Microelectr. Eng.* in press, cond-mat/0310195.
7. Yang Changyi, D. N. Jamieson, C. Pakes, S. Prawer, A. Dzurak, F. Stanley, P. Spizziri, L. Macks, E. Gauja, and R. G. Clark. *J. Appl. Phys.* **42**, 4124 (2003).
8. T. Shinnada, H. Koyama, C. Hinoshita, K. Imamura, and I. Ohdomari, *Jpn. J. Appl. Phys.* **41**, L287 (2002).
9. R. A. Baragiola, *Nucl. Instr. Meth. B* **78**, 223 (1993).
10. T. Schenkel, A. V. Barnes, M. A. Briere, A. Hamza, A. Schach von Wittenau, and D. Schneider, *Nucl. Instr. and Meth. B* **125**, 153 (1997).
11. F. Aumayr, H. Kurz, D. Schneider, M. A. Briere, J. W. McDonald, C. E. Cunningham, and H. P. Winter, *Phys. Rev. Lett* **71**, 1943 (1993).
12. J. Orloff, M. Utlaut, and L. Swenson, *High Resolution Focused Ion Beams: FIB and its Applications* (Kluwer, New York, 2003).
13. R. E. Marrs, *Nucl. Instr. Meth. B* **149**, 182 (1999).
14. T. Schenkel, A. Persaud, A. Kraemer, J. W. McDonald, J. P. Holder, A. V. Hamza, and D. H. Schneider, *Rev. Sci. Instr.* **73**, 663 (2002).
15. T. Schenkel, A. Persaud, S. J. Park, J. Meijer, J. R. Kingsley, J. W. McDonald, J. P. Holder, J. Bokor, and D. H. Schneider, *J. Vac. Sci. Technol. B* **20**, 2819 (2002).
16. M. Hattass, T. Schenkel, A. V. Hamza, A. V. Barnes, M. W. Newman, J. W. McDonald, T. R. Niedermayr, G. A. Machicoane, and D. H. Schneider, *Phys. Rev. Lett.* **82**, 4795 (1999).
17. B. E. Kane, N. S. McAlpine, A. S. Dzurak, and R. G. Clark, G. J. Milburn, He Bi Sun, and Howard Wiseman, *Phys. Rev. B* **61**, 2961 (2000).
18. I. W. Rangelow, F. Shi, P. Hudek, T. Gotszalk, P. B. Grabiec, and P. Dumania, *SPIE* **2879**, 56 (1996).
19. T. Schenkel, V. Radmilovic, E. A. Stach, S.-J. Park, and A. Persaud, *J. Vac. Sci. Technol. B* **21**, 2720 (2003).
20. R. Pedrak, Tzv. Ivanov, K. Ivanova, T. Gotszalk, N. Abedinov, and I. W. Rangelow, K. Edinger, E. Tomerov, T. Schenkel, and P. Hudek, *J. Vac. Sci. Technol. B* **21**, 3102 (2003).
21. A. Ural, P. B. Griffin, and J. D. Plummer, *J. Appl. Phys.* **85**, 6440 (1999).
22. J. Dabrowski, H.-J. Müssig, V. Zavodinsky, R. Baierle, and M. J. Caldas, *Phys. Rev. B* **65**, 245305 (2002).

23. see e.g., H. Ahmed, *J. Vac. Sci. Technol. B* **15**, 2101 (1997).
24. T.-C. Shen, C. Wang, G. C. Abeln, J. R. Tucker, J. W. Lyding, Ph. Avouris, and R. E. Walkup, *Science* **268**, 1590 (1995); T.-C. Shen, J.-Y. Ji, M. A. Zudov, R.-R. Du, J. S. Kline, and J. R. Tucker, *Appl. Phys. Lett.* **80**, 1580 (2002).
25. S. R. Schofield, N. J. Curson, M. Y. Simmons, F. J. Rue©, T. Hallam, L. Overbeck, and R. G. Clark, *Phys. Rev. Lett.* **91**, (2003).

Bang–Bang Refocusing of a Qubit Exposed to Telegraph Noise

Henryk Gutmann,[1] Frank K. Wilhelm,[1] William M. Kaminsky[2] and Seth Lloyd[3,4]

Solid state qubits promise the great advantage of being naturally scalable to large quantum computer architectures, but they also possess the significant disadvantage of being intrinsically exposed to many sources of noise in the macroscopic solid-state environment. With suitably chosen systems such as superconductors, many of sources of noise can be suppressed. However, imprecision in nanofabrication will inevitably induce defects and disorder, such as charged impurities in the device material or substrate. Such defects generically produce telegraph noise and can hence be modelled as bistable fluctuators. We demonstrate the possibility of the active suppression of such telegraph noise by bang–bang control through an exhaustive study of a qubit coupled to a single bistable fluctuator. We use a stochastic Schrödinger equation, which is solved both numerically and analytically. The resulting dynamics can be visualized as diffusion of a spin vector on the Bloch sphere. We find that bang–bang control suppresses the effect of a bistable fluctuator by a factor roughly equalling the ratio of the bang–bang period and the typical fluctuator period. Therefore, we show the bang–bang protocol works essentially as a high pass filter on the spectrum of such telegraph noise sources. This suggests how the influence of $1/f$-noise ubiquitous to the solid state world could be reduced, as it is typically generated by an ensemble of bistable fluctuators. Finally, we develop random walk models that estimate the level of noise suppression resulting from imperfect bang–bang operations, such as those that cannot be treated as δ-function impulses and those that have phase and axis errors.

KEY WORDS: Bistable fluctuator; telegraph noise; $1/f$-noise; quantum control; random walk.

PACS: 03.65.Yz; 03.67.Lx; 05.40.-a.

[1]Sektion Physik and CeNS, Ludwig-Maximilians-Universität, 80333 München, Germany.
[2]Department of Physics, Massachusetts Institute of Technology, Cambridge, MA 02139.
[3]Department of Mechanical Engineering, Massachusetts Institute of Technology, Cambridge, MA 02139.
[4]To whom correspondence should be addressed. E-mail: gutmann@theorie.physik.uni-muenchen.de

1. INTRODUCTION

In order to implement solid-state quantum information processing devices, the noise sources causing decoherence of their quantum states have to be carefully understood, controlled, and eliminated. This is a formidable task, as a solid-state environment generically couples a macroscopic number of degrees of freedom to any such device. Thus, a fundamental prerequisite for any design is that it must significantly decouple the quantum states used for computation from phonons and other quasiparticles in the underlying solid crystal. Examples of such designs are those employing discrete states in quantum dots[(1)] or superconductors with a gapped density of states.[(2–6)]

Most research going beyond this fundamental prerequisite has concentrated on decoupling devices from external noise sources such as electromagnetic noise generated by control and measurement apparatus.[(7)] On the other hand, there inevitably are internal noise sources because the fabrication of gates, tunnel junctions, and other functional components creates defects in the underlying crystal. Prominent examples of such defects are background charges in charge-based devices or cricital current fluctuations in flux-based devices.[(8,9)] A clear signature of such defects is telegraph noise in the case of a few defects or $1/f$-noise in the case of a larger ensemble.[(10)] With the growing success in engineering the electromagnetic environment, these defects are becoming more and more the key limiting sources of decoherence.

Such defects do not fall in the large class of noise sources that can be approximated well as a bosonic bath, and this fact complicates analysis. Whereas it is realistic to treat a bosonic bath in the tractable near-equilibrium thermodynamic limit where fluctuations are purely Gaussian,[(11–13)] localized noise sources with bounded spectra like the defects in which we are interested produce noise that is significantly non-Gaussian. Theories treating large ensembles of non-Gaussian noise sources have been presented.[(14,15)] However, with the ongoing improvement in nanofabrication technology, it is realistic to consider the case where non-Gaussian noise sources are reduced down to only a single one or a few per device. This is the case we treat here, and thus the defects find a more realistic representation as a small set of bistable fluctuators.[(16)] (henceforth abbreviated bfls). In principle, this approach can be extended to larger sets of bfls with a range of different mean switching times (e.g., an ensemble with an exponential distribution of switching times that produces $1/f$-noise.[(17–19)])

This report is organized as follows. Section 2 presents the model of a single bfl in the semiclassical limit, where it acts as a source of telegraph noise. Section 3 introduces an idealized open loop quantum control technique, quantum bang–bang control,[20–22] which is suitable for slowly fluctuating noise sources. Section 4 explains how we simulated the qubit dynamics under the influence of noise with and without bang–bang control by integrating of the corresponding time-dependent Schrödinger equation. As a measure of the decoherence, we analyze the deviations of the qubit's trajectory on the Bloch sphere from that of the noiseless case. These deviations take the form of a random walk around the noiseless-case trajectory. We therefore analyze the suppression of these deviations by comparing the variances of these random walks with and without bang–bang control. Both numerical and analytical solutions (the latter in the long-time or "diffusion" limit) are presented. Comparison of the numerical simulations to the analytical solutions shows excellent agreement. We then analyze how these results change when practical limitations are considered such as the fact that a bang–bang pulse cannot be an ideal δ-function impulse and the fact that the duration or polarization axis of the pulse may suffer from random fluctuations. We show at the end of subsec. 4.2 that within large margins bang–bang suppression of the bfl noise is not inhibited by having a finite, rather than infinitesimal, pulse length. However, in Sec. 5, we do find that duration and polarization axis errors in the bang–bang pulses can significantly affect the suppression of bfl noise. We present a point of optimum performance. Section 6 concludes with remarks on several recent publications concerning the suppression of telegraph or $1/f$-noise.

2. MODEL OF THE BISTABLE FLUCTUATOR IN ITS SEMICLASSICAL LIMIT

We describe the bfl-noise influenced evolution of the qubit in its semiclassical limit by using a stochastic Schrödinger equation[23,24] with the time-dependent effective Hamiltonian

$$H_q^{\text{eff}}(t) = H_{\text{q}} + H_{\text{noise}}(t) \tag{1}$$

$$H_{\text{q}} = \hbar\epsilon_{\text{q}}\hat{\sigma}_z^{\text{q}} + \hbar\Delta_{\text{q}}\hat{\sigma}_x^{\text{q}} \tag{2}$$

$$H_{\text{noise}}(t) = \hbar\alpha\ \hat{\sigma}_z^{\text{q}}\ \xi_{\text{bfl}}(t) \tag{3}$$

where ϵ_{q} and Δ_{q} define the free (noiseless) qubit dynamics. $\xi_{\text{bfl}}(t)$ denotes a function randomly switching between ± 1 (see, Fig. 1), which represents a telegraph noise signal. The switching events follow a symmetrical Poisson process, i.e., the probabilities of the bfl switching from $+1$ to -1 or -1 to $+1$ are the same and equal in time. The Poisson process is characterized by the mean time separation τ_{bfl} between two bfl flips. The coupling

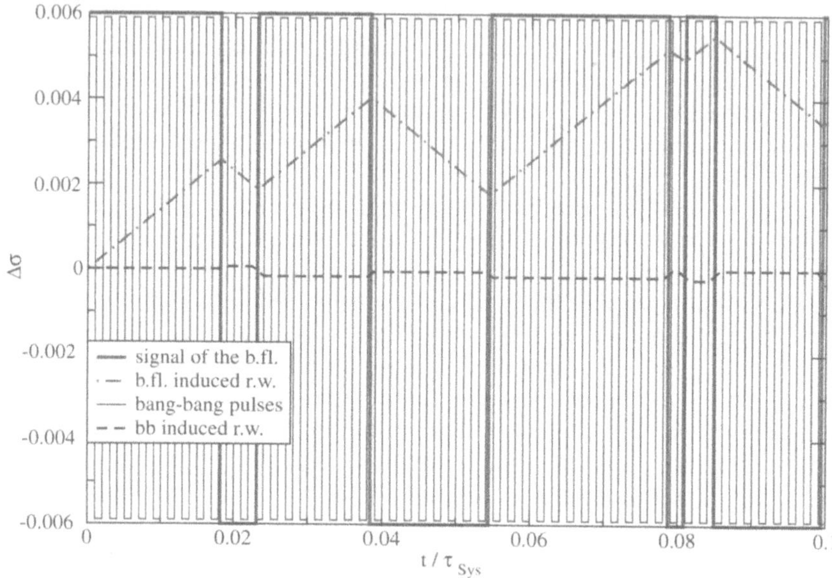

Fig. 1. Schematic plot of a typical Poisonian bfl noise signal and its resulting random walk behavior (in the limit of small deviations). The periodic fast switching step function represents a bang–bang pulse with a time scale ratio: $\tau_{\rm bfl}/\tau_{\rm bb} = 10$ and yields a quite smaller random walk step-length. $\tau_{\rm Sys} = \frac{\pi}{\sqrt{\epsilon_{\rm q}^2+\Delta_{\rm q}^2}}$ denotes the evolution period of the qubit in the noiseless case.

amplitude to the qubit in frequency units is α. The relation of this Hamiltonian to a microscopic model is explained in the Appendix.

Starting with an arbitrary initial state of the qubit, represented by some given point on the Bloch sphere, we can numerically integrate the corresponding stochastic differential equation and obtain the corresponding random walk on the Bloch sphere

$$\vec{\sigma}(t) = \mathrm{T}\exp\left(-i/\hbar \int_0^t H_{\rm q}^{\rm eff}(s)\,\mathrm{d}s\right)\vec{\sigma}(0) \tag{4}$$

with T denoting the usual time-ordering operator.

3. BANG–BANG CONTROL PROTOCOL

We propose to reduce the influence of the bfl-noise by applying to the qubit a continuous train of π-pulses along the σ_x-axis. This refocusing pulse scheme essentially corresponds to the standard quantum bang–bang procedure [20–22] or the Carr–Purcell–Gill–Meiboom echo technique from NMR.[25] For technical convenience, we consider the π-pulses to be of infinitesimal duration. This simplification is not crucial as will be detailed later in Subsec. 4.2 The pulses are assumed to be separated by a constant time interval $\tau_{\rm bb}$. The mean separation $\tau_{\rm bfl}$ between two bfl-flips is assumed to be much longer than $\tau_{\rm bb}$. For theoretical convenience, we also assume that $\tau_{\rm bfl}$ is shorter than the free precession period of the qubit. This too is not a crucial restriction. (It can always be overcome by changing to a co-precessing frame.)

Qualitatively, bang–bang control works as follows. Since $\tau_{\rm bb} \ll \tau_{\rm bfl}$, it is usually the case that the bfl does not flip during the time between two bang–bang pulses that flip the qubit. In this way, the bang–bang pulses average out the influence of $H_{\rm noise}(t)$. In fact, the refocusing scheme fully suppresses the σ_z-term of the static Hamiltonian (2) (compare Fig. 5); but this turns out to be no crucial obstacle to universal quantum computation as will be outlined later on. As one can visualize in Fig. 1, it is only when a bfl-flip occurs during a bang–bang period that the net influence of the bfl felt by the qubit is nonzero, and the qubit thus suffers some random deviation from its trajectory in the noiseless case. Taken together, these random deviations constitute a random walk around the noiseless case trajectory. While this walk is actually continuous, it can be modelled as a discrete walk with steps that are randomly distributed in time, one step for each bfl flip (see e.g. Ref. 26). The average step length is essentially the product of the noise coupling strength α and the mean time the bfl in its present state can influence the qubit. Without bang–bang control, this mean influence time is $\tau_{\rm bfl}$, whereas with bang–bang control, it is reduced to $\tau_{\rm bb}$. Therefore, both with and without bang–bang control, the random walk has the same time distribution of steps, but with bang–bang control the step size can be significantly reduced roughly by a factor of the ratio of time scales $\tau_{\rm bb}/\tau_{\rm bfl}$.

4. RANDOM WALK ON THE BLOCH SPHERE

Now we study this proposal quantitatively. We simulate these random walks both with and without bang–bang control by integrating both numerically and analytically the Schrödinger equation, Eq. (4), with the stochastic Hamiltonian of Eqs. (1–3). As generic conditions for the qubit dynamics, we choose $\epsilon_{\rm q} = \Delta_{\rm q} \equiv \Omega_0$. Without loss of generality, we set the qubit's initial state to be spin-up along the z-axis. If the qubit–bfl coupling α were zero, then the qubit would simply precess freely on the Bloch

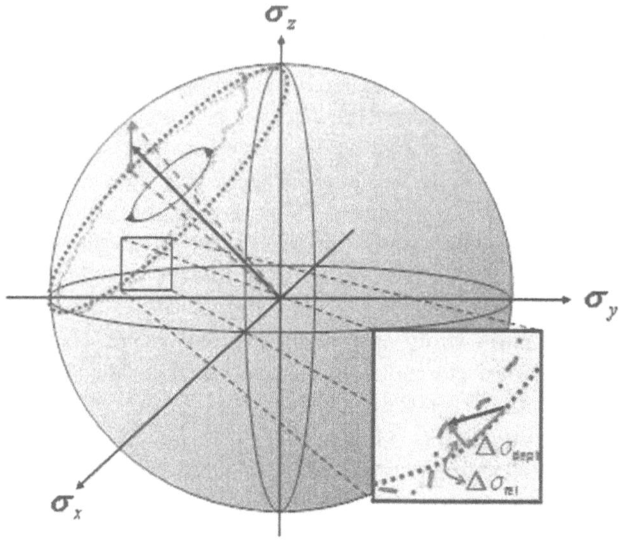

Fig. 2. Schematic plot of noisy qubit evolution generated by Poissonian telegraph noise. The resulting random walk (dot-dashed line) on the Bloch sphere is comprised both of deviations $\Delta\sigma_{\text{deph}}$ in parallel to the free precession trajectory (dotted line), which correspond to dephasing, and deviations $\Delta\sigma_{\text{rel}}$ perpendicular to it, which correspond to relaxation/excitation.

sphere around the rotation axis $\hat{\sigma}_x^{\text{q}}+\hat{\sigma}_z^{\text{q}}$ (the dotted line in Fig. 2). Hence, we expect for a sufficiently small coupling ($\alpha \ll \Omega_0$) only a slight deviation of the individual time evolution compared to the free evolution case (the dashed line in Fig. 2). For the coupling strength, we take $\alpha = 0.1\Omega_0$. All the following times and energies are given in units of the unperturbed system Hamiltonian, i.e., our time unit τ_{Sys} is given according the free precession time $\pi\tau_{\text{Sys}}/\sqrt{2}$, and our energy unit is given by $\Delta E = \sqrt{\epsilon_{\text{q}}^2+\Delta_{\text{q}}^2} = \sqrt{2}\Omega_0$. The time scale ratio is taken to be $\tau_{\text{bfl}}/\tau_{\text{bb}} = 10$ if not denoted otherwise.

This approach accounts for the essential features of our specific situation: the long correlation time of the external noise, essentially τ_{bfl}, its non-Gaussian statistics and its potentially large amplitude at low frequencies. These properties are crucial and are difficult, although not impossible, to take into account in standard master equation methods.

4.1. Numerical Simulations

We have numerically integrated Eq. (4) and averaged the deviations of the random walk evolution from the unperturbed trajectories for times up to $100\tau_{\mathrm{Sys}}$ over $N = 10^3$ realizations. Larger simulations have proven that convergence is already sufficient at this stage. We shall examine the root-mean-square (rms) deviations of this ensemble at given time points

$$\Delta\vec{\sigma}_{\mathrm{rms}}(t) = \sqrt{\frac{1}{N}\sum_{j=1}^{N}\left(\vec{\sigma}_j^{\mathrm{q}}(t) - \vec{\sigma}_{\mathrm{noisy},j}^{\mathrm{q}}(t)\right)^2} \tag{5}$$

with and without bang–bang control. In other approaches, such as those based on master equations, one separates dephasing and relaxation. Both are contained here in Eq. (5). We shall point out notable differences between these two channels. The deviation as a function of time is plotted in Fig. 3.

The total deviations on intermediate time scales are suppressed by a ratio on the order of 10. A detailed numerical analysis shows that *without* bang–bang suppression, the deviations parallel to the free precession

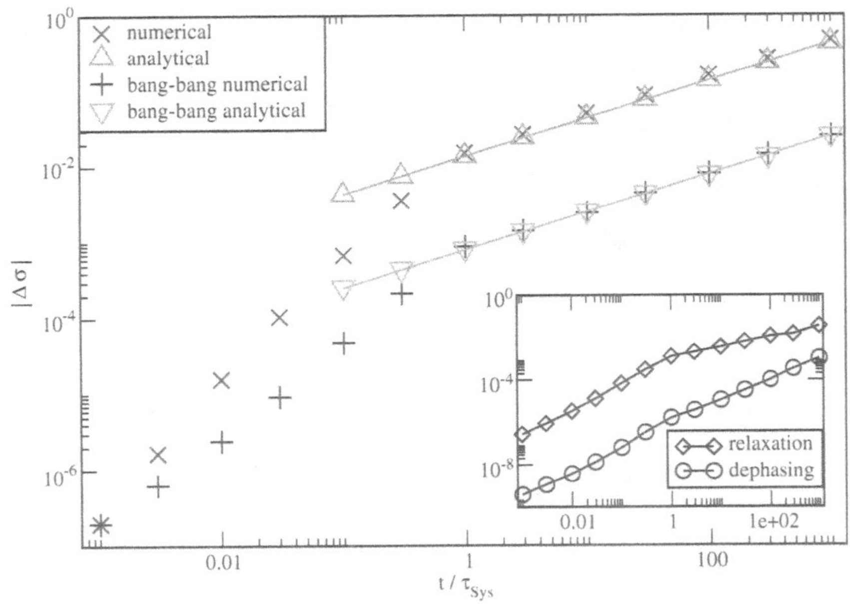

Fig. 3. Time evolution of the rms deviations for bfl-induced random walks with and without bang–bang control at a coupling constant $\alpha = 0.1$ and a typical flipping time scale $\tau_{\mathrm{bfl}} = 10^{-2}\tau_{\mathrm{Sys}}$. The separation between two bang–bang pulses is $\tau_{\mathrm{bb}} = 10^{-3}\tau_{\mathrm{Sys}}$. The straight lines are square-root fits of the analytical derived random walk model variances (plotted as triangles). Inset: Components of the deviations from the free precession trajectory that are parallel to it (dephasing) and perpendicular to it (relaxation/excitation) with bang–bang control.

trajectory (which correspond to dephasing) are of similar size to those perpendicular to free precession (which correspond to relaxation/excitation). In contrast, with bang–bang control, dephasing is almost totally absent as one can see in the inset of Fig. 3.

The main double-logarithmic plot of Fig. 3 shows that on short time scales ($t \lesssim 0.1\tau_{\rm Sys}$, which corresponds to $\lesssim 10$ random walk steps), deviations increase almost linearly in time. It is not until times on the order of $\tau_{\rm Sys}$ that the noise-induced deviations start to behave as typical classical random walks, increasing as a square-root in time.

4.2. Analytical Random Walk Models

We now develop analytical random walk models for our system. The random walk on the Bloch sphere is in general two-dimensional, consisting of both parallel and perpendicular deviations to the free evolution trajectory. Bang–bang control, as was seen in the above numerical results and as will be seen in the following analytical results, essentially reduces the random walk to one-dimension as only the perpendicular deviations remain significant. In the following, we restrict ourselves to the long-time (many random walk steps) regime.

We first calculate for both cases the probability distributions of the deviations after one bfl flip ("one-step deviations" in terms of the discrete random walk). The fluctuation of the period between $\tau_{\rm per}^{\pm}$ leads to dephasing, which can be evaluated at $\alpha \ll \epsilon_{\rm q}, \Delta_{\rm q}$ to

$$\Delta\vec{\sigma}_{\rm deph}^{\rm bfl} = 2\pi\cos\phi\left(\frac{1}{\tau_{\rm per}^{\pm}} - \frac{1}{\tau_{\rm per}}\right)\tau_{\rm bfl} \simeq \pm 2\frac{\Delta_{\rm q}\epsilon_{\rm q}}{\Delta_{\rm q}^2+\epsilon_{\rm q}^2}\alpha\tau_{\rm bfl} \qquad (6)$$

where the prefactor $\cos\phi = \frac{\Delta_{\rm q}}{\sqrt{\Delta_{\rm q}^2+\epsilon_{\rm q}^2}}$ takes the effective trajectory radius into account.

For the relaxation/excitation effect of the noise, one has to use the projection of the perturbation orthogonal to the free axis, using $\sin\eta = \alpha\Delta_{\rm q}/(\epsilon_{\rm q}^2+\Delta_{\rm q}^2)$. Furthermore this type of deviation also depends on the actual position of the spin on the Bloch sphere, e.g., there is no relaxation when the state is at one of the poles.

Averaging in rms-fashion over a full azimuthal cycle leads to a factor of $1/\sqrt{2}$. Moreover, the impact of relaxation/excitation is scaled down by an additional factor of $\cos\phi = \Delta_{\rm q}/\sqrt{\Delta_{\rm q}^2+\epsilon_{\rm q}^2}$ corresponding to the projection of the Bloch vector onto the precession axis, which furthermore decreases the deviation angle. In total, using $\tau_{\rm per}^{\pm} \simeq \tau_{\rm per}$ to first order in α, we find

$$\Delta\vec{\sigma}_{\rm rel}^{\rm bfl} = 2\pi\cos\phi\sin\eta\frac{1}{\sqrt{2}}\cos\phi\frac{\tau_{\rm bfl}}{\tau_{\rm per}^{\pm}} \simeq \sqrt{2}\frac{\Delta_{\rm q}^3}{(\epsilon_{\rm q}^2+\Delta_{\rm q}^2)^{3/2}}\alpha\tau_{\rm bfl}. \tag{7}$$

Our rms measure of the impact of the noise, Eq. (5), does not handle these two kinds of deviations separately, but rather adds them up to:

$$\begin{aligned}\Delta\vec{\sigma}_{\rm total}^{\rm bfl} &= \sqrt{\Delta\vec{\sigma}_{\rm deph}^{\rm bfl\ 2} + \Delta\vec{\sigma}_{\rm rel}^{\rm bfl 2}} = \sqrt{4\frac{\Delta_{\rm q}^2\epsilon_{\rm q}^2}{(\Delta_{\rm q}^2+\epsilon_{\rm q}^2)^2} + 2\frac{\Delta_{\rm q}^6}{(\epsilon_{\rm q}^2+\Delta_{\rm q}^2)^3}}\alpha\tau_{\rm bfl} \\ &= \frac{1}{(\Delta_{\rm q}^2+\epsilon_{\rm q}^2)^{3/2}}\sqrt{4(\Delta_{\rm q}^2+\epsilon_{\rm q}^2)\Delta_{\rm q}^2\epsilon_{\rm q}^2 + 2\Delta_{\rm q}^6}\alpha\tau_{\rm bfl}\end{aligned} \tag{8}$$

Our rms treatment disregard the different types of decoherence, dephasing and relaxation/excitation, corresponding to phase and bit-flip errors respectively. This is no crucial drawback but merely lies in the nature of our generic situation. If needed, both components can be isolated.

The derivation of the maximal one-step deviation for the bang–bang controlled situation has to be handled differently. The deviation resulting from a bfl-flip during a bang–bang pulse period is maximal if the step happens exactly at the moment of the second qubit spin-flip (i.e., in the middle of the bang–bang cycle). When this happens, the refocusing evolution has in its first half a drift, for example, to the "right" (compare to Fig. 5) and in the last half an equal aberration.

The resulting one-step deviation appears to be on the order of $2\alpha\tau_{\rm bb}$. However, this is scaled down by a factor of $1/\sqrt{2}$, as the impact of the aberration in x-direction is proportional to a factor of $\sin\chi$, where χ denotes the longitudinal angle of the present spin position on the Bloch sphere (see Fig. 5). This is because the σ_z-component of the noisy evolution does not influence the qubit, if it is near the $\sigma_z = \pm 1$-state and its influence is suppressed correspondingly in between. As we are mainly interested in mean aberrations after many random walk steps, we simply average the maximal one-step deviation over one precession period in the usual rms manner to obtain

$$\begin{aligned}\left\langle\Delta\vec{\sigma}_{\rm max}^{\rm bb}\right\rangle &= \sqrt{\frac{1}{2\pi}\int_0^{2\pi}\sin^2\chi\, 4\alpha^2\tau_{\rm bb}^2\, d\chi} \\ &= \sqrt{2}\alpha\tau_{\rm bb}.\end{aligned} \tag{9}$$

Obviously, this variance only contributes to relaxation.

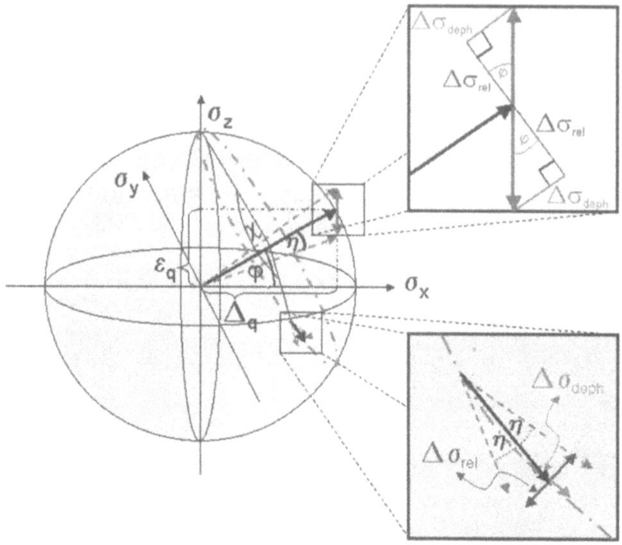

Fig. 4. Plot of a typical one-step deviation from the unperturbed qubit trajectory with generic values for ϵ_q and Δ_q. The fractions of the bfl fluctuations in $\hat{\sigma}_z$-direction have to be distinguished with respect to their effects on the qubit: those that yield dephasing deviations that are parallel to the free precession trajectory (proportional to $\sin\phi$) versus relaxation/excitation deviations that are perpendicular (proportional to $\sin\eta$). Both parts are additionally domineered by a factor of $\cos\phi$ due to the diminished radius of the trajectory starting from the initial state $\sigma_z = +1$. The impact of the relaxation/excitation generating part is furthermore depending on $\cos\phi$ as well as $\sin\chi$, the azimuth angle of the qubits present position.

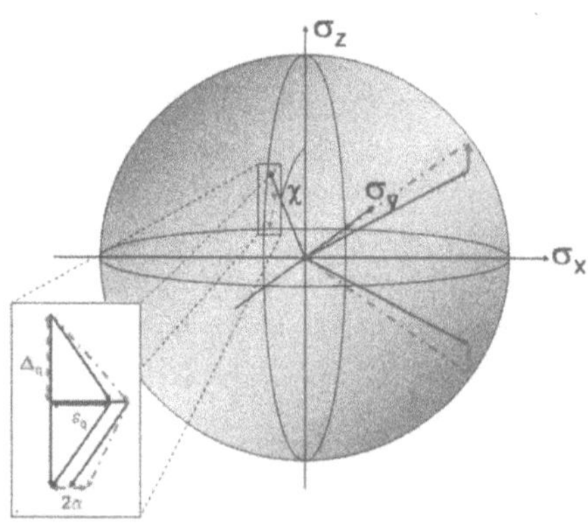

Fig. 5. Sketch of a maximal one-step deviation during a bang–bang modulated cycle, which appears if the bfl-state flips precisely at the intermediate bang–bang pulse time. The dephasing part of deviation evidently averages out, while a relaxating aberrance arise proportional to the noise-coupling constant α.

In the long time limit, we replace the fluctuating number of random walk steps for a given time Δt of noisy evolution by its expectation value $N_{\text{bfl}} = \Delta t/\tau_{\text{bfl}}$. This allows us to use the number of random walk steps as time parameter.[26] This simplification does not introduce significant error, as the relative number variation for Δt scales as $\frac{\sqrt{\Delta t/\tau_{\text{bfl}}}}{\Delta t/\tau_{\text{bfl}}} = \sqrt{\tau_{\text{bfl}}/\Delta t} \to 0$ in our preferred long-time limit. We encounter two different one-step-distributions, depending on whether the numeration of the step is an odd or even (corresponding to an "up" or "down" state of the bfl). For definiteness, we assume the bfl is initially in its "upper" state, which is of no influence on the long time limit as the memory to the initial state is already erased. The step-size distribution of the bfl model in our small deviation regime is given from Poisson statistics

$$\Phi^{\text{bfl}}_{\text{odd/even}}(x) = \frac{e^{\mp x/\beta}\theta(\pm x)}{\beta} \tag{10}$$

with $\beta = \frac{\sqrt{5}}{2}\alpha\tau_{\text{bfl}}$ the typical one-step deviation as calculated in Eq. (8). $\theta(x)$ denotes the Heaviside step function. We neglect the correlations between transverse and perpendicular deviations as we expect them to average out in the long-time limit.

For the bang–bang suppressed random walk, the flipping positions of the bfl-noise sign in the bang–bang time-slots are essentially randomly distributed as long as $\tau_{\text{bb}} \ll \tau_{\text{bfl}}$. That is why we find a constant step-size distribution between zero and a maximum divergence of $\gamma = \frac{2\alpha\tau_{\text{bb}}}{\sqrt{2}}$ (see Eq. (9)), namely

$$\Phi^{\text{bb}}_{\text{odd/even}}(x) = \frac{\theta(\pm x)\theta\,(\gamma \mp x)}{\gamma}. \tag{11}$$

By means of these one-step probability distributions, we are able to calculate via the convolution theorem the distributions for $2N_{\text{bfl}}$-step random walks. Specifically, they are the inverse Fourier transforms of the N_{bfl}-fold products of the Fourier transforms of the two-step distribution.[26] For the case without bang–bang control, we find

$$\begin{aligned}\Phi^{\text{bfl}}_{2\text{N}_{\text{bfl}}}(x) &= \mathscr{F}^{-1}\left[\left(\mathscr{F}\left[\Phi^{\text{bfl}}_{2}\right]\right)^{N_{\text{bfl}}}\right]\\ &= \int_{-\pi}^{\pi}\frac{dk}{2\pi\beta^{2N_{\text{bfl}}}}e^{-ikx}\left(\frac{1}{1-2\cos(k)e^{-1/\beta}+e^{-2/\beta}}\right)^{N_{\text{bfl}}}\end{aligned} \tag{12}$$

whereas for the case with bang–bang control, it is

$$\Phi^{\text{bb}}_{2\text{N}_{\text{bfl}}}(x) = \mathscr{F}^{-1}\left[\left(\mathscr{F}\left[\Phi^{\text{bb}}_{2}\right]\right)^{N_{\text{bfl}}}\right]$$

$$= \int_{-\pi}^{\pi} \frac{dk}{2\pi \gamma^{2N_{\rm bfl}}} e^{-ikx} \left(\frac{[1-\cos((\gamma+1)k)]}{[1-\cos(k)]} \right)^{N_{\rm bfl}} \tag{13}$$

with $\mathscr{F}$ and $\mathscr{F}^{-1}$ denoting the discrete Fourier transformation and its inverse, respectively.

Already for random walk step-numbers on the order of 10, the resulting distributions are almost Gaussian. Their standard deviations give the rms-deviations of the random walk models plotted in Fig. 3. As in the numerical simulations at long times, they grow as a square-root of the number of steps. As one can recognize, the underlying two-step distributions in the k-space (i.e., the functions in the large brackets of Eqs. (12) and (13)) are symmetric and differentiable around zero such that the above integrals can be evaluated analytically using the saddle point approximation (the small parameter is k, which is justified at least qualitatively in our bounded variable integral). We find for their variances in real space representation

$$\Delta\sigma_{\rm bfl}(N_{\rm bfl}) = \sqrt{N_{\rm bfl}}\beta = \sqrt{N_{\rm bfl}}\frac{\sqrt{5}}{2}\alpha\tau_{\rm bfl} \tag{14}$$

for the case without bang–bang control and

$$\Delta\sigma_{\rm bb}(N_{\rm bfl}) = \frac{\sqrt{N_{\rm bfl}}}{2}\gamma = \sqrt{\frac{N_{\rm bfl}}{2}}\alpha\tau_{\rm bb} \tag{15}$$

for the case with it. In the large-$N_{\rm bfl}$ limit, this model shows excellent agreement with the numerical simulations.

At first sight, treating bang–bang pulses as δ-function impulses appears to be an extraordinarily strong assumption, especially because in a physical implementation, the large bandwidth associated with very short pulses could excite other noise sources. However, this δ-function impulse approximation is only for technical simplification. In fact, going to the other extreme of a wide, continuous pulse of the form $\sin(\frac{\pi}{\tau_{bb}}t)$ would also refocus our bfl-noise over the course of its periods. Comparing the two-step deviation distributions arising from δ-function impulses versus continuous sine waves, one obtains for the δ-function case

$$\Phi_2^{\rm inf}(x) = \frac{|\gamma - x|}{\gamma}\theta(\gamma - x)\theta(\gamma + x) \tag{16}$$

and for the continuous sine wave case

$$\Phi_2^{\rm cont}(x) = \left\{ \left[\frac{\pi}{2\gamma} + \frac{\pi}{4\gamma}\cos\left(2\pi\frac{x}{\gamma}\right) \right] \left(1 - \frac{x}{\gamma}\right) + \frac{3}{16\gamma}\sin\left(2\pi\frac{x}{\gamma}\right) \right\} \theta(\gamma - x)\theta(\gamma + x) \tag{17}$$

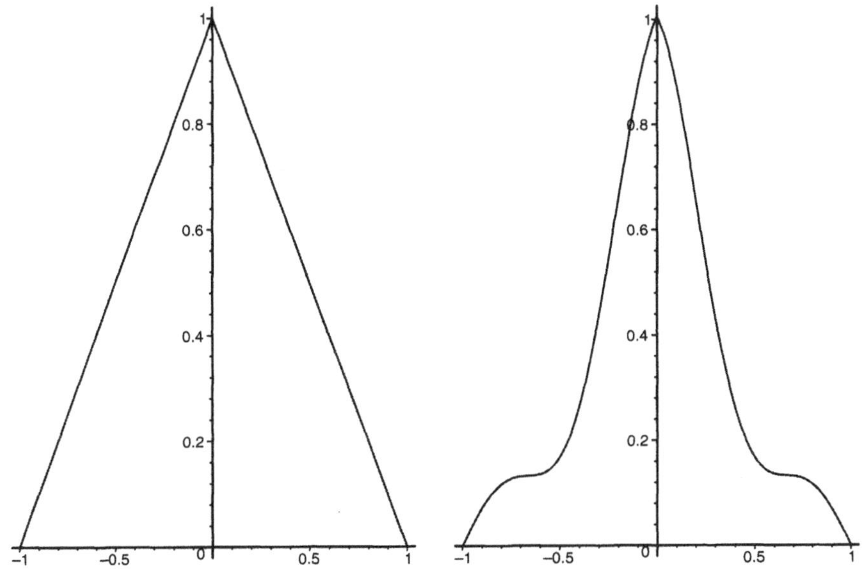

Fig. 6. Comparison of two-step distributions for the random walks with bang–bang control when the bang–bang pulses are taken to be δ-functions (left) versus a continuous sine wave $\sin(\frac{\pi}{\tau_{bb}}t)$ (right). For clarity, the y-axis is rescaled to the maximum values of the distributions, while the x-axis is given in units of γ.

These distributions are depicted in Fig. 6. One recognizes that in fact the distribution arising in the continuous sine wave case is narrower (and therefore indicates *more* effective noise suppression) than the δ-function impulse case, with the drawback of leaving less free evolution time for coherent operation.

4.3. Distributions of the Random Walks Deviation

Beyond predicting the variances of the random walks, our analysis also allows evaluation of their full probability distributions. We compare them to numerics with and without bang–bang compensation by use of simulations with 10^4 realizations at an evolution time $t_0 = \tau_{\mathrm{Sys}}$. The numerical histograms of the deviations with their respective one- and two-dimensional Gaussian fits are shown in Fig. 7.

We observe that not only the distribution obtained with bang–bang control is much narrower than the distribution obtained without it, but also that its shape is qualitatively different. The maximum of the bang–bang controlled distribution is at zero error. In contrast, the uncontrolled

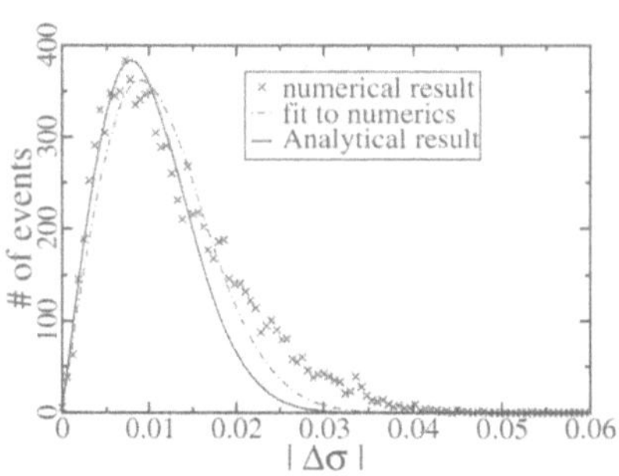

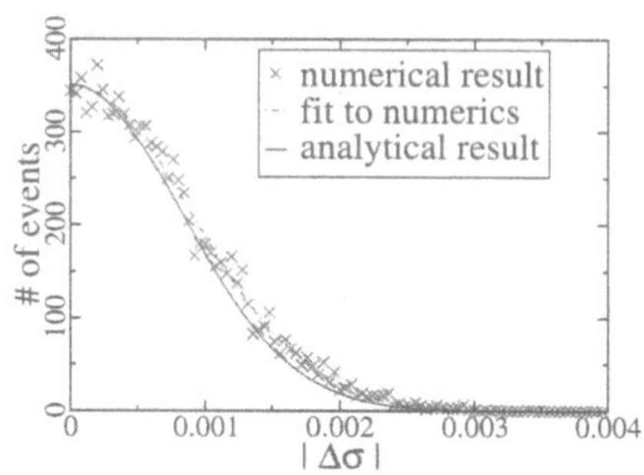

Fig. 7. Histograms of the deviation from free evolution both without bang–bang control (left) and with bang–bang control (right). Also plotted are fits to the expected two- and one-dimensional random walk statistics respectively associated with the uncontrolled and controlled cases. Numerical data were collected over 10^4 realizations at a fixed time $t_0 = \tau_{\rm Sys}$ defined such that $\tau_{\rm bfl} = 0.01\tau_{\rm Sys}$ and thus $N_{\rm bfl} = \tau_{\rm Sys}/\tau_{\rm bfl} = 100$ steps. (NB: The x-axis scale of the right graph depicting the bang–bang controlled case is 15 times smaller than that of the left graph depicting the uncontrolled case.)

distribution has its maximum at a finite error $|\Delta\sigma|_{\rm max} \approx 0.01$, and it has zero probability of zero error. This reflects the one-dimensional nature of the bang–bang controlled random walk in contrast to the two-dimensional nature of the uncontrolled random walk.

4.4. Bang–Bang Control Working as a High-Pass Filter

In order to measure the degree of noise suppression due to bang–bang control, we define the suppression factor $\mathcal{S}_{t_0}$ as follows for a given evolution time t_0

$$\mathcal{S}_{t_0}(\tau_{\rm bfl}/\tau_{\rm bb}) \equiv \frac{\Delta\vec{\sigma}^{\,\rm bfl}_{\rm rms}(t_0)}{\Delta\vec{\sigma}^{\,\rm bb}_{\rm rms}(t_0)}. \tag{18}$$

We now systematically study the dependence of $\mathcal{S}_{t_0}$ on $\tau_{\rm bfl}/\tau_{\rm bb}$ for a constant mean bfl switching rate $\tau_{\rm bfl} = 10^{-2}\tau_{\rm sys}$ at a fixed evolution time $t_0 = \tau_{\rm sys}$. The numerical data in Fig. 8 show that the suppression efficiency is linear in the bang–bang repetition rate, $S_{\tau_{\rm sys}} = \mu\tau_{\rm bfl}/\tau_{\rm bb}$. The numerically derived value of the coefficient, $\mu_{\rm numerical} \approx 1.679$, is in excellent agreement with the analytical result $\mu_{\rm analytical} = \sqrt{5/2} \simeq 1.581$ from our saddle point approximation, Eqs. (14) and (15).

This small discrepancy between the numerical and analytical results is due to the fact that the analytical calculations neglect correlations between the parallel and perpendicular components of the random walk. This leads to an underestimate of the rms-deviation $\Delta\vec{\sigma}^{\,\rm bfl}_{\rm rms}$ in the case without

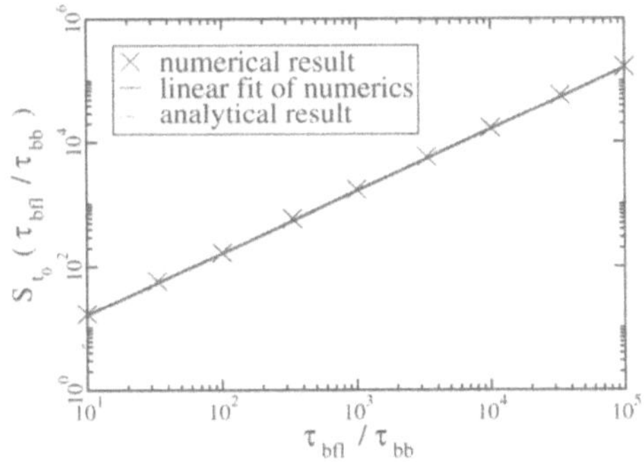

Fig. 8. The suppression factor $\mathcal{S}_{t_0}(\tau_{\text{bfl}}/\tau_{\text{bb}}) = \Delta\vec{\sigma}^{\text{bfl}}_{\text{rms}}(t_0)/\Delta\vec{\sigma}^{\text{bb}}_{\text{rms}}(t_0)$ evaluated for $t_0 = \tau_{\text{Sys}}$ as a function of the ratio of the mean switching time τ_{bfl} and the bang–bang pulse separation τ_{bb}.

bang–bang control (compare also to Fig. 7). Therefore, we have quantitatively proved our qualitative intuition: bang–bang control affects the bfl noise signal like a high-pass filter, an effect that one of the authors has generally predicted for dynamical decoupling techniques.[21]

5. LIMITATIONS DUE TO PULSE INACCURACIES

Thus far, we have tacitly assumed that one could apply perfect, zero-width π-pulses along exactly the $\hat{\sigma}_x$-axis of the Bloch sphere. We now take into account that the control pulses themselves typically will have slight fluctuations in their duration or polarization that interfere with the desired refocusing. As already shown at the end of Subsec. 4.2, the restriction of pulses to infinitesimal duration can be significantly relaxed. We now investigate to what extent the restriction to perfect pulses can be relaxed.

5.1. Two Generic Types of Bang–Bang Inaccuracies

We essentially analyze two generic types of errors that could occur in the control apparatus when trying to apply π-pulses in $\hat{\sigma}_x$-direction. One, the duration of each pulse could exhibit fluctuations, resulting in fluctuations in the rotation-angle around the desired value of π. Two, the polarization axis could suffer from directional deviations around the desired value of $\hat{\sigma}_x$. Assuming the statistical independence of each pulse error, we expect for both types of imperfections a random-walk-like behavior of increasing deviations compared to evolutions with perfect pulses.

5.1.1. One-dimensional pulse error (dephasing)

We make the quite general assumption that we may model the one-dimensional phase fluctuation of the imperfect bang–bang pulses $\phi_j(x)$ as a Gaussian distribution of the pulse durations and therefore of the rotation angles around their intended value π. This assumption should be valid for many physical situations, e.g., if the inaccuracy is due to electromagnetic noise in the pulse generator. The Gaussian is parameterized by its standard deviation $\delta\phi_0$ (see Fig. 9). Thus, the corresponding pulse angle aberration of the jth step is given by

$$\phi_j^{1\mathrm{d}}(x) = \frac{1}{\sqrt{2\pi}\,\delta\phi_0} e^{\left(-\frac{x^2}{2\delta\phi_0^2}\right)}. \tag{19}$$

Having assumed a Gaussian distribution, we can exactly evaluate the distributions of the N-step deviation $\Delta\Phi_N$ (which are usually given as N-fold time-convoluted integrals) as follows by use of the convolution theorem

$$\begin{aligned}\Phi_N^{1\mathrm{d}} &= \mathcal{F}^{-1}\left[\Pi_{j=1}^{N}\tilde{\phi}_j\right] \\ &= \frac{1}{\sqrt{2\pi N}\,\delta\phi_0} e^{\left(-\frac{x^2}{2N\delta\phi_0^2}\right)}\end{aligned} \tag{20}$$

with $\tilde{\phi}_j^{1\mathrm{d}} = \mathcal{F}[\phi_j]$ denoting the Fourier transform of $\phi_j^{1\mathrm{d}}$ and $\mathcal{F}^{-1}$ denoting the inverse Fourier transform.

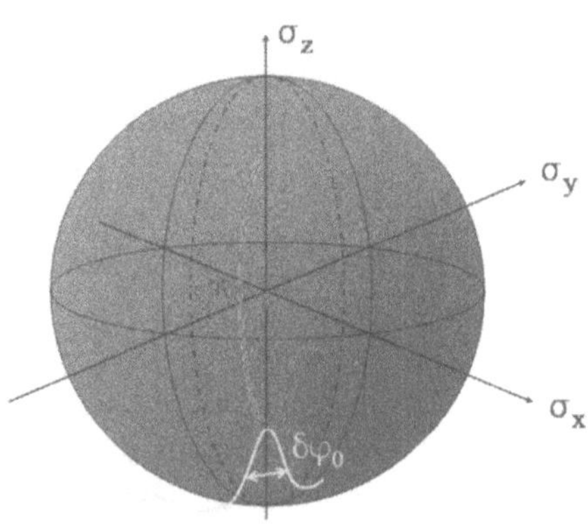

Fig. 9. Sketch of one-dimensional bang–bang aberration. The variations $\delta\phi_0$ of the rotation angle around the desired value of π leads to slight deviations in parallel to the permitted dynamical direction, thus generating dephasing.

Therefore the rms displacement in the random walk increases as a square-root in the number N of bang–bang pulses: $\delta\phi_N = \sqrt{N}\delta\phi_0$. Equivalently, the dephasing grows as square-root in time

$$\delta\phi(t) = \sqrt{t/\tau_{\rm bb}}\delta\phi_0 \tag{21}$$

on the time scale of our coarse-graining (which is here given as $\tau_{\rm bb}$).

5.1.2. Two-dimensional pulse error: dephasing and relaxation/excitation

A similar argument works when there are also fluctuations around the desired $\hat{\sigma}_x$ rotation axis. Each individual variation of the axis can be split into two components: (1) $\delta\phi_{\rm perp}$, which is perpendicular to the connecting vector between the $\hat{\sigma}_x$-axis and the qubit state $\vec{\sigma}(t)$ on the Bloch sphere, and (2) $\delta\phi_{\rm tan}$, which is transverse to it (see Fig. 10). To first order, the perpendicular part does not disturb the intended spin-flip.[27] However, the transverse part does cause a deviation from the ideal spin-flip in a direction toward or away from the previous qubit state. (Therefore, it produces relaxation or excitation, as its effect is orthogonal to the free $\hat{\sigma}_x$-evolution.) Consequently, in a statistical average we only have to consider $1/\sqrt{2}$ of the typical total mean $\delta\phi_0$ of the aberration. The effect of a π-rotation around an axis tilted by an angle $\delta\phi_{\rm tan}$ is a deviation $2\delta\phi_{\rm tan}$ from the trajectory of the perfect evolutions; thus we receive altogether a deviation on the order of $\sqrt{2}\delta\phi_0$.

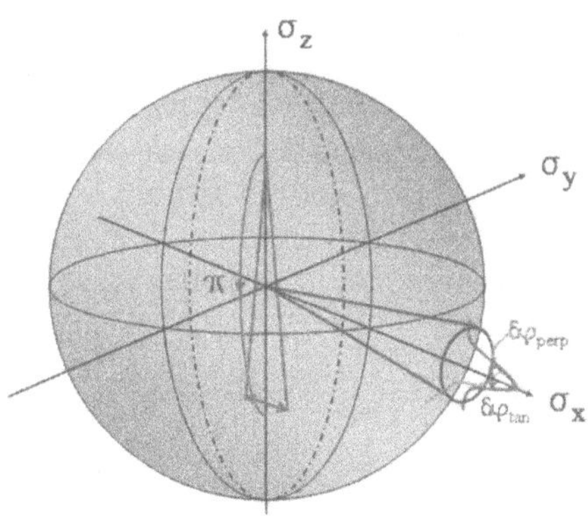

Fig. 10. Sketch of two-dimensional bang–bang aberration. To first order, variations $\delta\phi_{\rm perp}$ of the rotation axis perpendicular to the connection vector between σ_x and the qubit state (here for simplicity: $\sigma_z = +1$) do not influence the intended spin-flip, whereas the variations $\delta\phi_{\rm tan}$ along this line causes deviations on the Bloch sphere perpendicular to the permitted evolution trajectories (therefore producing relaxation or excitation).

Therefore, we obtain analogously to Eq. (19) for each single step distribution

$$\phi_j^{2d}(x) = \frac{1}{\sqrt{2\pi}\sqrt{2}\delta\phi_0} e^{\left(-\frac{x^2}{4\delta\phi_0^2}\right)}; \tag{22}$$

and analogously to Eq. (20) for the deviation after N steps

$$\Phi_N^{2d} = \frac{1}{\sqrt{2\pi}\sqrt{2N}\delta\phi_0} e^{\left(-\frac{x^2}{4N\delta\phi_0^2}\right)}. \tag{23}$$

Equivalently, in terms of the time t

$$\delta\phi(t) = \sqrt{2t/\tau_{\mathrm{bb}}} \cdot \delta\phi_0. \tag{24}$$

5.2. Numerical and Analytical Results

In the same manner as our previous integrations of a stochastic Schrödinger equation, we numerically simulate qubit dynamics under inaccurate pulses. In the first instance, we work without bfl-noise to verify our analytical random walk model. Later, we add the bfl-noise in order to study the competition between the two sources of error.

5.2.1. Random walk due to inaccurate bang–bang pulses only

We analyze deviations on the Bloch sphere between the noiseless case trajectories that occur when the bang–bang pulses are perfect and those when they are not. As per Eq. (5), we calculate the rms-deviation over ensembles of $N = 10^3$ realizations. As a representative time point, we once again choose $t_0 = \tau_{\mathrm{Sys}}$. This is because, as explained in the discussion surrounding Fig. 3, this time scale should exhibit neither short-time effects nor near-total decoherence. From Eqs. (21) and (24), it immediately follows that for the mean deviations at t_0 if there are phase errors

$$\Delta\sigma_{\mathrm{bb}}^{1d}(t_0) = \sqrt{N_{\mathrm{bb}}}\delta\phi_0 = \sqrt{\frac{t_0}{\tau_{\mathrm{bb}}}}\delta\phi_0, \tag{25}$$

and if there are axis errors

$$\Delta\sigma_{\mathrm{bb}}^{2d}(t_0) = \sqrt{2N_{\mathrm{bb}}}\delta\phi_0 = \sqrt{2\frac{t_0}{\tau_{\mathrm{bb}}}}\delta\phi_0. \tag{26}$$

As characteristic values for the mean accuracy of single pulses, we choose $\delta\phi_0$ in the range of 10^{-6} to 10^{-4}, which should be technologically feasible. As one can see in the double logarithmic plots of Fig. 11, the numerically determined evolutions follow the analytically expected square-root type random walk behavior.

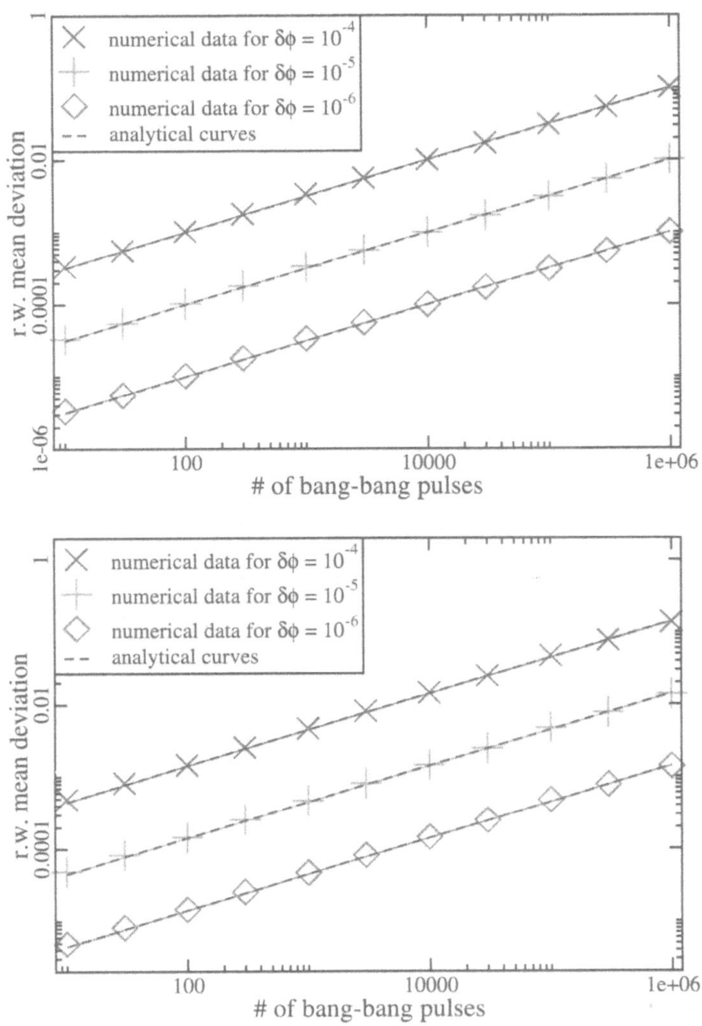

Fig. 11. Plot of the one- respectively two-dimensional imperfectly bang–bang pulsed evolution. Dashed lines are square-root fits of the numerical data, while the solid lines denotes the analytical calculations.

5.2.2. Random walk due to both inaccurate bb-pulses and bfl-noise

We now combine our imperfect bang–bang pulse operations with our former bfl-noise signal to discuss the applicability of our control scheme when "realistic" pulse generators are used. As before, we calculate the rms deviations at $t_0 = \tau_{\rm Sys}$ by averaging over 10^3 realizations. The bfl-parameters are those used previously: a coupling strength $\alpha = 0.1$ and an average switching time $\tau_{\rm bfl} = 0.01\tau_{\rm Sys}$. However, with the aim of determining the optimal bang–bang protocol in the presence of pulse imperfections, we now consider different pulse separation times $\tau_{\rm bb}/\tau_{\rm Sys}$ ranging from 10^{-5} to 10^{-2}.

We assume that the errors induced by the bfl and those induced by the pulse generator are statistically independent, and thus we sum together both sets of induced deviations in the usual rms-fashion. In comparison to the case of ideal bang–bang pulses, Eq. (15), we find here the average total deviations induced by both bfl telegraph noise and imperfect bang–bang pulses to be:

$$\begin{aligned}\Delta\sigma_{\rm tot}^{1d} &= \sqrt{\Delta\sigma_{\rm bfl}^2 + \Delta\sigma_{\rm bb}^{1d\,2}} \\ &= \sqrt{\frac{1}{2}N_{\rm bfl}\alpha^2\tau_{\rm bb}^2 + N_{\rm bb}\delta\phi_0^2} \\ &= \sqrt{\frac{1}{2}\alpha^2\tau_{\rm bb}^2\frac{t_0}{\tau_{\rm bfl}} + \delta\phi_0^2\frac{t_0}{\tau_{\rm bb}}} \end{aligned} \tag{27}$$

in the one-dimensional case where imperfect pulses only impart phase errors (due to imprecise pulse duration), and

$$\begin{aligned}\Delta\sigma_{\rm tot}^{2d} &= \sqrt{\Delta\sigma_{\rm bfl}^2 + \Delta\sigma_{\rm bb}^{2d\,2}} \\ &= \sqrt{\frac{1}{2}N_{\rm bfl}\alpha^2\tau_{\rm bb}^2 + 2N_{\rm bb}\delta\phi_0^2} \\ &= \sqrt{\frac{1}{2}\alpha^2\tau_{\rm bb}^2\frac{t_0}{\tau_{\rm bfl}} + 2\delta\phi_0^2\frac{t_0}{\tau_{\rm bb}}} \end{aligned} \tag{28}$$

in the two-dimensional case when imperfect pulses impart both phase and relaxation/excitation errors (due to imprecision in the pulses' polarization axes).

As Fig. 12 demonstrates, we observe a very good agreement between our numerical and analytical results. Such data make it possible to determine an optimal bang–bang separation time $\tau_{\rm bb}^{\rm opt}$. Specifically, this

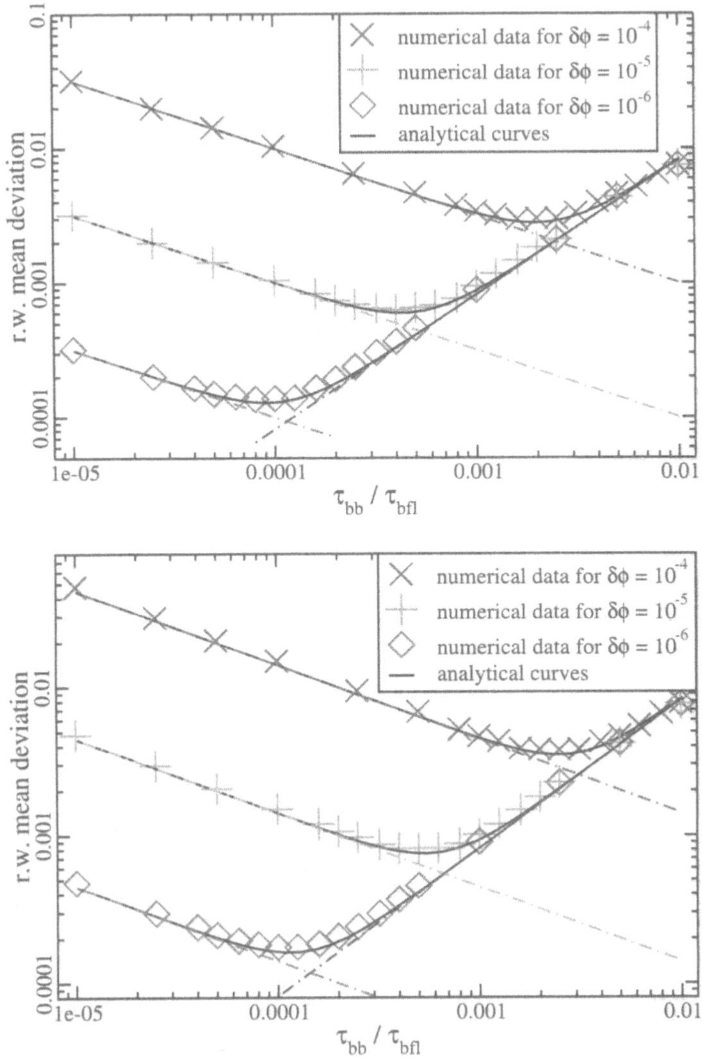

Fig. 12. Plot of the Bloch sphere rms deviations received by one-/two-dimensional inaccurately pulsed bang–bang compensation of the typical bfl-perturbation. Dashed lines describe the aberrances for pure faulty bang–bang (i.e., without bfl-noise), respectively the exactly compensated bfl-case (see Fig. 3), while solid lines denotes the deviations calculated by random walk analysis.

optimum can be derived by calculating the zero value of the derivative of Eqs. (27) and (28) with respect to τ_{bb}. We therefore conclude that the optimal period between bang–bang pulses is:

$$\tau_{\rm bb}^{1d} = \sqrt[3]{\tau_{\rm bfl}\frac{\delta\phi_0^2}{\alpha^2}} \tag{29}$$

for the one-dimensional case and

$$\tau_{\rm bb}^{2d} = \sqrt[3]{2\tau_{\rm bfl}\frac{\delta\phi_0^2}{\alpha^2}} \tag{30}$$

for the two-dimensional case. These optimal times respectively correspond to minimized variances at $t_0 = \tau_{\rm Sys}$ of

$$\Delta\sigma_{\rm opt}^{1d} = \sqrt{\frac{1}{2}+1}\frac{\alpha^{1/3}\delta\phi_0^{2/3}}{\tau_{\rm bfl}^{1/6}}\sqrt{t_0} \tag{31}$$

for the one-dimensional case of only imprecise pulse durations and

$$\Delta\sigma_{\rm opt}^{2d} = \sqrt{2^{-1/3}+2^{2/3}}\frac{\alpha^{1/3}\delta\phi_0^{2/3}}{\tau_{\rm bfl}^{1/6}}\sqrt{t_0} \tag{32}$$

for the two-dimensional case of both imprecise pulse durations and polarization axes.

6. CONCLUSION AND REMARKS

We have investigated the qubit errors that arise from the noise generated by a *single* bistable fluctuator (bfl) in its semiclassical limit, where it behaves as a telegraph noise source. We numerically integrated a corresponding stochastic Schrödinger equation, Eq. (4), as well as analytically solved (in the long-time limit) appropriate random walk models. As a characteristic measure of the resulting dephasing and relaxation effects, we used the rms deviation of noisy evolutions compared to noiseless ones. To suppress the effects of this noise, we presented a bang–bang pulse sequence analogous to the familiar spin–echo method. We claimed this pulse sequence to be capable of refocusing most of the bfl-noise induced aberrations. Both in the case without bang–bang control and the case with it, there was excellent agreement between our numerical and analytical results on the relevant intermediate time scales (i.e., times after a short initial phase where deviations grow linearly instead of as a square-root in time, but before the qubit becomes totally decohered).

In particular, we confirmed our preliminary qualitative picture that bang–bang control works as a high pass filter, suppressing qubit errors by a factor $\mu\tau_{\text{bfl}}/\tau_{\text{bb}}$ that is directly proportional to the ratio of the mean bfl switching time and the period between bang–bang pulses. The numerically and analytically calculated constants of proportionality $\mu_{\text{numerical}} \approx 1.679$ and $\mu_{\text{analytical}} = \sqrt{5/2} \simeq 1.581$ also matched to good accuracy. These results imply that the bang–bang procedure is an appropriate remedy against the $1/f$-noise that often is seen in solid-state environments. This is because bang–bang control exhibits maximal suppression of bfl telegraph noise, and $1/f$ noise generally arises from an ensemble of bfls. Finally, one has to be aware that also the static σ_z-term of the Hamiltonian is averaged out, and this generally reduces the degree of control on the qubit. But this is only a technical constraint, as one could imagine interchanging two different types of bang–bang pulses (e.g., along the x and the y-axis respectively) to admit corresponding quantum-gate operations.

We previously presented this basic idea in a short paper [28]. The present work extends that short paper by treating the effects of different types on non-ideal bang–bang pulses. Moreover, the analytical random walk method is outlined in much more detail, as this method should also be applicable to other problems that are difficult to treat in a master equation approach.

Meanwhile, several other extensions of Ref. 28 have been proposed by other research groups. Ref. 17 includes a larger number of fluctuators, described as semiclassical noise sources, but restricts itself to a single spin-echo cycle. Ref. 29 analyzes extensively the importance of higher, non-Gaussian cumulants and memory effects and arrives at a number of analytical results, but it does not treat the option of refocusing. Ref. 30 treats a full microscopic model and compares different variations of the bang–bang pulse sequence. Ref. 31 also treats a full microscopic model with potentially many fluctuators using a Lindblad-type approach and covers a wide range of ratios between the fluctuator and bang–bang pulse time scales. One of its main conclusions is that a Zeno effect is found in a parameter regime not covered by our work. Note that all of these other extensions of our work treat only the case of ideal bang–bang pulses.

ACKNOWLEDGEMENTS

We are indebted to G. Falci, I. Goychuk, T. P. Orlando and J. von Delft for helpful discussions. Our special thanks goes to A. Kaeck for pointing out the capability of our random walk model to describe deviations due to imperfect pulsing. HG and FKW also thank T. P. Orlando for his great hospitality during their stay at MIT. WMK gratefully acknowledges fellowship support from the Fannie and John Hertz Foundation.

This work was supported by a DAAD-NSF travel grant and by the National Security Agency (NSA) and Advanced Research and Development Activity (ARDO) under Army Research Office (ARO) contract P-43385-PH-QC as well as by the DFG through SFB 631.

APPENDIX A

We shall now connect the model of a *single* bfl as a telegraph noise source to a microscopic Hamiltonian. We start with the conventional Hamiltonian model of a single bfl, e.g., Refs. 14, 32 and 33. The original qubit is influenced by noise from another qubit, the bfl, which itself is coupled to a thermal environment by a bilinear spin-boson type interaction:

$$H = H_{\text{qubit}} + H_{\text{qubit,bfl}} + H_{\text{bfl}} + H_{\text{bfl,env}} + H_{\text{env}}. \tag{33}$$

where

$$H_{\text{qubit}} = \hbar\epsilon_{\text{q}}\hat{\sigma}_{\text{z}}^{\text{q}} + \hbar\Delta_{\text{q}}\hat{\sigma}_{\text{x}}^{\text{q}} \tag{34}$$

$$H_{\text{qubit,bfl}} = \hbar\alpha\hat{\sigma}_{\text{z}}^{\text{q}}\hat{\sigma}_{\text{z}}^{\text{bfl}} \tag{35}$$

$$H_{\text{bfl}} = \hbar\epsilon_{\text{bfl}}\hat{\sigma}_{\text{z}}^{\text{bfl}} + \hbar\Delta_{\text{bfl}}\hat{\sigma}_{\text{x}}^{\text{bfl}} \tag{36}$$

$$H_{\text{bfl,env}} = \hbar\lambda\hat{\sigma}_{\text{z}}^{\text{bfl}}\sum_j\left(\hat{a}_j^\dagger + \hat{a}_j\right) \tag{37}$$

$$H_{\text{env}} = \hbar\sum_j\omega_j\left(\hat{a}_j^\dagger\hat{a}_j + \text{id}/2\right) \tag{38}$$

The scalar α denotes the coupling strength between the original qubit and the bfl, while the scalar λ indicates the influence of the environmental heat bath on the bfl.

It is not obvious how to treat such a combined open quantum system[14,33]. The common approach of deriving a master equation for the reduced qubit system does not work, as it is not clear how to introduce an open quantum system "bfl" as the environment. Gassmann, *et al.* present four alternative approaches. [14] Their first approach is to derive a standard Markovian master equation for the combined open system "qubit + bfl" and trace out the parameters of the bfl afterwards. Their second approach is to consider the qubit as influenced by an effective bfl-bath environment by use of an Markovian and secular approximation in the limit of small α. Their third approach, which is both the most general and the most complicated, is to deduce a master equation by applying a non-Markovian weak-coupling perturbation ansatz in second order in α.

For our investigations, we prefer this last and most general approach: a stochastic treatment employing an appropriate randomly changing bfl-noise Hamiltonian term (compare also to Ref. 17). This choice is not only because of practical reasons (to make our numerics feasible), but also due to empirical considerations (see Refs. 8 and 9, where characteristics of telegraph noise were observed and attributed to bfls). Hence, we restrict our analysis to the limit $\lambda \gg \alpha$, i.e., the limit where the coupling of the bfl to the external environment is much larger than interaction between the bfl and the qubit. For convenience, we assume the bfl is in its high-temperature limit. (Note that this does not necessarily mean the qubit is also in a high-temperature regime for the qubit's energy scale might be much larger than that of the bfl.) We thus assume the bfl behaves like a classical (i.e., decohered) noise source, and we specifically describe the bfl's influence on the qubit with the following stochastic Hamiltonian:

$$H_{\text{qubit,bfl}} \overset{\text{semicl.}}{\longrightarrow} H^{\text{noise}}_{\text{qubit,bfl}}(t) = \hbar\alpha \; \hat{\sigma}^{\text{q}}_{\text{z}} \; \xi_{\text{bfl}}(t). \tag{39}$$

In the equation above, $\xi_{\text{bfl}}(t)$ is a random function of time representing the switching of the $\sigma_{\text{z}}^{\text{bfl}}$-value between ± 1. In our high-temperature limit, we assume $\xi_{\text{bfl}}(t)$ has symmetrical Poissonian statistics (*i.e.*, the probabilities of the bfl switching from $+1$ to -1 and from -1 to $+1$ are equal and constant over time). Such a symmetrical random process is readily described by just one parameter: the typical time separation τ_{bfl} between two bfl flips (see Fig. 1).

The high temperature limit is not a crucial constraint. Treating the strongly thermally coupled bfl in an intermediate temperature regime would only result in some asymmetrically switching $\xi_{\text{bfl}}(t)$. The typical switching times time $\tau_{\text{bfl}}^{\uparrow,\downarrow}$ for switching the bfl up and down respectively satisfy the detailed balance relation $\frac{\tau_{\text{bfl}}^{\downarrow}}{\tau_{\text{bfl}}^{\uparrow}} = e^{-\delta E_{\text{bfl}}/k_b T}$, where δE_{bfl} denotes the energy separation of the two bfl-states, and T the temperature of the heat bath which drives the switching of the bfl. The microscopic structure of the rates depends on details of the experiment. Typically, they will be golden rule rates containing the density of states of the heat bath and the matrix element of its coupling to the bfl. If that bath is made of harmonic oscillators with an ohmic spectral density, we, e.g., expect switching rater $\left(\tau_{\text{bfl}}^{\uparrow,\downarrow}\right)^{-1} = \pm\alpha_0 \delta E_{\text{bfl}}/(e^{\pm\delta E_{\text{bfl}}/k_b T} - 1)$, where α_0 is the dimensionless coupling strength to the Ohmic bath. This would essentially only lead to an additional drift of the qubit state, i.e., a random walk with a nonzero average value. Neither our analytical results nor our conclusions would otherwise change qualitatively. In fact, assuming the the bang–bang pulse cycles are sufficiently short relative both the typical $+1$ to -1 and -1 to $+1$

switching times of the bfl, bang–bang suppression of the bfl noise should not be diminished at all by the bfl's asymmetrical switching. We therefore obtain Eq. (1) as our starting point of the bfl-perturbed qubit dynamics.

REFERENCES

1. D. Loss and D. P. DiVincenzo, *Phys. Rev. A* **57**, 120 (1998).
2. D. Vion *et al.*, Science **296**, 886 (2002).
3. I. Chioresco, Y. Nakamura, C. J. P. M. Harmans, and J. E. *Mooij, Science* **299**, 1869 (2003).
4. J. M. Martinis, S.W. Nam, J. Aumentado, and C. Urbina, *Phys. Rev. Lett.* **89**, 117907 (2002).
5. Y. Nakamura, Y. A. Pashkin, and J. S. Tsai, *Nature* **398**, 786 (1999).
6. Yu. Makhlin, G. Schön, and A. Shnirman, *Rev. Mod. Phys.* **73**, 357 (2001).
7. C. van der Wal, F. Wilhelm, C. Harmans, and J. Mooij, *Eur. Phys. J.* **B 31**, 111 (2003).
8. R. Wakai and D. van Harlingen, *Phys. Rev. Lett.* **58**, 1687 (1987).
9. N. M. Zimmermann, J. L. Cobb, and A. F. Clark, *Phys. Rev. B* **56**, 7675 (1997).
10. M. B. Weissmann, *Rev. Mod. Phys.* **60**, 537 (1988).
11. A. J. Leggett *et al.*, *Rev. Mod. Phys.* **59**, 1 (1987).
12. Yu. Makhlin, G. Schön, and A. Shnirman, *Phys. Scr. T* **102**, 147 (2002).
13. U. Weiss, *Quantum Dissipative Systems* (World Scientific, Singapore, 2001).
14. H. Gassmann, F. Marquardt, and C. Bruder, *Phys. Rev. E* **66**, 041111 (2002).
15. N. Prokov'ef and P. Stamp, *Rep. Prog. Phys.* **63**, 669 (2000).
16. T. Ytakura and Y. Tokura, *Dephasing due to background charge fluctuations*, ISSP Int. Workshop "Quantum transport in mesoscopic scale and low dimensions", (2003).
17. Y. M. Galperin, B. L. Altshuler, and D. V. Shantsev, *Low-frequency noise as a source of dephasing of a qubit*, Proc. of NATO/Euresco Conf. "Fundamental problems of mesoscopic physics: Interaction and decoherence", Granada, Spain, (2003), NATO Science Series.
18. P. Dutta and P. M. Horn, *Rev. Mod. Phys.* **53**, 497 (1981).
19. J. M. Martinis *et al.*, *Phys. Rev.* B **67**, 094510 (2003).
20. S. Lloyd and L. Viola, *Phys. Rev. A* **58**, 2733 (1998).
21. S. Lloyd, E. Knill, and Viola, *Phys. Rev. Lett.* **82**, 2417 (1999).
22. S. Lloyd, E. Knill, and L. Viola, *Phys. Rev. Lett.* **83**, 4888 (1999).
23. L. Arnold, *stochastische Differentialgleichungen* (Oldenbourg, München, 1973).
24. N. G. van Kampen, *Stochastic Processes in Physics and Chemistry* (Elsevier, Amsterdam, 1997).
25. H. Carr and E. Purcell, *Phys. Rev.* **94**, 630 (1954).
26. G. H. Weiss, *Aspects and Applications of the Random Walk* (North-Holland, Amsterdam, 1994).
27. The attentive reader might object that a spin-flip around a different axis on the xy-equator doesn't commute with pure $\hat{\sigma}_x$-dynamics, but rather with something nearby. As we do not consider any $\hat{\sigma}_y$-components yet, we do not bother about the minimal distortion of pure $\hat{\sigma}_x$-manipulations, which can be estimated to be on the 2nd order of the aberration parameter $\delta\phi_0$ that we assumed to be very small anyway.
28. H. Gutmann, F.K. Wilhelm, W.M. Kaminsky, and S. Lloyd, cond-mat/0308107.
29. K. Rabenstein, V. A. Sverdlov, and D. V. Averin, JETP Vol. **79** is. 12, 783
30. L. Faoro and L. Viola, quant-ph/0312159.
31. G. Falci, A. D'Arrigo, A. Mastellone, and E. Paladino, *cond-mat*/0312442.
32. E. Paladino, L. Faoro, and G. Falci, *cond-mat*/0312411.
33. E. Paladino, L. Faoro, G. Falci, and R. Fazio, *Phys. Rev. Lett.* **88**, 228304 (2002).

Quantum Computing and Information Extraction for Dynamical Quantum Systems

Giuliano Benenti,[1,4] Giulio Casati[1,2] and Simone Montangero[3]

We discuss the simulation of complex dynamical systems on a quantum computer. We show that a quantum computer can be used to efficiently extract relevant physical information. It is possible to simulate the dynamical localization of classical chaos and extract the localization length with quadratic speed up with respect to any known classical computation. We can also compute with algebraic speed up the diffusion coefficient and the diffusion exponent, both in the regimes of Brownian and anomalous diffusion. Finally, we show that it is possible to extract the fidelity of the quantum motion, which measures the stability of the system under perturbations, with exponential speed up. The so-called quantum sawtooth map model is used as a test bench to illustrate these results.

KEY WORDS: Quantum computation; quantum simulation; information extraction; dynamical systems; complex systems; chaos.

PACS: 03.67.Lx; 05.45.Mt.

1. INTRODUCTION

One of the main applications of computers is the simulation of dynamical models describing the evolution of complex systems. From the viewpoint of quantum computation, quantum mechanical systems play a special role. Indeed, the simulation of quantum many-body problems on a classical

[1]Center for Nonlinear and Complex Systems, Università degli Studi dell'Insubria, Via Valleggio 11, 22100 Como, Italy and Istituto Nazionale per la Fisica della Materia, Unità di Como, Via Valleggio 11, 22100 Como, Italy. URL: http://www.unico.it/~dysco

[2]Istituto Nazionale per la Fisica Nucleare, Sezione di Milano, Via Celoria 16, 20133 Milano, Italy. E-mail: giulio.casati@uninsubria.it

[3]NEST-INFM & Scuola Normale Superiore, Piazza dei Cavalieri 7, 56126 Pisa, Italy, E-mail: monta@sns.it; URL: http://www.sns.it/~montangero

[4]To whom correspondence should be addressed. E-mail:giuliano.benenti@uninsubria.it

computer is a difficult task as the size of the Hilbert space grows exponentially with the number of particles. For instance, if we wish to simulate a chain of n spin-1/2 particles, the size of the Hilbert space is 2^n. Namely, the state of this system is determined by 2^n complex numbers. As observed by Feynman in the 1980s,[1] the growth in memory requirement is only linear on a quantum computer, which is itself a many-body quantum system. For example, to simulate n spin-1/2 particles we only need n qubits. Therefore, a quantum computer operating with only a few tens of qubits could outperform a classical computer. More recently, a few quantum efficient algorithms have been developed for various quantum systems, ranging from some many-body problems[2,3] to single-particle models of quantum chaos.[4–6]

Any quantum algorithm has to address the problem of efficiently extracting useful information from the quantum computer wave function. The result of the simulation of a quantum system is the wave function of this system, encoded in the n qubits of the quantum computer. The problem is that, in order to measure all $N = 2^n$ wave function coefficients by means of standard polarization measurements of the n qubits, one has to repeat the quantum simulation a number of times exponential in the number of qubits. This procedure would spoil any quantum algorithm, even in the case in which such algorithm could compute the wave function with an exponential gain with respect to any classical computation. Nevertheless, there are some important physical questions that can be answered in an efficient way, and we will discuss a few examples in this paper.

As a test bench to illustrate the power of quantum computation in the simulation of dynamical systems, we will discuss a quantum algorithm which efficiently simulates the quantum sawtooth map, a physical model with rich and complex dynamics.[6] This system is characterized by very different dynamical regimes, ranging from integrability to chaos, and from normal to anomalous diffusion; it also exhibits the phenomenon of dynamical localization of classical chaotic diffusion. We will show that some important physical quantities can be extracted efficiently by means of a quantum computer:

(i) the *localization length* of the system, which can be extracted with a *quadratic speed up* with respect to any known classical computation;[7]
(ii) the *diffusion coefficient* and the *diffusion exponent*, both in the regimes of normal (Brownian) and anomalous diffusion; in this case we obtain an *algebraic speed up*;
(iii) the *fidelity* of quantum motion, which characterizes the stability of the system under perturbations; for this quantity we achieve an *exponential speed up*.

The paper is organized as follows: the properties of the sawtooth map model are discussed in Sec. 2; our quantum algorithm simulating the quantum dynamics of this model in Sec. 3; the quantum computation of the localized regime and the extraction of the localization length in Sec. 4; the quantum simulation of the phenomena of normal and anomalous diffusion and the computation of the diffusion coefficient and diffusion exponent in Sec. 5; the quantum computation of the fidelity of quantum motion in Sec. 6; our conclusions are summarized in Sec. 7.

2. THE SAWTOOTH MAP

The sawtooth map is a prototype model in the studies of classical and quantum-dynamical systems and exhibits a rich variety of interesting physical phenomena, from complete chaos to complete integrability, normal and anomalous diffusion, dynamical localization, and cantori localization. Furthermore, the sawtooth map gives a good approximation to the motion of a particle bouncing inside a stadium billiard (which is a well-known model of classical and quantum chaos).

The sawtooth map belongs to the class of periodically driven dynamical systems, governed by the Hamiltonian

$$H(\theta, I; \tau) = \frac{I^2}{2} + V(\theta) \sum_{j=-\infty}^{+\infty} \delta(\tau - jT), \tag{1}$$

where (I, θ) are conjugate action-angle variables $(0 \le \theta < 2\pi)$. This Hamiltonian is the sum of two terms, $H(\theta, I; \tau) = H_0(I) + U(\theta; \tau)$, where $H_0(I) = I^2/2$ is just the kinetic energy of a free rotator (a particle moving on a circle parametrized by the coordinate θ), while

$$U(\theta; \tau) = V(\theta) \sum_j \delta(\tau - jT), \tag{2}$$

represents a force acting on the particle that is switched on and off instantaneously at time intervals T. Therefore, we say that the dynamics described by Hamiltonian Eq. (1) is *kicked*. The corresponding Hamiltonian equations of motion are

$$\begin{aligned} \dot{I} &= -\frac{\partial H}{\partial \theta} = -\frac{dV(\theta)}{d\theta} \sum_{j=-\infty}^{+\infty} \delta(\tau - jT), \\ \dot{\theta} &= \frac{\partial H}{\partial I} = I. \end{aligned} \tag{3}$$

These equations can be easily integrated and one finds that the evolution from time lT^- (prior to the l-th kick) to time $(l+1)T^-$ (prior to the $(l+1)$-th kick) is described by the map

$$\begin{aligned} \bar{I} &= I + F(\theta), \\ \bar{\theta} &= \theta + T\bar{I}, \end{aligned} \tag{4}$$

where $F(\theta) = -\mathrm{d}V(\theta)/\mathrm{d}\theta$ is the force acting on the particle.

In the following, we will consider the special case $V(\theta) = -k(\theta - \pi)^2/2$. This map is called the *sawtooth map*, since the force $F(\theta) = -\mathrm{d}V(\theta)/\mathrm{d}\theta = k(\theta - \pi)$ has a sawtooth shape, with a discontinuity at $\theta = 0$. By rescaling $I \to J = TI$, the classical dynamics is seen to depend only on the parameter $K = kT$. Indeed, in terms of the variables (J, θ) map (4) becomes

$$\begin{aligned} \bar{J} &= J + K(\theta - \pi), \\ \bar{\theta} &= \theta + \bar{J}. \end{aligned} \tag{5}$$

The sawtooth map exhibits sensitive dependence on initial conditions, which is the distinctive feature of classical chaos: any small error is amplified exponentially in time. In other words, two nearby trajectories separate exponentially, with a rate given by the maximum Lyapunov exponent λ, defined as

$$\lambda = \lim_{|t| \to \infty} \frac{1}{t} \ln\left(\frac{\delta(t)}{\delta(0)}\right), \tag{6}$$

where the discrete time $t = \tau/T$ measures the number of map iterations and $\delta(t) = \sqrt{[\delta J(t)]^2 + [\delta\theta(t)]^2}$. To compute $\delta J(t)$ and $\delta\theta(t)$, we differentiate map (5), obtaining

$$\begin{bmatrix} \delta\bar{J} \\ \delta\bar{\theta} \end{bmatrix} = M \begin{bmatrix} \delta J \\ \delta\theta \end{bmatrix} = \begin{bmatrix} 1 & K \\ 1 & 1+K \end{bmatrix} \begin{bmatrix} \delta J \\ \delta\theta \end{bmatrix}. \tag{7}$$

The iteration of map (7) gives $\delta J(t)$ and $\delta\theta(t)$ as a function of $\delta J(0)$ and $\delta\theta(0)$ [$\delta J(0)$ and $\delta\theta(0)$ represent a change of the initial conditions]. The stability matrix M has eigenvalues $\mu_\pm = \frac{1}{2}(2 + K \pm \sqrt{K^2 + 4K})$, which do not depend on the coordinates J and θ and are complex conjugate for $-4 \le K \le 0$ and real for $K < -4$ and $K > 0$. Thus, the classical motion is stable for $-4 \le K \le 0$ and completely chaotic for $K < -4$ and $K > 0$. For $K > 0$, $\delta(t) \propto (\mu_+)^t$ asymptotycally in t, and therefore the maximum Lyapunov exponent is $\lambda = \ln \mu_+$. Similarly, we obtain $\lambda = \ln|\mu_-|$ for $K < -4$. In the stable region $-4 \le K \le 0$, $\lambda = 0$.

The sawtooth map can be studied on the cylinder $[J \in (-\infty, +\infty)]$, or on a torus of sinite size ($-\pi L \leq J < \pi L$, where L is an integer, to assure that no discontinuities are introduced in the second equation of (5) when J is taken modulus $2\pi L$). Although the sawtooth map is a deterministic system, for $K > 0$ and $K < -4$ the motion along the momentum direction is in practice indistinguishable from a random walk. Thus, one has normal diffusion in the action (momentum) variable and the evolution of the distribution function $f(J,t)$ is governed by a Fokker–Planck equation:

$$\frac{\partial f}{\partial t} = \frac{\partial}{\partial J}\left(\frac{1}{2} D \frac{\partial f}{\partial J}\right). \tag{8}$$

The diffusion coefficient D is defined by

$$D = \lim_{t \to \infty} \frac{\langle (\Delta J(t))^2 \rangle}{t}, \tag{9}$$

where $\Delta J \equiv J - \langle J \rangle$, and $\langle \ldots \rangle$ denotes the average over an ensemble of trajectories. If at time $t = 0$ we take a phase space distribution with initial momentum J_0 and random phases $0 \leq \theta < 2\pi$, then the solution of the Fokker–Planck equation (8) is given by

$$f(J,t) = \frac{1}{\sqrt{2\pi D t}} \exp\left[-\frac{(J - J_0)^2}{2Dt}\right]. \tag{10}$$

The width $\sqrt{\langle (\Delta J(t))^2 \rangle}$ of this Gaussian distribution grows in time, according to

$$\langle (\Delta J(t))^2 \rangle \approx D(K)\, t\,. \tag{11}$$

For $K > 1$, the diffusion coefficient is well approximated by the random phase approximation, in which we assume that there are no correlations between the angles (phases) θ at different times. Hence, we have

$$D(K) \approx \langle (\Delta J_1)^2 \rangle = \frac{1}{2\pi} \int_0^{2\pi} d\theta\, (\Delta J_1)^2 = \frac{1}{2\pi} \int_0^{2\pi} d\theta\, K^2 (\theta - \pi)^2 = \frac{\pi^2}{3} K^2, \tag{12}$$

where $\Delta J_1 = \bar{J} - J$ is the change in action after a single map step. For $0 < K < 1$ diffusion is slowed, due to the sticking of trajectories close to broken tori (known as cantori), and we have $D(K) \approx 3.3\, K^{5/2}$ (this regime is discussed in Ref. 8). For $-4 < K < 0$ the motion is stable, the phase space has a complex structure of elliptic islands down to smaller and smaller scales, and one can observe anomalous diffusion, that is, $\langle(\Delta J)^2\rangle \propto t^{\alpha}$, with $\alpha \neq 1$ (for instance, $\alpha = 0.57$ when $K = -0.1$, see Fig. 4). The cases $K = -1, -2, -3$ are integrable.

The quantum version of the sawtooth map is obtained by means of the usual quantization rules, $\theta \to \hat{\theta}$ and $I \to \hat{I} = -i\partial/\partial\theta$ (we set $\hbar = 1$). The quantum evolution in one map iteration is described by a unitary operator $\hat{U}$, called the Floquet operator, acting on the wave function ψ:

$$\bar{\psi} = \hat{U}\,\psi = \exp\left[-i\int_{lT^-}^{(l+1)T^-} d\tau\, H(\hat{\theta}, \hat{I}; \tau)\right]\psi, \tag{13}$$

where H is Hamiltonian (1). Since the potential $V(\theta)$ is switched on only at discrete times lT, it is straightforward to obtain

$$\bar{\psi} = e^{-iT\hat{I}^2/2}\, e^{-iV(\hat{\theta})}\,\psi = e^{-iT\hat{I}^2/2}\, e^{ik(\hat{\theta}-\pi\hat{\mathbb{1}})^2/2}\,\psi, \tag{14}$$

where $\hat{\mathbb{1}}$ denotes the identity operator. It is important to emphasize that, while the classical sawtooth map depends only on the rescaled parameter $K = kT$, the corresponding quantum evolution (14) depends on k and T separately. The effective Planck constant is given by $\hbar_{\text{eff}} = T$. Indeed, if we consider the operator $\hat{J} = T\hat{I}$ ($\hat{J}$ is the quantization of the classical rescaled action J), we have

$$[\hat{\theta}, \hat{J}] = T[\hat{\theta}, \hat{I}] = iT = i\hbar_{\text{eff}}. \tag{15}$$

The classical limit $\hbar_{\text{eff}} \to 0$ is obtained by taking $k \to \infty$ and $T \to 0$, while keeping $K = kT$ constant.

3. QUANTUM COMPUTING OF THE QUANTUM SAWTOOTH MAP

In the following, we describe an exponentially efficient quantum algorithm for simulation of the map (14).[6] It is based on the forward/backward quantum Fourier transform between momentum and angle bases. Such an approach is convenient since the operator $\hat{U}$, introduced in Eq. (13), is the product of two operators, $\hat{U}_k = e^{ik(\hat{\theta}-\pi\hat{\mathbb{1}})^2/2}$ and $\hat{U}_T = e^{-iT\hat{I}^2/2}$, diagonal in the θ and I representations, respectively. This quantum algorithm requires the following steps for one map iteration:

1. We apply $\hat{U}_k$ to the wave function $\psi(\theta)$. In order to decompose the operator $\hat{U}_k$ into one- and two-qubit gates, we first of all write θ in binary notation:

$$\theta = 2\pi \sum_{j=1}^{n} \alpha_j 2^{-j}, \tag{16}$$

with $\alpha_j \in \{0, 1\}$. Here n is the number of qubits, so that the total number of levels in the quantum sawtooth map is $N = 2^n$. From this expansion, we obtain

$$(\theta - \pi)^2 = 4\pi^2 \sum_{j_1, j_2=1}^{n} \left(\frac{\alpha_{j_1}}{2^{j_1}} - \frac{1}{2n}\right)\left(\frac{\alpha_{j_2}}{2^{j_2}} - \frac{1}{2n}\right), \tag{17}$$

that is

$$\begin{aligned}(\hat{\theta} - \pi\hat{\mathbb{1}})^2 = 4\pi^2 \sum_{j_1, j_2=1}^{n} & \hat{\mathbb{1}}_1 \otimes \cdots \otimes \hat{\mathbb{1}}_{j_1-1} \otimes \hat{O}_{j_1} \otimes \hat{\mathbb{1}}_{j_1+1} \\ & \otimes \cdots \otimes \hat{\mathbb{1}}_{j_2-1} \otimes \hat{O}_{j_2} \otimes \hat{\mathbb{1}}_{j_2+1} \otimes \cdots \otimes \hat{\mathbb{1}}_{j_n},\end{aligned} \tag{18}$$

where $\hat{\mathbb{1}}_j$ is the identity operator for the qubit j and the one-qubit operators $\hat{O}_{j_1}$ and $\hat{O}_{j_2}$ act on qubits j_1 and j_2, respectively. We have

$$\hat{O}_j = \frac{1}{2^j} \frac{\hat{\mathbb{1}}_j - (\hat{\sigma}_z)_j}{2} - \frac{1}{2n}\hat{\mathbb{1}}_j, \tag{19}$$

where $(\hat{\sigma}_z)_j$ denotes the Pauli operator $\hat{\sigma}_z$ for the qubit j. Note that the operator $\hat{O}_j$ is diagonal in the computational basis $\{|0\rangle, |1\rangle\}$. We can insert Eq. (18) into the unitary operator $\hat{U}_k$, obtaining the decomposition

$$\begin{aligned}e^{ik(\hat{\theta} - \pi\hat{\mathbb{1}})^2/2} = \prod_{j_1, j_2=1}^{n} & \exp\Big[i2\pi^2 k \Big(\hat{\mathbb{1}}_1 \otimes \cdots \otimes \hat{\mathbb{1}}_{j_1-1} \otimes \hat{O}_{j_1} \otimes \hat{\mathbb{1}}_{j_1+1} \\ & \otimes \cdots \otimes \hat{\mathbb{1}}_{j_2-1} \otimes \hat{O}_{j_2} \otimes \hat{\mathbb{1}}_{j_2+1} \otimes \cdots \otimes \hat{\mathbb{1}}_{j_n}\Big)\Big],\end{aligned} \tag{20}$$

which is the product of n^2 two-qubit gates (controlled phase-shift gates), each acting non-trivially only on the qubits j_1 and j_2. In the computational basis $\{|\alpha_{j_1}\alpha_{j_2}\rangle = |00\rangle, |01\rangle, |10\rangle, |11\rangle\}$ each two-qubit gate can be written as $\exp(i2\pi^2 k D_{j_1, j_2})$, where D_{j_1, j_2} is a diagonal matrix:

$$D_{j_1,j_2} = \begin{bmatrix} \frac{1}{4n^2} & 0 & 0 & 0 \\ 0 & -\frac{1}{2n}\left(\frac{1}{2^{j_2}} - \frac{1}{2n}\right) & 0 & 0 \\ 0 & 0 & -\frac{1}{2n}\left(\frac{1}{2^{j_1}} - \frac{1}{2n}\right) & 0 \\ 0 & 0 & 0 & \left(\frac{1}{2^{j_1}} - \frac{1}{2n}\right)\left(\frac{1}{2^{j_2}} - \frac{1}{2n}\right) \end{bmatrix}. \tag{21}$$

Note that decomposition (20) of $\hat{U}_k$ is specific to the sawtooth map.

2. The change from the θ to the I representation is obtained by means of the quantum Fourier transform, which requires n Hadamard gates and $\frac{1}{2}n(n-1)$ controlled phase-shift gates (see, e.g., Ref. 9).

3. In the I representation, the operator $\hat{U}_T$ has essentially the same form as the operator $\hat{U}_k$ in the θ representation, and therefore it can be decomposed into n^2 controlled phase-shift gates, similarly to Eq. (20).

4. We return to the initial θ representation by application of the inverse quantum Fourier transform.

Thus, overall, this quantum algorithm requires $3n^2 + n$ gates per map iteration ($3n^2 - n$ controlled phase-shifts and $2n$ Hadamard gates). This number is to be compared with the $O(n2^n)$ operations required by a classical computer to simulate one map iteration by means of a fast Fourier transform. Thus, the quantum simulation of the quantum sawtooth map dynamics is exponentially faster than any known classical algorithm. Note that the resources required to the quantum computer to simulate the evolution of the sawtooth map are only logarithmic in the system size N. Of course, there remains the problem of extracting useful information from the quantum-computer wave function. This will be discussed in the subsequent sections.

4. QUANTUM COMPUTING OF DYNAMICAL LOCALIZATION

Dynamical localization is one of the most interesting phenomena that characterize the quantum behavior of classically chaotic systems: quantum interference effects suppress chaotic diffusion in momentum, leading to exponentially localized wave functions. This phenomenon was first found and studied in the quantum kicked-rotator model[(10)] and has profound analogies with Anderson localization of electronic transport in disordered materials.[(11)] Dynamical localization has been observed experimentally in the microwave ionization of Rydberg atoms[(12)] and in experiments with cold atoms.[(13)]

Dynamical localization can be studied in the sawtooth map model. In this case, map (14) is studied on the cylinder $[I \in (-\infty, +\infty)]$, which is cut-off at a finite number N of levels due to the finite quantum (or classical) computer memory. Similarly to other models of quantum chaos, quantum interference in the sawtooth map leads to suppression of classical chaotic diffusion after a *break time* $t^\star$. For $t > t^\star$, while the classical distribution goes on diffusing, the quantum distribution reaches a steady state which *decays exponentially* over the momentum eigenbasis:

$$W_m \equiv \left|\langle m|\psi\rangle\right|^2 \approx \frac{1}{\ell} \exp\left[-\frac{2|m-m_0|}{\ell}\right], \tag{22a}$$

with m_0 the initial value of the momentum (the index m singles out the eigenstates of $\hat{I}$, that is, $\hat{I}|m\rangle = m|m\rangle$).[5] Therefore, for $t > t^\star$ only $\sqrt{\langle(\Delta m)^2\rangle} \sim \ell$ levels are populated.

An estimate of $t^\star$ and ℓ can be obtained by means of the following argument.[(15)] The localized wave packet has significant projection over about ℓ basis states, both in the basis of the momentum eigenstates and in the basis of the eigenstates of the Floquet operator $\hat{U}$ defined by Eq. (13). This operator is unitary and therefore its eigenvalues can be written as $\exp(i\lambda_i)$, and the so-called *quasienenergies* λ_i are in the interval $[0, 2\pi[$. Thus, the mean level spacing between "significant" quasienergy eigenstates is $\Delta E \approx 2\pi/\ell$. The Heisenberg principle tells us that the minimum time required to the dynamics to resolve this energy spacing is given by

$$t^\star \approx 1/\Delta E \approx \ell. \tag{23}$$

---

[5] Strictly speaking, the asymptotic tails of the localized wave functions decay, for the sawtooth map model, as a power law:

$$W_m \propto \frac{1}{|m-m_0|^4}. \tag{22b}$$

This happens due to the discontinuity in the kicking force $F(\theta) = k(\theta - \pi)$, when the angle variable $\theta = 0$. For this reason the matrix elements of the evolution operator $\hat{U}$ [defined by Eq. (13)] decay as a power law in the momentum eigenbasis: $U_{mm'} = \langle m|\hat{U}|m'\rangle \sim 1/|m-m'|^\alpha$, with $\alpha = 2$. This case was investigated for random matrices, where it was shown that eigenfunctions are also algebraically localized with the same exponent α.[(14)] However, the localization picture is not very sensitive to the behavior of the tails of the wave function. Indeed, a rough estimate of the crossover between the exponential decay Eq. (22a) and the power law decay Eq. (22b) is given by their crossing point,

$$m_c \sim \frac{3}{2}\ell \log \ell, \quad W_m(m_c) \sim \frac{1}{\ell^4 \log \ell}. \tag{22c}$$

This implies that by increasing ℓ the exponential localization is pushed to larger momentum windows and down to smaller probabilities.

This is the break time after which the quantum feature of the dynamics becomes apparent. Diffusion up to time $t^\star$ involves a number of levels given by

$$\sqrt{\langle(\Delta m)^2\rangle} \approx \sqrt{D_m t^\star} \approx \ell, \tag{24}$$

where $D_m = D/T^2$ is the classical diffusion coefficient, measured in number of levels. The relations (23) and (24) imply

$$t^\star \approx \ell \approx D_m. \tag{25}$$

Therefore, the quantum localization length ℓ for the average probability distribution is approximately equal to the classical diffusion coefficient. For the sawtooth map,

$$\ell \approx D_m \approx (\pi^2/3)k^2. \tag{26}$$

Note that the quantum localization can take place on a finite system only if ℓ is smaller than the system size N.

In Fig. 1 (taken from Ref. 7), we show that exponential localization, obtained using our quantum algorithm, can already be clearly seen with $n = 6$ qubits. It is important to stress that in a quantum computer the memory capabilities grow exponentially with the number of qubits (the number of levels N is equal to 2^n). Therefore, already

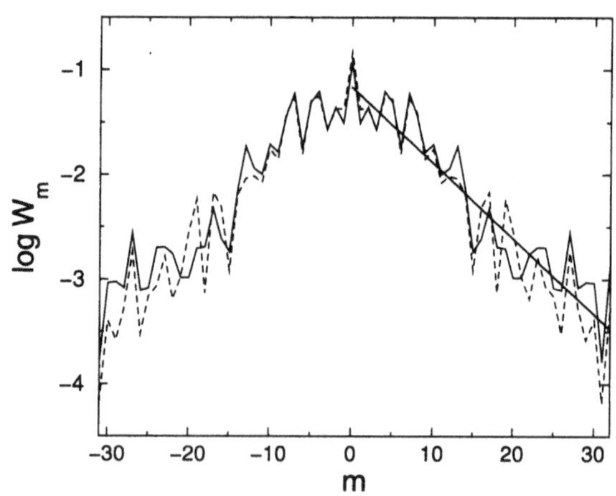

Fig. 1. The probability distribution over the momentum basis for the sawtooth map with $n = 6$ qubits, $k = \sqrt{3}$, $K = \sqrt{2}$, and initial momentum $m_0 = 0$; the time average is taken in the intervals $10 \leq t \leq 20$ (full curve) and $290 \leq t \leq 300$ (dashed curve). The straight line fit, $W_m \propto \exp(-2|m|/\ell)$, gives a localization length $\ell \approx 12$. Note that the logarithm is base ten.

with less than 40 qubits, one could make simulations inaccessible to today's supercomputers. Figure 1 shows that the exponentially localized distribution, appearing at $t \approx t^\star$, is *frozen* in time, apart from quantum fluctuations, which we partially smooth out by averaging over a few map steps. The freezing of the localized distribution can be seen from comparison of the probability distributions taken immediately after $t^\star$ (the full curve in Fig. 1) and at a much larger time $t = 300 \approx 25t^\star$ (the dashed curve in the same figure). Here the localization length is $\ell \approx 12$, and classical diffusion is suppressed after a break time $t^\star \approx \ell \approx D_m$, in agreement with estimates (25) and (26) [the classical diffusion coefficient is $D_m \approx (\pi^2/3)k^2 \approx 9.9$]. This quantum computation up to times of the order of ℓ requires a number $N_g \approx 3n^2\ell \sim 10^3$ of one- or two-qubit quantum gates.

In Fig. 2, we show a quantum computation that might be performed already with a three-qubit quantum processor. It is possible to compare two very different regimes, namely the localized and the ergodic regime, by varying only the value of the quantum parameter k, while keeping the classical parameter K constant. In both cases the wave function is stationary (apart from quantum fluctuations), as can be seen from the comparison of the wave function patterns at different times. The difference between the two cases is striking. Notice that, in this example, the localization length $\ell < 1$ and one can explain the results of this simulation using

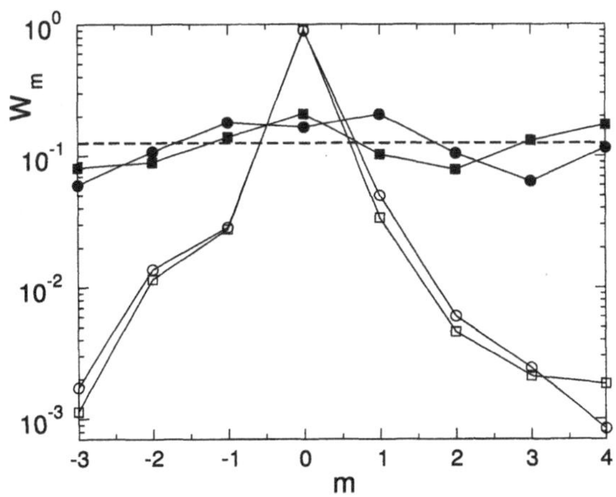

Fig. 2. The probability distribution over the momentum basis for the sawtooth map with $n = 3$ qubits, $k = K/T = KN/2\pi L$, and initial momentum $m_0 = 0$: ergodic regime at $L = 1$ (full symbols) and localized regime at $L = 5$ (empty symbols). Circles (squares) represent the wave function after $t = 3$ ($t = 50$) time steps. The dashed line represents an equally weighted wave function. To smooth the results, we average over ten (one hundred) different values of $K \in [1.4, 1.5]$ for the localized (ergodic) case.

perturbation theory. Indeed, we have $k \sim 0.35 < 1$, and therefore we can treat the kick $\hat{U}_k$ as a perturbation of the free-evolution operatore $\hat{U}_T$. The case shown in Fig. 2 is interesting since it involves only $n=3$ qubits and a few tens on quantum gates. Therefore this quantum computation seems to be accessible or close to the present capabilities of NMR-based[16,17] and ion-trap[18] quantum processors.

We now discuss how to extract the relevant information (the value of the localization length) from a quantum computer simulating the sawtooth-map dynamics. The localization length can be measured by running the algorithm repeatedly up to time $t > t^\star$. Each run is followed by a standard projective measurement on the computational (momentum) basis. Since the wave function at time t can be written as

$$|\psi(t)\rangle = \sum_m \hat{\psi}(m,t)\,|m\rangle\,, \tag{27}$$

with $|m\rangle$ momentum eigenstates, such a measurement gives outcome $\bar{m}$ with probability

$$W_{\bar{m}} = \left|\langle \bar{m}|\psi(t)\rangle\right|^2 = \left|\hat{\psi}(\bar{m},t)\right|^2. \tag{28}$$

A first series of measurements would allow us to give a rough estimate of the variance $\langle(\Delta m)^2\rangle$ of the distribution W_m. In turn, $\sqrt{\langle(\Delta m)^2\rangle}$ gives a first estimate of the localization length ℓ. After this, we can store the results of the measurements in histogram bins of width $\delta m \propto \ell \approx \sqrt{\langle(\Delta m)^2\rangle}$. Finally, the localization length is extracted from a fit of the exponential decay of this coarse-grained distribution over the momentum basis. Elementary statistical theory tells us that, in this way, the localization length can be obtained with accuracy ν after the order of $1/\nu^2$ computer runs. It is interesting to note that it is sufficient to perform a coarse-grained measurement to generate a coarse-grained distribution. This means that it will be sufficient to measure the most significant qubits, and ignore those that would give a measurement accuracy below the coarse graining δm. Thus, the number of runs and measurements is independent of ℓ.

In Fig. 3, we report a simulation of the measurement process. In the upper figure we compare the exact probabilities given by the wave function with the result of a complete measurement of all qubits and the result of a coarse-grained measurement. The histograms are built from the same number of computational runs, each followed by a projective measurement. The coarse-grained measurement does not resolve the thinnest structures of the exact wave function. However, it is still possible to extract a good estimate of the localization length ℓ from a fit of the exponential decay of the probability distribution W_m. In the lower figure we compare the localization lengths, extracted from the complete and the coarse-grained measurements, as a function of the number N_M of projective measurements. Two distinct behaviors are clearly distinguishable: the localization length computed from the complete measurement of all qubits converges slowly to the exact value for the localization length, since a large number of projective measurement is required in order to resolve the exponentially decaying tails. On the contrary, the coarse-grained measurements

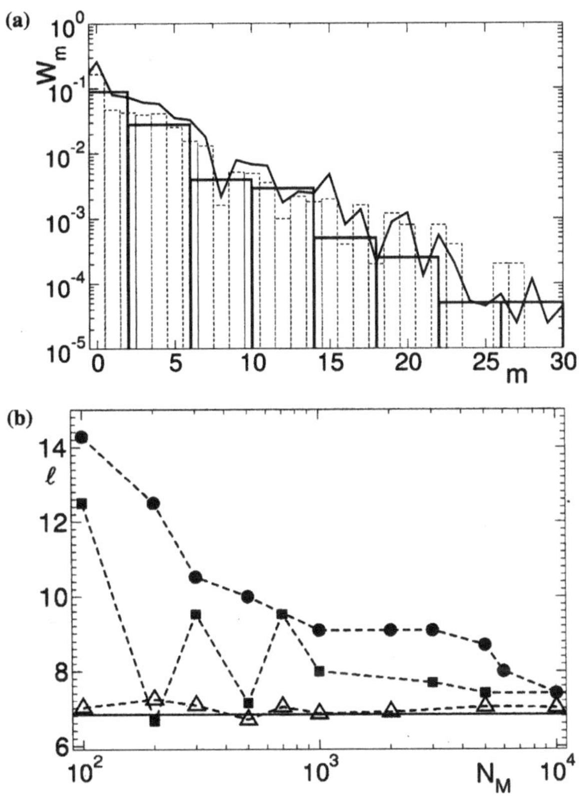

Fig. 3. Top: Simulation of a measurement experiment for the quantum sawtooth map at $n = 6$, $K = \sqrt{2}$, $T = 2\pi L/N$, $L = 10$, $t = 50$. The thick line is the exact wave function, the thin dashed (thick full) histogram represents the result of N_M runs, each followed by a projective measurement of all (all except two) qubits. In both cases $N_M = 5 \times 10^3$; Bottom: The estimated localization length ℓ as a function of the number N_M of projective measurements. We estimate ℓ by fitting the probability decay for the complete (circles) and coarse-grained (squares) measurements. Triangles give 2ξ, with the inverse partecipation ratio computed from the coarse-grained probability distribution. The straight line is the theoretical result $\ell \sim 6.8$, obtained from Eq. (26).

approach the exact value after a much smaller number of measurements, even though the fluctuations as a function of the number of measurements are quite large.

It is possible to give a better estimate of the localization length by computing the inverse participation ratio

$$\xi = \frac{1}{\sum_m W_m^2}. \tag{29}$$

The inverse participation ratio determines the number of basis states significantly populated by the wave function and gives an estimate of the localization length of the system. We have $1 \leq \xi \leq N$, with the limiting cases $\xi = 1$ and $\xi = N$ corresponding to a wave function delta-peaked ($W_m = \delta_{m,m_0}$) or uniformly spread ($W_m = 1/N$). In the localized regime, $\xi \approx \ell/2$. We stress that the inverse participation ratio is almost insensitive to the behavior of exponentially small tails of the wave function. Thus, the estimate $\ell \approx 2\xi$ is quite accurate already with a small number of coarse-grained mesurements (see Fig. 3).

We now come to the crucial point, of estimating the gain of quantum computation of the localization length with respect to classical computation. First of all, we recall that it is necessary to make about $t^\star = O(\ell)$ map iterations to obtain the localized distribution, see Eq. (25). This is true, both for the present quantum algorithm and for classical computation. It is reasonable to use a basis size $N = O(\ell)$ to detect localization (say, N equal to a few times the localization length). In such a situation, a classical computer requires $O(\ell^2 \log \ell)$ operations to extract the localization length, while a quantum computer would require $O(\ell(\log \ell)^2)$ elementary gates. Indeed, both classical and quantum computers need to perform $t \approx t^\star = O(\ell) = O(N)$ map iterations. Therefore, the quantum computer provides a *quadratic speed up* in computing the localization length. We point out that the number of runs and measurements necessary to extract the localization length ℓ does not depend on ℓ itself and, therefore, on the system size $N = O(\ell)$. Hence, this number does not affect the quadratic speed up provided by the quantum computer. It is also interesting to notice that, even though the speed up is only quadratic, the advantage of the quantum computer in memory resources remains exponential. This point becomes crucial in many-dimensional or many-body physical models, in which, using a classical computer, it is very hard, if not impossible, to merely establish if the system is localized or not. As the results of this section can be extended to this kind of problems, quantum computers promise to become a useful tool for their investigation.

As we shall see in Sec.6, the quantum computation can provide an *exponential gain* (with respect to any known classical computation) in problems that require the simulation of dynamics up to a time t which is independent of the number of qubits. In this case, provided that we can extract the relevant information in a number of measurements polynomial in the number of qubits, one should compare $O(t(\log N)^2)$ elementary gates (quantum computation) with $O(tN \log N)$ elementary gates (classical computation).

5. QUANTUM COMPUTING OF BROWNIAN AND ANOMALOUS DIFFUSION

As we have discussed in Sec. 2, the classical sawtooth map is characterized by different diffusive behaviors in the chaotic and semi-integrable regimes. Quantum computers could help us to study these different regimes by simulating the map in the deep semiclassical region $\hbar_{\text{eff}} \to 0$. Let us first show that a quantum computer would be useful in computing the Brownian diffusion coefficient D_m. For this purpose, we can repeat several times the quantum simulation of the sawtooth map up to a given time t, ending each run with a standard projective measurement in the momentum basis. This allows us to compute, up to statistical errors, $\langle(\Delta m)^2\rangle$. The diffusion coefficient is then obtained from Eq. (24) as $D_m \approx \langle(\Delta m)^2\rangle / t$. Therefore a computation of the diffusion coefficient up to time t significantly involves the order of $\sqrt{D_m t}$ momentum eigenstates (other levels are only weakly populated for times smaller than t and can be neglected). Thus, a basis of dimension $N = O(t^{1/2})$ is sufficient for this computation. To estimate the speed up of quantum computation, one should compare $O(t(\log N)^2) = O(N^2(\log N)^2)$ elementary gates (quantum computation) with $O(tN \log N) = O(N^3 \log(N))$ elementary gates (classical computation). This gives an *algebraic speed up*.

We note that similar computations could be done in the regime of anomalous diffusion, in which $\langle(\Delta J)^2\rangle = T^2\langle(\Delta m)^2\rangle \propto t^\alpha$, to evaluate the exponent α, a quantity of great physical interest. Such a regime is quite complex in the sawtooth map: Fig. 4 shows, for the classical map, the dependence of the exponent α as a function of K. As can be seen from this figure, the map explores subdiffusive ($\alpha < 1$) and superdiffusive ($\alpha > 1$) regions, up to ballistic diffusion ($\alpha = 2$). As required by the principle of quantum to classical correspondence, the quantum sawtooth map follows this behavior in the deep semiclassical regime $\hbar_{\text{eff}} \ll 1$, up to some time scale which diverges when $\hbar_{\text{eff}} \to 0$. It is important to point out that $\hbar_{\text{eff}}$ drops to zero exponentially with the number of qubits ($\hbar_{\text{eff}} \propto 1/N = 1/2^n$), and therefore the deep semiclassical region can be reached with a small number of qubits. For large $\hbar_{\text{eff}}$, one can also study how diffusion is mod-

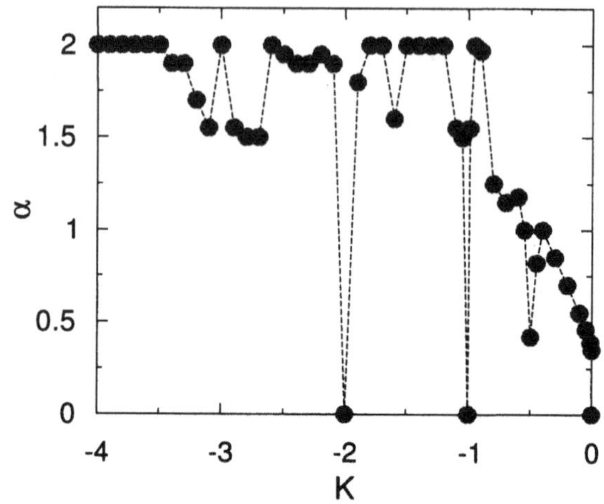

Fig. 4. Left: Exponent α of the anomalous diffusion $[\langle(\Delta J)^2\rangle \propto t^{\alpha}]$ as a function of K for the classical sawtooth map in the semi-integrable regime.

ified by important quantum phenomena, like quantum tunneling, localization, and quantum resonances.

A quantum computer could help us in obtaining the exponent α of the anomalous diffusion. In this case, since $\langle(\Delta m)^2\rangle \propto t^{\alpha}$, a rough estimate of the size of the basis required for the computation up to time t is $N = O(t^{\alpha/2})$. Hence, we must compare $O(t(\log N)^2) = O(N^{2/\alpha}(\log N)^2)$ elementary gates (quantum computation) with $O(tN \log N) = O(N^{(\alpha+2)/\alpha}$ $\log(N))$ elementary gates (classical computation). The speed up is again algebraic.

6. QUANTUM COMPUTING OF THE FIDELITY OF QUANTUM MOTION

The simulation of quantum dynamics up to a time t which is independent of the number of qubits is useful, for instance, to measure dynamical correlation functions of the form

$$C(t) \equiv \langle\psi|\,\hat{A}^{\dagger}(t)\,\hat{B}(0)\,|\psi\rangle = \langle\psi|\,(\hat{U}^{\dagger})^{t}\,\hat{A}^{\dagger}(0)\,\hat{U}^{t}\,\hat{B}(0)\,|\psi\rangle, \tag{30}$$

where $\hat{U}$ is the time-evolution operator. Similarly, we can efficiently compute the fidelity of quantum motion, which is a quantity of central interest in the study of the stability of a system under perturbations (see, e.g., Refs. 19–26 and references therein). The fidelity $f(t)$ (also called the

Loschmidt echo), measures the accuracy with which a quantum state can be recovered by inverting, at time t, the dynamics with a perturbed Hamiltonian. It is defined as

$$f(t) = \langle\psi| (\hat{U}_\epsilon^\dagger)^t \hat{U}^t |\psi\rangle . \tag{31}$$

Here the wave vector $|\psi\rangle$ evolves forward in time with the time-evolution operator U up to time t, and then evolves backward in time with a perturbed time-evolution operator U_ϵ. If the operators $\hat{U}$ and $\hat{U}_\epsilon$ can be simulated efficiently on a quantum computer, as is the case in most physically interesting situations, then the fidelity of quantum motion can be evaluated with exponential speed up with respect to known classical computations. The same conclusion is valid for the correlation functions (30).

The fidelity can be efficiently evaluated on a quantum computer, with the only requirement of an ancilla qubit, using the *scattering circuit* drawn in Fig. 5.[27,28] This circuit has various important applications in quantum computing, including quantum state tomography and quantum spectroscopy.[28] The circuit ends up with the measurement of the ancilla qubit, and we have

$$\langle\sigma_z\rangle = \mathrm{Re}[\mathrm{Tr}(\hat{W}\rho)], \qquad \langle\sigma_y\rangle = \mathrm{Im}[\mathrm{Tr}(\hat{W}\rho)], \tag{32}$$

where $\langle\sigma_z\rangle$ and $\langle\sigma_y\rangle$ are the expectation values of the Pauli spin operators $\hat{\sigma}_z$ and $\hat{\sigma}_y$ for the ancilla qubit, and $\hat{W}$ is a unitary operator. These two expectation values can be obtained (up to statistical errors) if one runs several times the scattering circuit. If we set $\rho = |\psi\rangle\langle\psi|$ and $\hat{W} = (\hat{U}_\epsilon^\dagger)^t \hat{U}^t$, it is easy to see that $f(t) = |\mathrm{Tr}(\hat{W}\rho)|^2 = \langle\sigma_z\rangle^2 + \langle\sigma_y\rangle^2$. For this reason, provided that the quantum algorithm which implements $\hat{U}$ is efficient, the fidelity can be efficiently computed by means of the circuit described in Fig. 5. This applies to a large class of problems and, in particular, to the quantum sawtooth map. We note that another possible way to efficiently measure the fidelity has been proposed in Ref. 25.

It is interesting to discuss the time scale of the fidelity decay in chaotic systems.[20–24] To be concrete, let us consider the quantum sawtooth map model and choose to perturb the kicking strength: $k \to k' =$

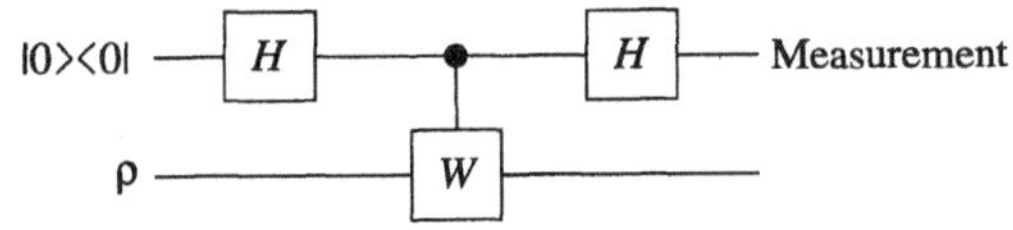

Fig. 5. Scattering circuit. The top line denotes the ancilla qubit, the bottom line a set of n qubits, H the Hadamard gate, and W a unitary transformation.

$k + \sigma$, corresponding to a classical perturbation $K = kT \to K' = K + \epsilon$, with $\epsilon \equiv \sigma T \ll K$. In this case, the typical transition matrix elements $V_{jk} = \langle u_j | \hat{V} | u_k \rangle$ of the perturbation operator $\hat{V} = -[\sigma(\hat{\theta} - \pi\hat{1})^2/2] \sum_j \delta(\tau - jT)$ between the eigenstates $\{|u_j\rangle\}$ of the Floquet operator $\hat{U}$ are of the order of $V \sim \sigma/\sqrt{N}$ ($|u_j\rangle$ is an eigenstate of $\hat{U}$ corresponding to the quasienergy λ_j if $\hat{U}|u_j\rangle = \exp(i\lambda_j)|u_j\rangle$). This energy scale has to be compared with the typical energy separation $\Delta E \sim 1/N$ between the unperturbed quasienergies. It can be shown[20–24] that there exist three regimes for the fidelity decay:

(1) for $\sigma < \sigma_p \approx 1/\sqrt{N}$, corresponding to $V < \Delta E$, the fidelity decay can be calculated in perturbation theory. This gives a Gaussian decay: $f(t) \sim \exp(-V^2 t^2)$;

(2) for $\sigma_p < \sigma < \sigma_c \sim 1$, the fidelity decay is exponential: $f(t) \sim \exp(-\Gamma t)$, with rate $\Gamma \sim V^2/\Delta E \sim \sigma^2$ given by the Fermi golden rule;

(3) in the semiclassical regime $\sigma > \sigma_c \sim 1$, the fidelity decay is again exponential, but with a perturbation independent rate $\Gamma = \lambda$, where λ is the Lyapunov exponent of classical chaotic dynamics.

As an example, we show in Fig. 6 the crossover between the Fermi golden rule and the Lyapunov regime. It is interesting to note that the Lyapunov exponent in this case could be extracted already with $n = 10$ qubits and a few kicks.

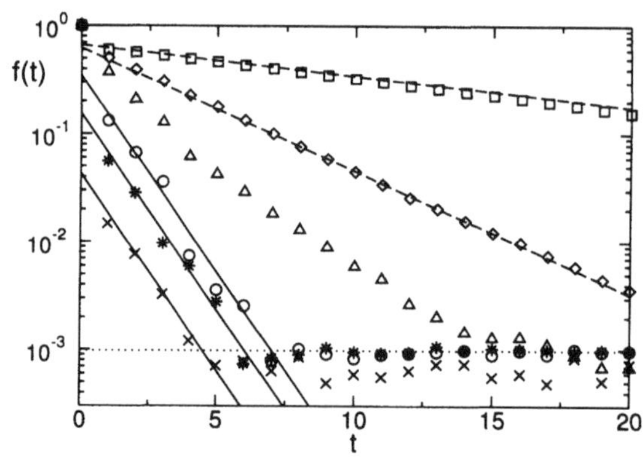

Fig. 6. Fidelity decay for the sawtooth map with $n = 10$ qubits, $K = 0.75$, $L = 1$, $\epsilon = 10^{-3}$ (squares), 2×10^{-3} (diamonds), 3×10^{-3} (triangles), 10^{-2} (circles), 2×10^{-2} (stars), and 3×10^{-2} (crosses). Dashed lines correspond to the Fermi Golden rule decay: $f(t) \sim \exp(-A\sigma^2 t)$, where $A \approx 2.4$. Full lines show the Lyapunov decay: $f(t) \sim \exp(-\lambda t)$, with the maximum Lyapunov exponent $\lambda = \ln[(2 + K + \sqrt{K^2 + 4K})/2] \approx 0.84$. Note that the crossover between the Fermi golden rule and the Lyapunov regime takes place at $\epsilon \sim 5 \times 10^{-3}$, corresponding to $\sigma \sim 1$. The dotted line shows the fidelity saturation value $f_\infty = 1/N$. We choose as initial wave function a momentum eigenstate and data are obtained after averaging over 100 different initial conditions.

7. CONCLUSION

In this paper, we have discussed relevant physical examples of efficient information extraction in the quantum computation of a dynamical system. We have shown that a quantum computer with a small number of qubits can efficiently simulate the quantum localization effects, simulate both the Brownian and anomalous diffusion in the deep semiclassical regime, and compute the fidelity of quantum motion. We would like to stress that the simulation of complex dynamical systems is accessible to the first generation of quantum computers with less than 10 qubits. Therefore, we believe that quantum algorithms for dynamical systems deserve further studies, since they are the ideal software for the first quantum processors. Furthermore, we emphasize that the quantum computation of quantities like dynamical localization or fidelity is a demanding testing ground for quantum computers. In the first case, we want to simulate dynamical localization, a purely quantum phenomena which is quite fragile in the presence of noise;[(29,30)] in the latter case, fidelity is computed as a result of a sophisticated many-qubit Ramsey-type interference experiment. Therefore, the computation of these quantities appears to be a relevant test for quantum processors operating in the presence of decoherence and imperfection effects.

It is interesting to discuss why the speed up in computing the localization length and the diffusion coefficient is not exponential, even though a single step of the dynamical evolution can be simulated with exponential speed up. The ultimate reason is that, in diffusive processes, the spread of information is "slow". More precisely, the number N of quantum levels which are significant to describe the system's dynamics grows only algebraically in time. Therefore, in order to use N levels, dynamics must be iterated up to a time scale which is algebraic in N, that is, exponential in the number of qubits.

Finally, we would like to stress that the results discussed in this paper have a general validity, beyond the sawtooth map model. The localization length, the diffusion coefficient and the fidelity can be extracted, with the same speed up (algebraic or exponential), also in all other one-dimensional driven dynamical systems modeled by Eq. (1), provided that the driving force is sufficiently regular to be computed efficiently (this is the case, for instance, when it can be expanded in Taylor series). Moreover, our results can be extended to many-dimensional and many-particle systems, for which the computation of the quantities discussed in this paper by means of a classical computer turns out to be a prohibitive task.

ACKNOWLEDGEMENTS

This work was supported in part by the EC contracts IST-FET EDIQIP and RTN QTRANS, the NSA and ARDA under ARO contract No. DAAD19-02-1-0086, and the PRIN 2002 "Fault tolerance, control and stability in quantum information processing".

REFERENCES

1. R. P. Feynman, *Int. J. Theor. Phys.* **21**, 467 (1982).
2. S. Lloyd, *Science* **273**, 1073 (1996).
3. G. Ortiz, J. E. Gubernatis, E. Knill, and R. Laflamme, *Phys. Rev. A* **64**, 022319 (2001).
4. R. Schack, *Phys. Rev. A* **57**, 1634 (1998).
5. B. Georgeot and D. L. Shepelyansky, *Phys. Rev. Lett.* **86**, 2890 (2001).
6. G. Benenti, G. Casati, S. Montangero, and D. L. Shepelyansky, *Phys. Rev. Lett.* **87**, 227901 (2001).
7. G. Benenti, G. Casati, S. Montangero, and D. L. Shepelyansky, *Phys. Rev. A* **67**, 052312 (2003).
8. I. Dana, N. W. Murray, and I. C. Percival, *Phys. Rev. Lett.* **62**, 233 (1989).
9. M. A. Nielsen and I. L. Chuang, *Quantum Computation and Quantum Information* (Cambridge University Press, Cambridge, 2000).
10. G.Casati, B. V. Chirikov, J. Ford, and F. M. Izrailev, *Lecture Notes Phys.* **93**, 334 (1979); for a review see, e.g., F.M. Izrailev, *Phys. Rep.* **196**, 299 (1990).
11. S. Fishman, D. R. Grempel, and R. E. Prange, *Phys. Rev. Lett.* **49**, 509 (1982).
12. P. M. Koch and K. A. H. van Leeuwen, *Phys. Rep.* **255**, 289 (1995), and references therein.
13. F. L. Moore, J. C. Robinson, C. F. Barucha, B. Sundaram, and M. G. Raizen, *Phys. Rev. Lett.* **75**, 4598 (1995); H. Ammann, R. Gray, I. Shvarchuck, and N. Christensen, *ibid.* **80**, 4111 (1998); D. A. Steck, W. H. Oskay, and M. G. Raizen, *ibid.* **88**, 120406 (2002).
14. A. D. Mirlin, Y. V. Fyodorov, F. -M. Dittes, J. Quezada, and T. H. Seligman, *Phys. Rev. E* **54**, 3221 (1996).
15. B. V. Chirikov, F. M. Izrailev, and D.L. Shepelyansky, *Sov. Sci. Rev.* C **2**, 209 (1981).
16. Y. S. Weinstein, S. Lloyd, J. Emerson, and D. G. Cory *Phys. Rev. Lett.* **89**, 157902 (2002).
17. L. M. K. Vandersypen, M. Steffen, G. Breyta, C. S. Yannoni, M. H. Sherwood, and I. L. Chuang, *Nature* **414**, 883 (2001).
18. S. Gulde, M. Riebe, G. P. T. Lancaster, C. Becher, J. Eschner, H. Häffner, F. Schmidt-Kaler, I. L. Chuang, and R. Blatt, *Nature* **421**, 48 (2003).
19. A. Peres, *Phys. Rev. A* **30**, 1610 (1984).
20. R. A. Jalabert and H. M. Pastawski, *Phys. Rev. Lett.* **86**, 2490 (2001).
21. Ph. Jacquod, P. G. Silvestrov, and C. W. J. Beenakker, *Phys. Rev. E* **64**, 055203(R) (2001).
22. N. R. Cerruti and S. Tomsovic, *Phys. Rev. Lett.* **88**, 054103 (2002).
23. G. Benenti and G. Casati, *Phys. Rev. E* **65**, 066205 (2002).
24. T. Prosen and M. Žnidarič, *J. Phys. A* **35**, 1455 (2002).
25. J. Emerson, Y. S. Weinstein, S. Lloyd, and D. Cory, *Phys. Rev. Lett.* **89**, 284102 (2002).

26. F. M. Cucchietti, D. A. R. Dalvit, J. P. Paz, and W. H. Zurek, *Phys. Rev. Lett.* **91**, 210403 (2003).
27. S. A. Gardiner, J. I. Cirac, and P. Zoller, *Phys. Rev. Lett.* **79**, 4790 (1997).
28. C. Miquel, J. P. Paz, M. Saraceno, E. Knill, R. Laflamme, and C. Negrevergne, *Nature* **418**, 59 (2002).
29. E. Ott, T. M. Antonsen, Jr., and J. D. Hanson, *Phys. Rev. Lett.* **53**, 2187 (1984).
30. P. H. Song and D. L. Shepelyansky, *Phys. Rev. Lett.* **86**, 2162 (2001).

One-Dimensional Continuous-Time Quantum Walks

D. ben-Avraham,[1] E. M. Bollt,[2] and C. Tamon[3,4]

We survey the equations of continuous-time quantum walks on simple one-dimensional lattices, which include the finite and infinite lines and the finite cycle, and compare them with the classical continuous-time Markov chains. The focus of our expository article is on analyzing these processes using the Laplace transform on the stochastic recurrences. The resulting time evolution equations, classical vs. quantum, are strikingly similar in form, although dissimilar in behavior. We also provide comparisons with analyses performed using spectral methods.

KEY WORDS: Quantum walks; continuous time; Laplace transform.

PACS: 03.67.Lx.

1. INTRODUCTION

The theory of Markov chains on countable structures is an important area in Mathematics and Physics. A quantum analog of continuous-time Markov chains on the infinite line is well studied in Physics (for example, it can be found in Ref. 12, Chapters 13 and 16). More recently, it was studied by Aharonov *et al.*(2) and by Farhi and Gutmann.(11) The latter work placed the problem in the context of quantum algorithms for search problems on graphs. Here, the symmetric stochastic matrix of the graph is viewed as a Hamiltonian of the quantum process. Using Schrödinger's equation with this Hamiltonian, we obtain a quantum walk on the underlying graph, instead of a classical random walk.

[1]Department of Physics, Clarkson University. E-mail: qd00@clarkson.edu
[2]Department of Mathematics and Computer Science, and Department of Physics, Clarkson University. E-mail: bolltem@clarkson.edu
[3]Department of Mathematics and Computer Science, and Center for Quantum Device Technology, Clarkson University. E-mail: tino@clarkson.edu
[4]To whom correspondence should be addressed.

Recent works on continuous-time quantum walks on finite graphs include the analyses of mixing and hitting times on the n-cube,[16,14] of mixing times on circulant graphs and Cayley graphs of the symmetric group,[6,13] and of hitting times on path-like graphs.[8,9] Most of these are structural results based on spectral analysis of the underlying graphs, such as the n-cube, circulant and Cayley graphs, and (weighted) paths. For example, Moore and Russell[16] proved that the mixing time of a quantum walk on the n-cube is asymptotically faster than a classical random walk; Kempe[14] proved that the hitting time for vertices on opposite ends of the n-cube is exponentially faster than in a classical random walk. Ahmadi *et al.*[6] and Gerhardt and Watrous[13] proved that circulants and the Cayley graph of the symmetric group lack the uniform mixing property found in classical random walks.

A recent work of Childs *et al.*[9] gave intriguing evidence that continuous-time quantum walk is a powerful method for designing new quantum algorithms. They analyzed diffusion processes on one-dimensional structures (finite path and infinite line) using spectral methods. Another work by Childs and Goldstone[10] explored the application of continuous-time quantum walks to perform Grover search on spatial lattices.

There is an alternate theory of *discrete* quantum walks on graphs, which we will not discuss here. This alternate model was studied in Aharonov *et al.*[4] and Ambainis *et al.*,[5] but had appeared earlier in work by Meyer.[15] The work by Ambainis *et al.*[5] had also focused on one-dimensional lattices. Recently, Ambainis[1] developed an optimal (discrete) quantum walk algorithm for the fundamental problem of Element Distinctness. This offers another idea for developing quantum algorithms.

We survey and (re)derive equations for the continuous-time classical and quantum walks on one-dimensional lattices using the Laplace transform that works directly with the recurrences. The Laplace transform is a well-known tool in stochastic processes (see Ref. 7) and it might offer a useful alternative to the Fourier transform in certain settings.

1.1. Stochastic Walks on Graphs

Let $G=(V,E)$ be a simple (no self-loops), countable, undirected graph with adjacency matrix A. Let D be a diagonal matrix whose jth entry is the degree of the jth vertex of G. The Laplacian of G is defined as $H=A-D$. Suppose that $P(t)$ is a time-dependent probability distribution of a stochastic (particle) process on G. The classical evolution of the continuous-time walk is given by the Kolmogorov equation

$$P'(t)=HP(t). \tag{1}$$

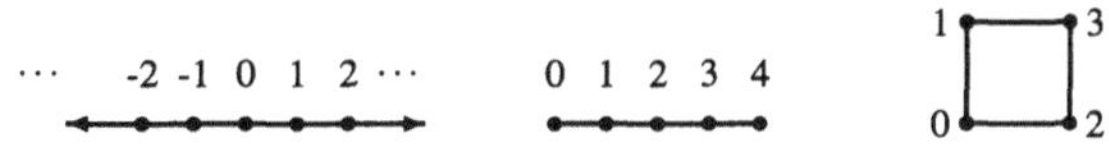

Fig. 1. Examples of some one-dimensional lattices. From left to right: $\mathbb{Z}$, $\mathbb{P}_4$, $\mathbb{Z}_4$.

Graph	**Classical walk** $P_j(t) =$ probability on vertex j at time t	**Quantum walk** $\psi_j(t) =$ amplitude on vertex j at time t
$\mathbb{Z}$	$e^{-t}I_{\lvert j\rvert}(t)$	$(-i)^{\lvert j\rvert}J_{\lvert j\rvert}(t)$
$\mathbb{Z}_N$	$\sum_{\alpha\equiv\pm j(\text{mod} N)} e^{-t}I_\alpha(t)$	$\sum_{\alpha\equiv\pm j(\text{mod} N)} (-i)^\alpha J_\alpha(t)$
$\mathbb{P}_N$	$\sum_{\alpha\equiv\pm j(\text{mod } 2N)} e^{-t}I_\alpha(t)$	$\sum_{\alpha\equiv\pm j(\text{mod } 2N)} (-i)^\alpha J_\alpha(t)$

Fig. 2. The equations of the continuous-time classical vs. quantum walks on the infinite line, finite cycle, and the finite line, assuming the particle starts at position 0.

The solution of this equation, modulo some conditions, is $P(t)=e^{tH}P(0)$, which can be solved by diagonalizing the symmetric matrix H. This *spectral* approach requires full knowledge of the spectrum of H.

A quantum analog of this classical walk uses the Schrödinger equation in place of the Kolmogorov equation. Let $\psi : V(G) \to \mathbb{C}$ be the time-independent amplitude of the quantum process on G. Then, the wave evolution is

$$i\hbar\frac{\mathrm{d}}{\mathrm{d}t}\psi(t) = H\psi(t). \tag{2}$$

Assuming $\hbar = 1$ for simplicity, the solution of this equation is $\psi(t) = e^{-itH}\psi(0)$, which, again, is solvable via spectral techniques. The classical behavior of this quantum process is given by the probability distribution $P(t)$ whose j th entry is $P_j(t)=|\psi_j(t)|^2$, where $\psi_j(t)=\langle j|\psi(t)\rangle$. The *average probability* of vertex j is defined as $\overline{P}(j)=\lim_{T\to\infty}\frac{1}{T}\int_0^T P_j(t)\mathrm{d}t$ (see Ref. 4) (Fig. 1).

The table in Fig. 2 shows the known equations for continuous-time stochastic walks on the infinite (integer) line $\mathbb{Z}=\{\ldots,-2,-1,0,1,2,\ldots\}$, the finite cycle $\mathbb{Z}_N = \{0,\ldots,N-1\}$ on N vertices, and the finite path $\mathbb{P}_N = \{0,\ldots,N\}$ on $N+1$ vertices, in terms of the two kinds of Bessel functions $I(\cdot)$ and $J(\cdot)$. We assume here that the particle is initially at 0. The plots in Figs. 3 and 4 show the dissimilar behavior of the classical versus quantum walks.

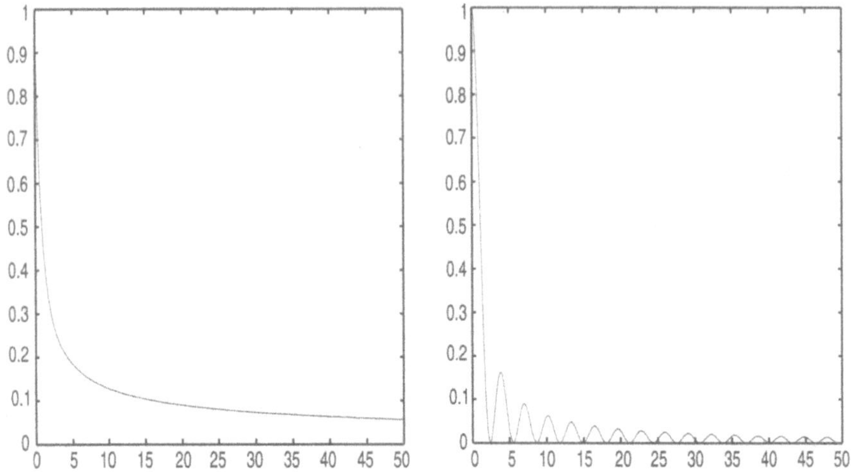

Fig. 3. Stochastic walks on the infinite line $\mathbb{Z}$: (a) plot of $P_0(t)$ in the continuous-time random walks for $t \in [0, 50]$. (b) plot of $|\psi_0(t)|^2$ in a continuous-time quantum walk for $t \in [0, 50]$. Both processes exhibit exponential decay, but with the quantum walk showing an oscillatory behavior.

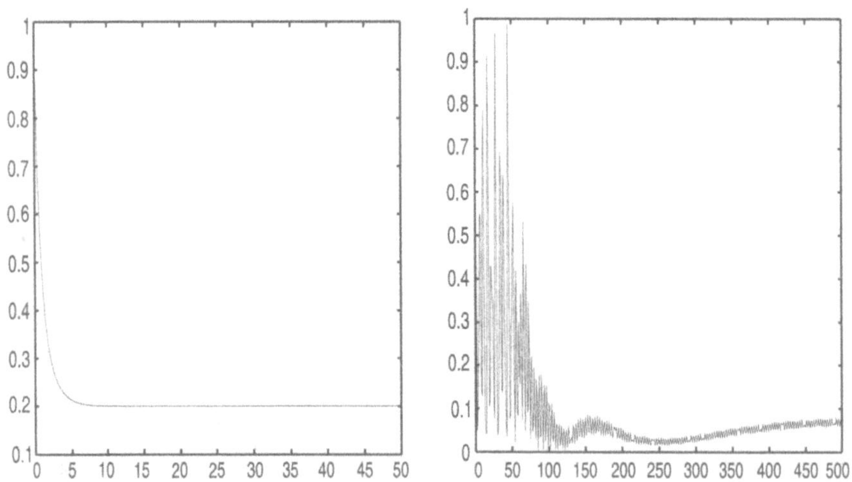

Fig. 4. Stochastic walks on the finite cycle $\mathbb{Z}_5$, each approximated using 21 terms: (a) plot of $P_0(t)$ in the continuous-time random walks for $t \in [0, 50]$. (b) plot of $|\psi_0(t)|^2$ in the continuous-time quantum walk for $t \in [0, 500]$. The classical walk settles quickly to 1/5, while the quantum walk exhibit a short-term chaotic behavior and a long-term oscillatory behavior below 0.1.

1.2. Laplace Transform

The Laplace transform of a time-dependent function $P(t)$, denoted $\hat{P}(s)=\mathcal{L}\{P(t)\}$, is defined as

$$\mathcal{L}\{P(t)\}=\int_0^{\infty} e^{-st}P(t)\,dt. \tag{3}$$

The only basic properties of the Laplace transform which we will need are (see Ref. 3):

- Linearity: $\mathcal{L}\{aP(t)+bQ(t)\}=a\hat{P}(s)+b\hat{Q}(s)$
- Derivative: $\mathcal{L}\{P'(t)\}=s\hat{P}(s)-P(0)$
- Shifting: $\mathcal{L}\{e^{at}P(t)\}=\hat{P}(s-a)$

The relevant inverse Laplace transform involving the Bessel functions are (for $\nu>-1$):

$$\hat{P}(s)=\frac{(s-\sqrt{s^2-a^2})^{\nu}}{\sqrt{s^2-a^2}} \iff P(t)=a^{\nu}I_{\nu}(at), \quad \text{(Eq. (29.3.59) in Ref. 3)} \tag{4}$$

$$\hat{P}(s)=\frac{(\sqrt{s^2+a^2}-s)^{\nu}}{\sqrt{s^2+a^2}} \iff P(t)=a^{\nu}J_{\nu}(at). \quad \text{(Eq. (29.3.56) in Ref. 3)} \tag{5}$$

2. THE INFINITE LINE

2.1. Classical Process

The Kolmogorov equation for the infinite line is

$$P_j'(t)=\tfrac{1}{2}P_{j-1}(t)-P_j(t)+\tfrac{1}{2}P_{j+1}(t), \tag{6}$$

with initial value $P_j(0)=\delta_{0,j}$. The Laplace transform of Eq. (6) is

$$\hat{P}_{j+1}(s)-2(s+1)\hat{P}_j(s)+\hat{P}_{j-1}(s)=-P_j(0). \tag{7}$$

The solution of $q^2-2(s+1)q+1$ is $q_{\pm}=(s+1)\pm\sqrt{(s+1)^2-1}$. A natural guess of the solution is

$$\hat{P}_j(s)=\begin{cases} Aq_+^j & \text{if } j<0 \\ Aq_-^j & \text{if } j>0. \end{cases} \tag{8}$$

When $j=0$, we get $A=(1+s-q_-)^{-1}$. Thus, for $j\in\mathbb{Z}$,

$$\hat{P}_j(s)=\frac{q_-^{|j|}}{(1+s-q_-)}=\frac{[(s+1)-\sqrt{(s+1)^2-1}]^{|j|}}{\sqrt{(s+1)^2-1}}. \tag{9}$$

Using the inverse Laplace transform Eq. (4), after shifting $S=s+1$, we get

$$P_j(t)=\mathrm{e}^{-t}I_{|j|}(t). \tag{10}$$

This is a probability function, since $\mathrm{e}^{t/2(z+1/z)}=\sum_{k=-\infty}^{\infty} z^k I_k(t)$, if $z\neq 0$ (see Eq. (9.6.33) in Ref. 3).

2.2. Quantum Process

The Schrödinger equation for the infinite line is

$$\mathrm{i}\psi_j'(t)=\tfrac{1}{2}\psi_{j-1}(t)+\tfrac{1}{2}\psi_{j+1}(t). \tag{11}$$

The Laplace transform of Eq. (11) is

$$\hat{\psi}_{j+1}(s)-2\mathrm{i}(s\hat{\psi}_j(s)-\psi_j(0))+\hat{\psi}_{j-1}(s)=0. \tag{12}$$

The solutions of $q^2-2\mathrm{i}sq+1$ are $q_{\pm}=\mathrm{i}(s\pm\sqrt{s^2+1})$, where $q_+q_-=1$. A guess for the solution is

$$\hat{\psi}_j(s)=\begin{cases} Aq_+^j & \text{if } j\leq 0\\ Aq_-^j & \text{if } j>0.\end{cases} \tag{13}$$

When $j=0$, we get $A=(s+\mathrm{i}q_-)^{-1}$. Thus,

$$\hat{\psi}_j(s)=\frac{q_-^{|j|}}{(s+\mathrm{i}q_-)}=(-\mathrm{i})^{|j|}\frac{(\sqrt{s^2+1}-s)^{|j|}}{\sqrt{s^2+1}}. \tag{14}$$

The inverse Laplace transform Eq. (5) yields, for $j\in\mathbb{Z}$,

$$\psi_j(t)=(-\mathrm{i})^{|j|}J_{|j|}(t). \tag{15}$$

This is a probability function, since $1=J_0^2(z)+2\sum_{k=1}^{\infty}J_k^2(z)$ (see Eq. (9.1.76) in Ref. 3).

2.3. Spectral Analysis

Let H be a Hamiltonian defined as $\langle j|H|k\rangle = \frac{1}{2}$ if $j = k \pm 1$, and 0 otherwise. For each $p \in [-\pi, \pi]$, define $|p\rangle$ so that

$$\langle j|p\rangle = \frac{1}{\sqrt{2\pi}} e^{ipj}. \tag{16}$$

The eigenvalue equation $H|p\rangle = \lambda_p|p\rangle$ has the solution $\lambda_p = \cos(p)$. Thus, the amplitude of position j when the particle starts at position 0 is

$$\begin{aligned}\langle j|e^{-iHt}|0\rangle &= \frac{1}{2\pi}\int_{-\pi}^{\pi} e^{ipj} e^{-it\cos(p)}\, dp \\ &= (-i)^j J_j(t). \qquad \text{(see Eq. (9.1.21) in Ref. 3)}\end{aligned} \tag{17}$$

Childs *et al.*[9] gave a more generalized analysis along these lines.

3. THE FINITE CYCLE

3.1. Classical Process

If A is the adjacency matrix of the finite cycle, let $H = \frac{1}{2}A - I$ be its Laplacian matrix. The Kolmogorov equation for the finite cycle is

$$P_j'(t) = \tfrac{1}{2}P_{j-1}(t) - P_j(t) + \tfrac{1}{2}P_{j+1}(t). \tag{18}$$

Applying the Laplace transform to Eq. (18), we get

$$(s+1)\hat{P}_j(s) - P_j(0) = \tfrac{1}{2}\hat{P}_{j-1}(s) + \tfrac{1}{2}\hat{P}_{j+1}(s). \tag{19}$$

For convenience, define the extra condition $\hat{P}_{-1}(s) = \hat{P}_{N-1}(s) + 2$, so that $\hat{P}_{j+1}(s) - 2(s+1)\hat{P}_j(s) + \hat{P}_{j-1}(s) = 0$ holds for $j \in \mathbb{Z}_N$. The cycle condition is $\hat{P}_N(s) = \hat{P}_0(s)$. We guess the solution to be

$$\hat{P}_j(s) = Aq_+^j + Bq_-^j, \tag{20}$$

where $q_\pm$ is the solution to $x^2 - 2(s+1)x + 1 = 0$, i.e., $q_\pm = (s+1) \pm \sqrt{(s+1)^2 - 1}$, with $q_+q_- = 1$. Using the cycle condition, we get

$$Aq_+^N + Bq_-^N = A + B \implies A(q_+^N - 1) = B(1 - q_-^N) \implies B = Aq_+^N. \tag{21}$$

Using the extra condition and Eq. (21), we get $A=2((q_+-q_-)(q_+^N-1))^{-1}$. Thus, for $j\in\mathbb{Z}_N$,

$$\begin{aligned}\hat{P}_j(s) &= Aq_+^j+Bq_-^j=A(q_+^j+q_+^{N-j})\\ &=\frac{2}{(q_+-q_-)}\frac{(q_-^j+q_-^{N-j})}{(1-q_-^N)}=\frac{2}{(q_+-q_-)}\sum_{k=0}^{\infty}\left(q_-^{kN+j}+q_-^{(k+1)N-j}\right)\\ &=\sum_{k=0}^{\infty}\left[\frac{((s+1)-\sqrt{(s+1)^2-1})^{kN+j}}{\sqrt{(s+1)^2-1}}+\frac{((s+1)-\sqrt{(s+1)^2-1})^{(k+1)N-j}}{\sqrt{(s+1)^2-1}}\right].\end{aligned}$$

The inverse Laplace transform Eq. (4), after shifting, yields, for $j\in\mathbb{Z}_N$,

$$P_j(t)=\sum_{k=0}^{\infty}\mathrm{e}^{-t}\left[I_{kN+j}(t)+I_{(k+1)N-j}(t)\right]=\sum_{\alpha\equiv\pm j(\mathrm{mod}N)}\mathrm{e}^{-t}I_\alpha(t). \tag{22}$$

3.2. Quantum Process

Since the finite cycle is a regular graph, instead of the Laplacian, we use the adjacency matrix directly. In a continuous-time quantum walk, this simply introduces an irrelevant phase factor in the final expression. The Schrödinger equation, in this case, is

$$\mathrm{i}\psi_j'(t)=\tfrac{1}{2}\psi_{j-1}(t)+\tfrac{1}{2}\psi_{j+1}(t). \tag{23}$$

The Laplace transform of Eq. (23) is

$$\hat{\psi}_{j+1}(s)-2\mathrm{i}(s\hat{\psi}_j(s)-\psi_j(0))+\hat{\psi}_{j-1}(s)=0. \tag{24}$$

The cycle boundary condition is $\hat{\psi}_N(s)=\hat{\psi}_0(s)$. For convenience, define

$$\hat{\psi}_{-1}(s)=\hat{\psi}_{N-1}(s)=2\mathrm{i}. \tag{25}$$

The solutions of $q^2-2\mathrm{i}sq+1$ are $q_\pm=i(s\pm\sqrt{s^2+1})$, with $q_+q_-=1$. A solution guess, for $j\in\mathbb{Z}_N$, is

$$\hat{\psi}_j(s)=Aq_+^j+Bq_-^j. \tag{26}$$

The cycle boundary condition yields $B = Aq_+^N$. By Eq. (25), we get $A = 2\mathrm{i}((q_+ - q_-)(q_+^N - 1))^{-1}$. Thus, for $j \in \mathbb{Z}_N$,

$$\begin{aligned}\hat{\psi}_j(s) &= Aq_+^j + Bq_-^j = A(q_+^j + q_+^{N-j})\\ &= \frac{2\mathrm{i}}{(q_+ - q_-)}\frac{(q_-^j + q_-^{N-j})}{(1-q_-^N)} = \frac{2\mathrm{i}}{(q_+ - q_-)}\sum_{k=0}^{\infty}\left(q_-^{kN+j} + q_-^{(k+1)N-j}\right)\\ &= \sum_{k=0}^{\infty}\left[\frac{((-\mathrm{i})(\sqrt{s^2+1}-s))^{kN+j}}{\sqrt{s^2+1}} + \frac{((-\mathrm{i})(\sqrt{s^2+1}-s)^{(k+1)N-j}}{\sqrt{s^2+1}}\right].\end{aligned}$$

The inverse Laplace transform Eq. (5) gives, for $j \in \mathbb{Z}_N$,

$$\begin{aligned}\psi_j(t) &= \sum_{k=0}^{\infty}\left[(-\mathrm{i})^{kN+j} J_{kN+j}(t) + (-\mathrm{i})^{(k+1)N-j} J_{(k+1)N-j}(t)\right]\\ &= \sum_{\alpha \equiv \pm j (\mathrm{mod} N)} (-\mathrm{i})^{\alpha} J_{\alpha}(t). \qquad (27)\end{aligned}$$

3.3. Spectral Analysis

The normalized adjacency matrix H of $\mathbb{Z}_N$ is the circulant matrix

$$H = \begin{pmatrix} 0 & 1/2 & 0 & \dots & 0 & 1/2\\ 1/2 & 0 & 1/2 & \dots & 0 & 0\\ 0 & 1/2 & 0 & \dots & 0 & 0\\ \vdots & \vdots & \vdots & \vdots & & \vdots\\ 1/2 & 0 & 0 & \dots & 1/2 & 0 \end{pmatrix}. \qquad (28)$$

It is well known that all $N \times N$ circulant matrices are unitarily diagonalized by the Fourier matrix $F = \frac{1}{\sqrt{N}}V(\omega_N)$, where $\omega_N = \mathrm{e}^{2\pi\mathrm{i}/N}$ and $V(\omega_N)$ is the Vandermonde matrix defined as $V(\omega_N)[j,k] = \omega_N^{jk}$, for $j,k \in \{0,1,\dots,N-1\}$. The eigenvalues of H are $\lambda_j = \frac{1}{2}(\omega_N^j + \omega_N^{j(N-1)}) = \cos(2\pi j/N)$, for $j = 0,1,\dots,N-1$. Thus, the wave amplitude at vertex j at time t is

$$\psi_j(t) = \frac{1}{N}\sum_{k=0}^{N-1}\mathrm{e}^{-\mathrm{i}t\cos(2\pi k/N)}\omega_N^{jk}. \qquad (29)$$

From earlier analysis, we get the following Bessel equation

$$\frac{1}{N}\sum_{k=0}^{N-1}\mathrm{e}^{-\mathrm{i}t\cos(2\pi k/N)}\mathrm{e}^{2\pi\mathrm{i}jk/N} = \sum_{\alpha\equiv\pm j(\mathrm{mod}\ N)}(-\mathrm{i})^{\alpha}J_{\alpha}(t). \qquad (30)$$

It is an open question if there exists a time $t \in \mathbb{R}^+$ such that for all $j \in \mathbb{Z}_N$ we have $|\psi_j(t)|^2 = 1/N$, i.e., uniformity is achieved at time t. For $N = 2, 3, 4$, it is known that *instantaneous exact uniform mixing* is achieved (see Refs. 6 and 16).

4. THE FINITE PATH

4.1. Classical Process

Let A be the *normalized* adjacency matrix of the finite path, where A is a stochastic matrix where the probability transitions are proportional to the degrees of the vertices. Let $H = A - I$ be its Laplacian. Then, the Kolmogorov equation, in this case, is

$$P_j'(t) = \tfrac{1}{2} P_{j-1}(t) - P_j(t) + \tfrac{1}{2} P_{j+1}(t), \tag{31}$$

for $0 < j < N$, with initial condition $P_j(0) = \delta_{j,0}$ and boundary conditions

$$P_0'(t) = P_1(t) - P_0(t), \qquad P_N'(t) = P_{N-1}(t) - P_N(t). \tag{32}$$

The Laplace transform of Eq. (31) is

$$\hat{P}_{j+1}(s) - 2(s+1)\hat{P}_j(s) + \hat{P}_{j-1}(s) = 0, \qquad 0 < j < N, \tag{33}$$

and two boundary equations $(1+s)\hat{P}_0(s) - 1 = \hat{P}_1(s)$, and $(1+s)\hat{P}_N(s) = \hat{P}_{N-1}(s)$. A guess of the solution is

$$\hat{P}_j(s) = Aq_+^j + Bq_-^j, \qquad 0 \le j \le N, \tag{34}$$

where $q_\pm = (s+1) \pm \sqrt{(s+1)^2 - 1}$. The boundary equations give $B - A = 2/(q_+ - q_-)$ and $A = Bq_-^{2N}$. Combining these last two equations, we get

$$A = \frac{2}{(q_+ - q_-)} \frac{q_-^{2N}}{(1 - q_-^{2N})}. \tag{35}$$

Thus, for $j = 0, 1, \ldots, N$,

$$\begin{aligned}
\hat{P}_j(s) &= Aq_+^j + Bq_-^j = A(q_+^j + q_+^{2N-j}) = \frac{2}{(q_+ - q_-)} \frac{(q_-^j + q_-^{2N-j})}{(1 - q_-^{2N})} \\
&= \frac{2}{(q_+ - q_-)} \sum_{k=0}^{\infty} (q_-^{2Nk+j} + q_-^{2N(k+1)-j}).
\end{aligned}$$

The inverse Laplace transform Eq. (4), after shifting, yields, for $j = 0, 1, \ldots, N$,

$$P_j(t) = \sum_{k=0}^{\infty} \mathrm{e}^{-t}\left[I_{2Nk+j}(t) + I_{2N(k+1)-j}(t)\right] = \sum_{\alpha \equiv \pm j \,(\mathrm{mod}\ 2N)} \mathrm{e}^{-t} I_\alpha(t). \tag{36}$$

4.2. Quantum Process

The Schrödinger equation for the finite path is

$$\mathrm{i}\psi_j'(t) = \tfrac{1}{2}\psi_{j-1}(t) + \tfrac{1}{2}\psi_{j+1}(t), \tag{37}$$

for $0 < j < N$, with initial condition $\psi_j(0) = \delta_{0,j}$ and boundary conditions

$$\mathrm{i}\psi_0'(t) = \psi_1(t), \quad \mathrm{i}\psi_N'(t) = \psi_{N-1}(t). \tag{38}$$

The Laplace transform of Eq. (37) is

$$\hat{\psi}_{j+1}(s) - 2\mathrm{i}s\hat{\psi}_j(s) + \hat{\psi}_{j-1}(s) = 0, \quad 0 < j < N, \tag{39}$$

and two boundary equations $is\hat{\psi}_0(s) - \mathrm{i} = \hat{\psi}_1(s)$, and $is\hat{\psi}_N(s) = \hat{\psi}_{N-1}(s)$. The solutions of $q^2 - 2\mathrm{i}sq + 1$ are $q_\pm = i(s \pm \sqrt{s^2+1})$. A guess of the solution is

$$\hat{\psi}_j(s) = Aq_+^j + Bq_-^j, \quad 0 \le j \le N. \tag{40}$$

From the boundary equations, we get $B - A = 2\mathrm{i}/(q_+ - q_-)$ and $B = Aq_+^{2N}$. Thus,

$$A = \frac{2\mathrm{i}}{(q_+ - q_-)} \frac{q_-^{2N}}{(1 - q_-^{2N})}. \tag{41}$$

For $j = 0, 1, \ldots, N$,

$$\begin{aligned}
\hat{\psi}_j(s) &= Aq_+^j + Bq_-^j = A(q_+^j + q_+^{2N-j}) \\
&= \frac{2\mathrm{i}}{(q_+ - q_-)} \frac{(q_-^j + q_-^{2N-j})}{(1 - q_-^{2N})} = \frac{2\mathrm{i}}{(q_+ - q_-)} \sum_{k=0}^{\infty} \left(q_-^{2Nk+j} + q_-^{2(k+1)N-j}\right) \\
&= \sum_{k=0}^{\infty} \left[\frac{((-\mathrm{i})(\sqrt{s^2+1} - s))^{2Nk+j}}{\sqrt{s^2+1}} + \frac{((-\mathrm{i})(\sqrt{s^2+1} - s))^{2(k+1)N-j}}{\sqrt{s^2+1}} \right].
\end{aligned}$$

The inverse Laplace transform Eq. (5) yields, for $j = 0, 1, \ldots, N$,

$$\begin{aligned}\psi_j(t) &= \sum_{k=0}^{\infty}\left[(-\mathrm{i})^{2Nk+j} J_{2Nk+j}(t) + (-\mathrm{i})^{2N(k+1)-j} J_{2N(k+1)-j}(t)\right] \\ &= \sum_{\alpha\equiv\pm j \,(\mathrm{mod}\ 2N)} (-\mathrm{i})^{\alpha} J_{\alpha}(t). \end{aligned} \tag{42}$$

4.3. Spectral Analysis

The spectrum of a path on n vertices is given by Spitzer.[17] For $j \in \{0, 1, \ldots, N\}$, the eigenvalue λ_j and its eigenvector v_j are given by

$$\lambda_j = \cos\left(\frac{(j+1)\pi}{N+2}\right), \quad v_j(\ell) = \sqrt{\frac{2}{N+2}} \sin\left(\frac{(j+1)\pi}{N+2}(\ell+1)\right). \tag{43}$$

The probability of measuring vertex 0 at time t is given by

$$P_0(t) = \frac{4}{(N+2)^2} \sum_{j,k} \sin^2\left(\frac{(j+1)\pi}{N+2}\right) \sin^2\left(\frac{(k+1)\pi}{N+2}\right) \mathrm{e}^{-\mathrm{i}t(\lambda_j - \lambda_k)}. \tag{44}$$

Since all eigenvalues are distinct, the *average probability* of measuring the starting vertex 0 is

$$\begin{aligned}\overline{P}(0) &= \frac{4}{(N+2)^2} \sum_{j,k} \sin^2\left(\frac{(j+1)\pi}{N+2}\right) \sin^2\left(\frac{(k+1)\pi}{N+2}\right) \\ &\quad\times \lim_{T\to\infty} \frac{1}{T} \int_0^T \mathrm{e}^{-\mathrm{i}t(\lambda_j - \lambda_k)}\, \mathrm{d}t \\ &= \frac{4}{(N+2)^2} \sum_{j} \sin^4\left(\frac{(j+1)\pi}{N+2}\right). \end{aligned}$$

Equating this with Eq. (42), we obtain a Bessel-like equation:

$$\lim_{T\to\infty} \frac{1}{T} \int_0^T \left| \sum_{a\equiv 0 (\mathrm{mod}\ 2N)} (-\mathrm{i})^a J_a(t) \right|^2 \mathrm{d}t = \frac{4}{(N+2)^2} \sum_{k=0}^{N} \sin^4\left(\frac{(k+1)\pi}{N+2}\right) \tag{45}$$

5. CONCLUSIONS

This expository survey reviews equations for the continuous-time quantum walks on one-dimensional lattices. The focus was on analysis based on the Laplace transform which works directly with the stochastic recurrences. It would be interesting to extend this analysis to higher-dimensional or to regular graph-theoretic settings. Another interesting direction is to consider lattices with defects and weighted graphs.[9]

ACKNOWLEDGMENTS

Authors thank the anonymous referees for pointing out several omissions and for helpful comments that improve the presentation of this manuscript. Supported by NSF grant DMR-0121146, DMS-0404778 and PHY-0140094.

REFERENCES

1. A. Ambainis, Quantum walk algorithm for element distinctness, quant-ph/0311001.
2. Y. Aharonov, L. Davidovich, and N. Zagury, *Phys. Rev. Lett.* **48(2)**, 1687 (1993)
3. M. Abramowitz and I. Stegun, *Handbook of Mathematical Functions With Formulas, Graphs, and Mathematical Tables* (Dover, 1974).
4. D. Aharonov, A. Ambainis, J. Kempe, and U. Vazirani, Quantum Walks on Graphs, in *Proceedings of 33rd ACM Annual Symposium Theory of Computing* (ACM Press, 2001), pp. 50–59.
5. A. Ambainis, E. Bach, A. Nayak, A. Viswanath, and J. Watrous, One-dimensional Quantum Walks, in *Proceedings of 33rd ACM Annual Symposium Theory of Computing* (ACM Press, 2001), pp. 37–49.
6. A. Ahmadi, R. Belk, C. Tamon, and C. Wendler, *Quantum Inform. Comput.* **3(6)**, 611 (2003).
7. K. L. Chung, *Markov Chains with Stationary Transition Probabilities*, 2nd edn. (Springer, 1967).
8. A. Childs, E. Farhi, and S. Gutmann, *Quantum Inform. Process.* 1, 35 (2002).
9. A. Childs, E. Deotto, R. Cleve, E. Farhi, S. Gutmann, and D. Spielman, Exponential algorithmic speedup by quantum walk, in *Proceedings of 35th ACM Annual Symposium Theory of Computing* (ACM Press, 2003), pp. 59–68.
10. A. Childs, and J. Goldstone, quant-ph/0306054 (2003).
11. E. Farhi, and S. Gutmann, *Phys. Rev. A* **58**, 915 (1998).
12. R. Feynman, R. Leighton, and M. Sands, *The Feynman Lectures on Physics*, Vol. III (Addison-Wesley, 1965).
13. H. Gerhardt and J. Watrous, Continuous-time quantum walks on the symmetric group, in *Proceedings 7th International Workshop on Randomization and Approximation Techniques in Computer Science*, LNCS 2764 (Springer, 2003) pp. 290–301.
14. J. Kempe, Quantum Random walks hit exponentially faster, in *Proceedings 7th International Workshop Randomization and Approximation Techniques in Computer Science*, LNCS 2764 (Springer, 2003), pp. 354–369.

15. D. Meyer, *J. Stat. Phys.*, **85**, 551 (1996).
16. C. Moore and A. Russell, Quantum walks on the hypercube, in *Proceedings 6th International Workshop on Randomization and Approximation Techniques in Computer Science*, LNCS 2483 (Springer, 2002), pp. 164–178.
17. F. Spitzer, *Principles of Random Walk*, 2nd edn. (Springer, 1976).

Kolibri 360 GmbH
Ferdinand-Jühlke-Straße 7
99095 Erfurt, Deutschland
produktsicherheit@kolibri360.de